Climate-Environmental Responses and Countermeasure Study on Atmospheric Component Changes

大气组成变化及其影响与对策研究

主　编（Chief Editors）

石广玉（SHI Guangyu）

符淙斌（FU Congbin）

副主编（Associate Editor）

陈　彬（CHEN Bin）

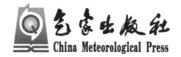

气象出版社
China Meteorological Press

内 容 简 介

本书介绍了大气科学与环境领域基于国际科技合作项目的一系列最新科研和技术开发成果，包括大气组成变化及其气候效应，陆地、海洋和湖泊等生态系统，传染病病原体等对气候变化的响应和从节能和减排等方面开发的应对气候变化的新技术等。本书可供大气科学、环境科学与技术、生态科学与技术等领域的专家、教师、学生等参考。

图书在版编目(CIP)数据

大气组成变化及其影响与对策研究 / 石广玉，符淙斌主编.
—北京：气象出版社，2014.4

ISBN 978-7-5029-5916-6

Ⅰ. ①大… Ⅱ. ①石… ②符… Ⅲ. ①气候变化—研究 Ⅳ. ①P467

中国版本图书馆 CIP 数据核字(2014)第 068321 号

Daqi Zucheng Bianhua jiqi Yingxiang yu Duice Yanjiu

大气组成变化及其影响与对策研究

主编 石广玉 符淙斌

副主编 陈 彬

出版发行：	气象出版社		
地 址：	北京市海淀区中关村南大街 46 号	**邮政编码：**	100081
总 编 室：	010-68407112	**发 行 部：**	010-68409198
网 址：	http://www.cmp.cma.gov.cn	**E-mail：**	qxcbs@cma.gov.cn
责任编辑：	李太宇	**终 审：**	黄润恒
封面设计：	博雅思企划	**责任技编：**	吴庭芳
印 刷：	北京京华虎彩印刷有限公司		
开 本：	787 mm×1092 mm 1/16	**印 张：**	18
字 数：	461 千字		
版 次：	2014 年 6 月第一版	**印 次：**	2014 年 6 月第一次印刷
定 价：	90.00 元		

本书编委会

主　编　石广玉　符淙斌

副主编　陈　彬

编　委　（按姓氏拼音排序）

艾丽坤　陈云浩　贾仲君　王东云

王建峰　王　伟　杨　军　张延荣

赵　晟　朱建国

前　言

气候变化特别是全球变暖已经成为世界各国高度关注的问题，中国的情况亦不例外。气候变化的原因可以归为自然和人为两大类。政府间气候变化专门委员会第四次评估报告(IPCC AR4)认为，对大部分观测到的20世纪中叶以来的全球温度上升来说，很可能是人为大气CO_2等温室气体浓度增加引起的，主要源于煤、石油和天然气等化石燃料的燃烧，而主要的碳汇则是海洋和生态系统。与此同时，大气气溶胶(颗粒物)的浓度也在增加，这种污染除了对空气质量、交通运输和人体健康具有重要影响外，还具有不可忽视的气候效应。大气组成变化将产生一系列的气候、环境和生态后果。为了应对气候变化，世界各国正在采取应对措施。其中，最重要的是节能减排(提高能源使用效率，减少温室气体排放)和增加碳汇(增加碳的吸收)。中国目前已经超过美国成为全球最大的温室气体排放国，受到的环境，外交压力空前巨大。发展是中国的第一要务，但是，气候变化问题与13亿中国人息息相关。因此，探讨全球变暖的原因、影响及其对策，无疑是国家的急需。书中介绍的项目成果是通过国际合作，发挥中日两国的科技优势取得的，具体内容有：探讨大气组成变化及其气候效应；揭示陆地、海洋和湖泊等生态系统以及传染病病原体等对气候变化的响应；从节能和减排等多方面开发应对气候变化的新技术。项目成果将为东北亚地区以致全球大气环境与气候变化研究带来新的启示与促进，为减小气候变化研究的不确定性做出贡献，为我国制定长期气候变化应对方案和实现减排目标提供科学依据和信息数据。

本书得到国家国际科技合作专项(编号：2010DFA22770、2013DFG22820)和国家自然科学基金项目(41130104)共同资助。

编者

2014年1月

目　录

第一部分　大气组成与气候变化的相关研究与进展

Part One　The Related Study and Development on the Atmospheric Components and Climate Change

第二部分 气候变化影响的相关研究与进展

Part Two The Related Study and Development on the Responses to Climate Change

第三部分 气候变化对策的相关研究与进展

Part Three The Related Study and Development on the Countermeasures to the Climate Change

第一部分

大气组成与气候变化的相关研究与进展

Part One

The Related Study and Development on the Atmospheric Components and Climate Change

IDENTIFICATION OF CULTURABLE BIOAEROSOLS COLLECTED OVER DRYLAND IN NORTHWEST CHINA: OBSERVATION USING A TETHERED BALLOON①

CHEN Bin[1], Fumihisa Kobayashi[2]②, Maromu Yamada[3], Yang-Hoon Kim[4], Yasunobu Iwasaka[5] and SHI Guangyu[1]

[1] State Key Laboratory of Numerical Modeling for Atmospheric Sciences and Geophysical Fluid Dynamics, Institute of Atmospheric Physics, Chinese Academy of Sciences, Beijing 100029, China

[2] School of Natural System, College of Science & Engineering, Kanazawa University, Kakuma-machi, Kanazawa, Ishikawa 920-1192, Japan

[3] Center for Innovation, Kanazawa University, Kakuma-machi, Kanazawa, Ishikawa 920-1192, Japan

[4] Department of Microbiology, Chungbuk National University, 410 Sungbong-Ro, Heungduk-Gu, Cheongju 361-763, South Korea

[5] Frontier Science Organization, Kanazawa University, Kakuma-machi, Kanazawa, Ishikawa 920-1192, Japan

Abstract

Dryland ecosystems are important biological component, especially in the fields of the biogeography and extrenophiles. For the research of atomospheric bioaerosols over dryland, we carried out the direct sampling using a tethered balloon over Dunhuang City, China. Bioaerosols were collected using a tethered balloon with a bioaerosol collector at 820 m above the ground (1960 m above the sea level) in the night on August 16, 2007. The bioaerosols were cultivated immediately after the collection at Dunhuang Meteorological Station. Two strains of molds were isolated using the nutrient agar medium. About 400-bp 18S rRNA partial sequences were amplified by PCR and determined. The results of a homology search by 18S rRNA sequences of isolates in DNA databases (GenBank, DDBJ, and EMBL) and an observation of the form revealed that two bioaerosols in the convective mixed layer over Dunhuang City were *Cladosporium* sp. and *Aspergillus* sp.

Key words: atmospheric bioaerosols, dryland ecosystem, direct sampling, tethered balloon, molds

1 Introduction

Drylands are extensive, covering 30% of the Earth's land surface and 50% of Africa

① The paper published in *Asian Journal of Atmospheric Environment*, Vol. 5-3, pp. 172-180, September 2011.

② Corresponding author. Tel.: +81-76-234-4820, E-mail: fumihisa@t.kanazawa-u.ac.jp

(Sankaran *et al.*, 2005; Scholes and Archer, 1997). Furthermore, dryland ecosystems support a large fraction of the human population and most pastoralist societies (Trenton *et al.*, 2010; Sankaran *et al.*, 2005). Soil microbiology of dryland has been researched in fields of biogeography and extrenophiles (Friedmann *et al.*, 1993; Friedmann, 1982; Friedmann and Kibler, 1980; Pielou, 1979). Friedmann and Kibler (1980) repoted that the nitrogen source for endolithic microorganisms in deserts is abioticlly fixed nitrogen produced by atmosphericelectric discharges (lightning or aurorase), conveyed to the rock by atmospheric precipitation. About dryland in the West China, Kenzaka *et al.* (2010) reported that the phylogenetic analysis of arid soils of the Loess Plateau, China, revealed that most phylotypes had low similarity with known strains in various phyla.

Atmospheric bioaerosols over the dryland play an important role in the dryland ecosystem because they deposit on and raise from the dryland soil by an atmospheric phenomena, *i.e.* provide and lost to the soil ecosystem. Wang *et al.* (2010) examined the seasonal variations of airborne bacteria in the Mogao Grottoes, Dunhuang, China. Their data were interesting, but it was difficult to lead to atmospheric bioaerozol because the sampler was mounted 1.5 m above ground level. We investigated the atmospheric mineral particles over Beijing using balloon-borne measurement (Chen *et al.*, 2010 a; 2010 b). Using our tethered-balloon technique, we can collect atmospheric bioaerosols in the convective mixed layer over dryland.

In this study, for the research of atomospheric bioaerosols over the dryland, Dunhuang in Northwest China, we tried to collect them using a tethered balloon and bioaerosol sampler. The separated culture and identifications of the atomospheric bioaerosols were reported.

2 Experimental

2.1 Location and Sampling Date

The sampling of the atmospheric bioaerosols using tethered balloon was made at Dunhuang (40°00′N, 94°30′E, 1140 m above sea level), China, which is on the east side of the Tarim Basin (Taklamakan desert). Dunhuang is a very interesting location for atmospheric bioaerosol measurements over the dryland because the particulate matter originated from Taklimakan desert is frequently transported through combination of westerly wind and local circulations. The sampling was carried out around noon (13:20—14:20 Beijing Standard Time) on August 17, 2007. The weather of the sampling date was cloudy.

2.2 Direct Sampling Method and Atmospheric Observations

The aerosol particle collection and the atmospheric observations were carried out using a tethered balloon which can lift instruments up to 1000 m with maximum payload of 10 kg. As shown in Fig. 1, a bioaerosol sampler, an optical particle counter (OPC; KR-12A, Rion Co., Ltd.), and a thermo-hygrometer (EX-501, EMPEX Instruments, Inc.) were mounted on the tethered balloon. The altitudes of the balloon were monitored by the on-load global positioning system (GPS), and the data were transferred to the operating room on the ground by radio signal.

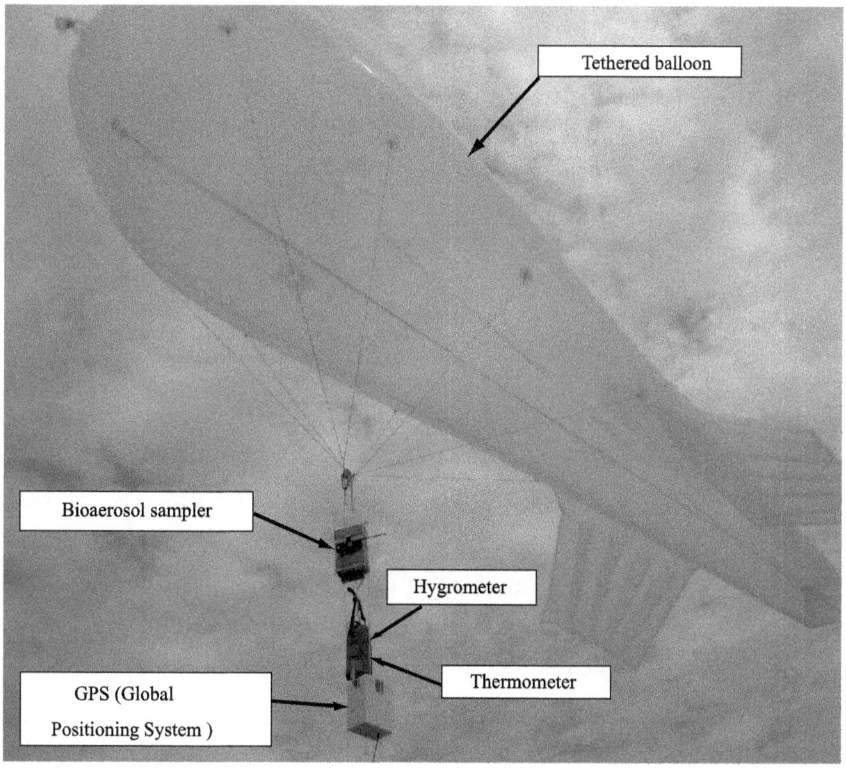

Fig. 1　The tethered balloon and bioaerosol sampler

In this study, the bioaerosols were collected on 0.45 μm pore-size membrane filters with the bioaerosol sampler. The filter was set into a filter holder (In-Line Filter Holder, 47 mm; Millipore Co., Ltd.) in the sampler under sterile condition after those were autoclaved. The sampling was started at 820 m above the ground (1960 m above the sea level) using a remote controller with a radio transmitter. The sampling volume of the air was estimated about 0.8 m^3 with a 60-minute sampling duration and the flow rate of 13.5 L/min by an air pump. To avoid contamination during non-sampling period, the inlet and outlet of the filter holder were closed by the shutter valves. The mechanical details of the sampler were described by Iwasaka *et al*. (2009).

In addition to the direct sampling of the bioaerosols, the profiles of the temperature and the relative humidity were measured by the thermo-hygrometer. The OPC measured aerosol number concentrations with the diameters of the particles larger than 0.3, 0.5, 0.7, 1.0, 2.0 and 5.0 μm. These data were used for understanding the atmospheric condition during the observation and speculating the origin of the collected aerosols.

2.3 Separated Culture and Identification

The atmospheric bioaerosols were cultivated immediately after the collection in the clean booth at Dunhuang Meteorogical Station. The filter sample put on the plate contains the Nutrient agar medium (Difco BD Co. Ltd.). The microorganisms were observed using an optical microscope (E2T-C, Nikon Co., Ltd.). The DNA was extracted from the isolates on the pate using cell wall lytic enzyme, lysozyme, and proteinase K (Sigma—Aldrich). 18S rDNA for eukaryote was amplified by polymerase chain reaction (PCR) using primer F2 (5'-TGGTTGATCCTGCCAGAGG-3') and R1 (5'-GGCTACCTTG TTACG ACTT-3'). PCR reaction mixture (vol. 20 μL) included the following: 4 μL of 5×Buffer, 1.6 μL of 10×dNTP (2.5 mM each, dATP, dCTP, GTP, dTTP), 0.2 μL of each primers (20 mM), 12.8 μL of sterile deionized H_2O, 1 U of Prime STAR DNA polymerase (TAKARA BIO INC. Co., Ltd.), 1 μL of DNA (~30 ng). The thermal cycler (Dice, TAKARA BIO INC. Co., Ltd.) was used under the following conditions for amplification: initial 2 min denatureation at 98℃; 35 cycle-10 s denaturation at 98℃, 10 s annealing at 54℃, and 1.5 min extension at 72℃; final 3 min extension at 72℃.

The DNA sequencing of cloned rDNA was determined by agenetic analyzer (Applied Biosystems Co., Ltd.), and the related species of the isolates were searched by BLAST analysis (http://www.ncbi.nlm.nih.gov/BLAST/) to DNA databases (GenBank/EMBL/DDBJ). These sequence data of isolates have been submitted to the DDBJ database under accession numbers AB455104 and AB455105.

3 Results and Discussion

3.1 Atmospheric Observations

During the observation, Dunhuang was overcast with stratocumulus which was formed in the boundary layer. Due to the cloudy weather, the temperature and the relative humidity (R. H.) near the ground at the observation time were relatively lower (about 22℃) and higher (about 65%) than usual summer daytime, respectively. The observation was carried out in moderate wind conditions. Dust events were not reported in Dunhuang and even in the upwind areas before the observation.

The balloon took 17 minutes to reach the sampling height with the ascent rate of about 36 m/min, measuring temperature, R. H. and particle concentration simultaneously.

Fig. 2 shows the vertical profiles of the temperature and the R. H. during the balloon flight in addition to the average and standard deviation of those during the bioaerosol sampling. The humidity gradually increased from 63%−70% near the ground surface to nearly 100% at the sampling height. The relative humidity gradually decreased to 70% during the sampling.

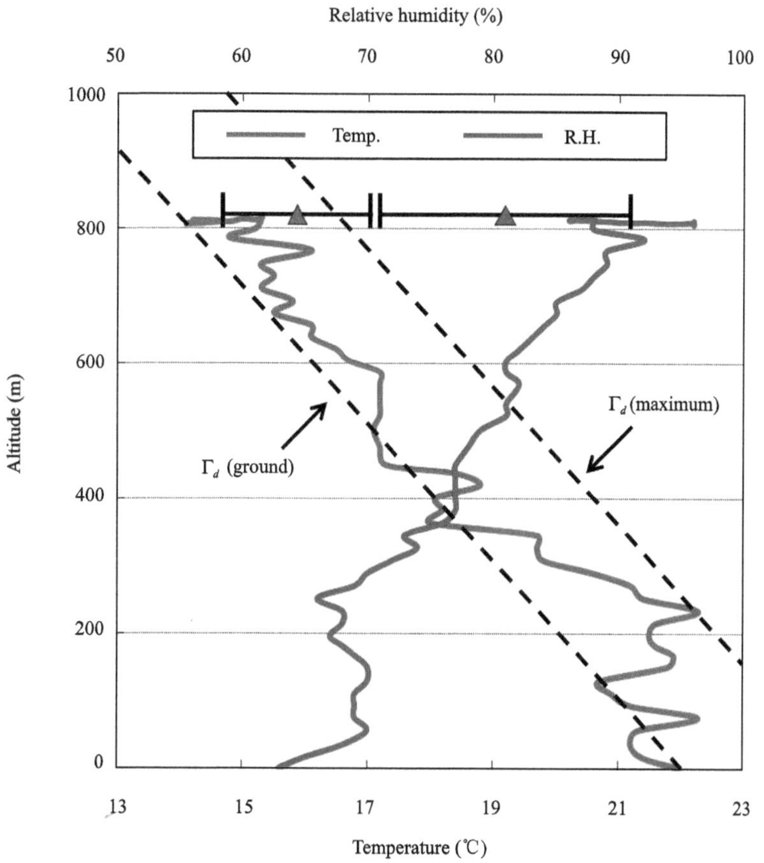

Fig. 2　Vertical profiles of temperature and R. H. during the balloon flight. Symbols (▲) and error bars at the altitude of 820 m indicate average values and standard deviations during the bioaerosol sampling, respectively. Broken lines show the dry adiabatic lapse rate (Γ_d=0.976℃/100 m).

To make inference about the vertical distribution of turbulence, the dry adiabatic lapse rate (Γ_d = 0.976℃/100 m) was drawn by the broken lines in Fig. 2. Comparing the temperature on the ground to that at the sampling height, the air seems to have decreased along with Γ_d. However, considering the temperature profile in detail, the environment temperature between 150 and 350 m was clearly higher than the Γ_d(ground) line drawn based on the ground temperature. This means that the air parcel around 300 m has a potential to be thermodynamically lifted along with the Γ_d(maximum) line until it hits the environment temperature again. Instead, the upper air parcels will descend according to those Γ_d and fill in the space. During the sampling the temperature at the sampling height gradually increased to the value on the Γ_d (maximum) line, suggesting that the warm air parcel were moved by convective flow.

Fig. 3 shows the vertical profiles of the number concentrations of the particles at the size ranges of 0.3—0.5 μm, 2.0—5.0 μm and >5.0 μm. The particle concentrations were high below several tens meters, which correspond well with the lower temperature than Γ_d (ground), suggesting the direct influence of the particle emissions just from the ground near the balloon-launching site. There were no clear differences of particle concentrations between 50 and 750 m. However, at the heights around 800 m, the concentration of the particles larger than 2.0 μm suddenly increased, but the smaller particles in the size range of 0.3—0.5 μm did not change considerably. Such concentration changes are frequently observed within clouds. Concerning the high R. H. around 800 m, the balloon should have encountered the stratocumulus in the boundary layer. After passing the high concentration event of coarse particles, the concentrations immediately decreased to almost same value as that observed below 750 m, and the concentration did not change very much during the sampling. It is suggested that the particle concentrations from 50 m to the sampling height were homogenized by convective mixing in the boundary layer and the sample collected would reflect the local air quality in the boundary layer above the surface atmosphere (below the altitude of about 50 m). The aerosols lifted up to the sampling level will be locally diffused by horizontal wind which is stronger than that near the ground surface.

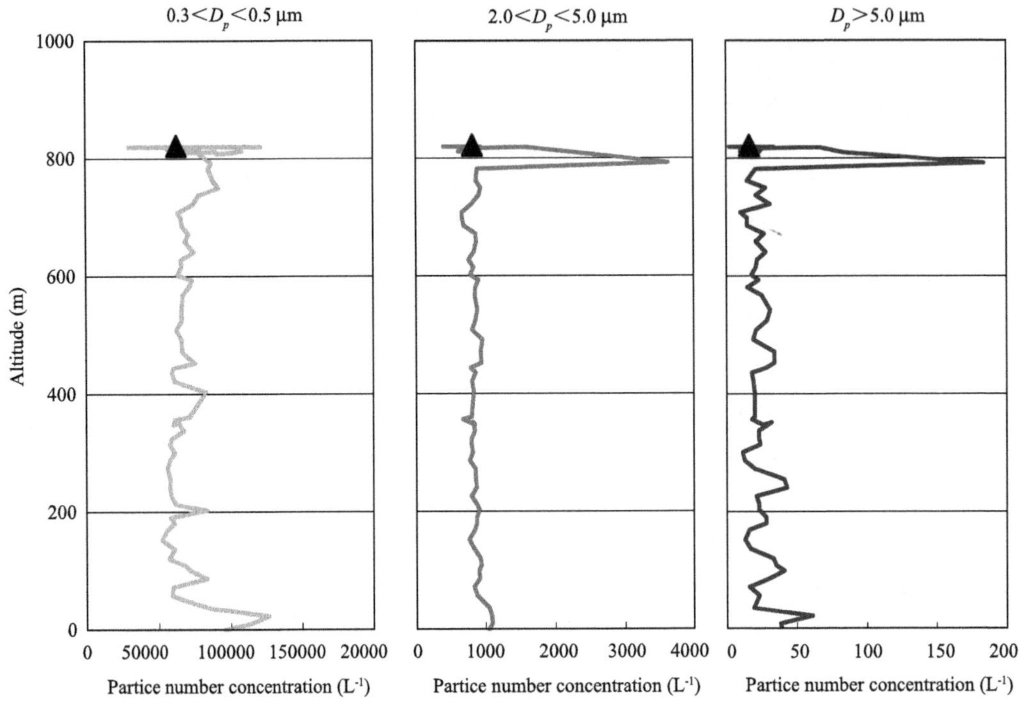

Fig. 3 Vertical profiles of particle number concentrations (L^{-1}) in size ranges of 0.3—0.5 μm, 2.0—5.0 μm and >5.0 μm. Symbols (▲) indicate average values of particle concentration during the bioaerosol sampling

According to previous observations using balloon-borne OPC in Dunhuang, the concentrations of the coarse mode particles (diameter > 1.0 μm) which were mostly composed of mineral dust remained high from the ground to 4 km a.s.l. in several summer cases without dust events (background condition) (Iwasaka *et al.* 2007). The convective flow over dryland in summer is more active than that over other locations such as marine or vegetated areas because of the large temperature differences between daytime and nighttime. Accordingly, the strong convection lifts and mixes the airmass containing the particulate matter to the higher altitude. The aerosols lifted up to the top of the boundary layer have longer life time and the distances of the aerosol dispersion would be long. Moreover, a fraction of the aerosols overshot through the capping inversion on the top of the mixed layer might be transported to wide areas by the prevailing westerlies in the free troposphere.

3.2 Separated Culture and Identification

We collected the bioaerosols directly at the altitude about 820 m above the ground (1 960 m above the sea level) under the atmospheric conditions as shown in Figs. 2 and 3. Two colonies grew overnight on the plate containing Nutrient agar medium. One colony was named BADHUN 0701 and the other one was named BADHUN 0702. Fig. 4 shows photographs of BADHUN 0701 (a) and BADHUN 0702 (b) on the plate after isolation. Two strains were observed to be of a kind of mold. BADHUN 0701 strain was black colony. In the case of BADHUN 0702 strain, the center of colony was brown and the circumference was white. These strains were observed by microscope. Fig. 5 (a) and (b) show the microphotograph of mycelium (a) and conidia (b) of BADHU 0701 strain. The mycelium was branched, 2-4 μm wide. The conidia were shaped, about 6 to 7 and 3 to 4 μm in diameter. Fig. 5 (c), (d), and (e) show the microphotograph of mycelium (c), conidiophores (d), and conidia (e) of BADHU 0702 strain. The mycelium was narrow and twisted, about 1 μm wide. The conidiophores were straight and small head-like swelling. The conidia were sphere, about 2 to 3 μm in diameter.

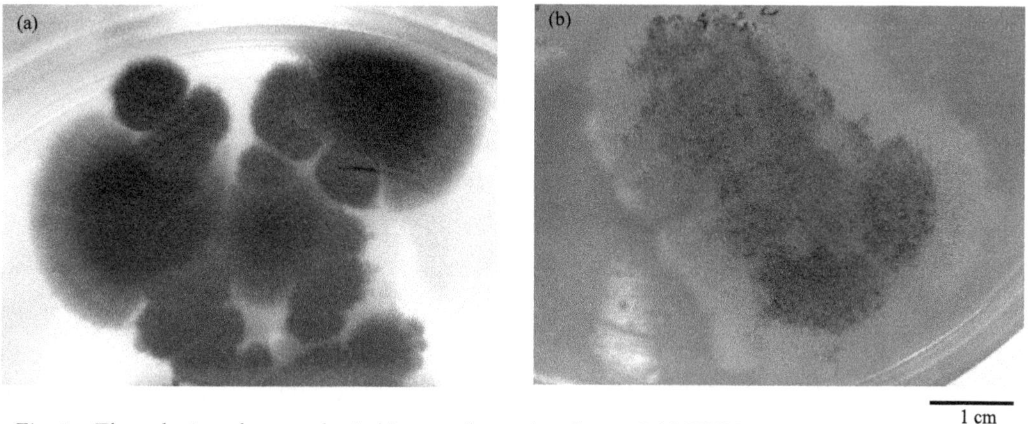

Fig. 4　The colonies of atmospheric bioaerosols on the plates: BADHUN 0701 (a) and BADHUN 0702 (b) strains. Scale bar shows 1 cm.

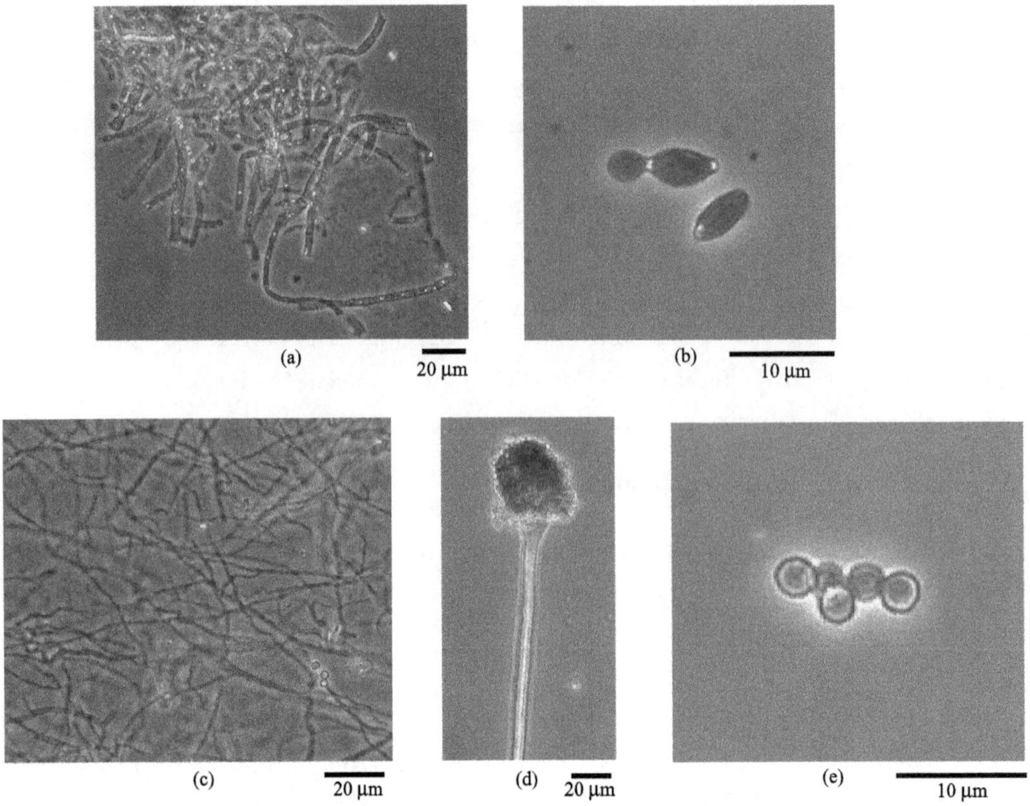

Fig. 5 Microphotograph of mycelium (a), conidia (b) of BADHU 0701 strain, mycelium (c), conidiophores (d), and conidia (e) of BADH 0702 strain. Scale bars show 20 μm in (a), (c), and (d); 10 μm in (b) and (e).

The 18S rDNA sequences of the fungus strains BADHU 0701 and 0702 were determined for identification under accession number AB455104 and AB455105, respectively. The similarities between its 18S rDNA sequence and DNA sequences in the GenBank, DDBJ, and EMBL were researched using a homology search program package, BLAST (Altschul *et al.*, 1997). The partial DNA sequence data of BADHU 0701 strain, 320 base pairs in length, were closely related to *Cladosporium cladosporioides* (DQ678004, 100.0%), *Cladosporium cladosporioides* (AF548071, 100.0%), and *Cladosporium cladosprioides* (AF548070, 100.0%), as shown in Fig. 6. Though BADHU 0701 strain was closely related to *Cladosoprium cladosperioides* from results of similarities, it belonged to the species of *Cladosporium* because it could not decide *Cladosporium cladosporioides* from the microphotograph data, especially microphotograph of hyphae (Fig. 5). BADHU 0701 was found *Cladosporium* sp. The partial DNA sequence data of BADHU 0702 strain, 396 base pairs in length, were closely related to *Aspergillus versicolor* (EU263603, 99.0%), *Aspergillus* sp. (DQ810193, 99.0%), *Aspergillus versicolor* (AF548069, 99.0%), *Aspergillus versicolor* (AF548068, 99.0%), *Aspergillus silvaticus* (AF548067, 99.0%), *Aspergillus versicolor* (AB008411, 99.0%), *Aspergillus ustus* (AB008410, 99.0%), *Aspergillus ustus* (AB002072, 99.0%), and *Aspergillus sparsus* (AB002066, 99.0%), as shown in Fig. 7. From the data of similarities, the colony form, micrographics, BADHU 0702 strain was belong to species of *Aspergillus* sp. (Figs. 4, 5, 7).

```
BADHUN 0701 strain             (AB455104)     1 GGGGCATCAGTATTCAATCGTCAGAGGTGAAATTCTTGGA 40
Cladosporium cladosporioides   (DQ678004)   917 GGGGCATCAGTATTCAATCGTCAGAGGTGAAATTCTTGGA 956
Cladosporium cladosporioides   (AF548071)   799 GGGGCATCAGTATTCAATCGTCAGAGGTGAAATTCTTGGA 838
Cladosporium cladosporioides   (AF548070)   799 GGGGCATCAGTATTCAATCGTCAGAGGTGAAATTCTTGGA 838
                                                ****************************************

  41 TTGATTGAAGACTAACTACTGCGAAAGCATTTGCCAAGGATGTTTTCATTAATCAGTGAACGAAAGTTAGGGGATCGAAG 120
 957 TTGATTGAAGACTAACTACTGCGAAAGCATTTGCCAAGGATGTTTTCATTAATCAGTGAACGAAAGTTAGGGGATCGAAG 1036
 839 TTGATTGAAGACTAACTACTGCGAAAGCATTTGCCAAGGATGTTTTCATTAATCAGTGAACGAAAGTTAGGGGATCGAAG 918
 839 TTGATTGAAGACTAACTACTGCGAAAGCATTTGCCAAGGATGTTTTCATTAATCAGTGAACGAAAGTTAGGGGATCGAAG 918
     ********************************************************************************

 111 ACGATCAGATACCGTCGTAGTCTTAACCATAAACTATGCCGACTAGGGATCGGACGGTGTTAGTATTTTGACCCGTTCGG 190
1027 ACGATCAGATACCGTCGTAGTCTTAACCATAAACTATGCCGACTAGGGATCGGACGGTGTTAGTATTTTGACCCGTTCGG 1106
 909 ACGATCAGATACCGTCGTAGTCTTAACCATAAACTATGCCGACTAGGGATCGGACGGTGTTAGTATTTTGACCCGTTCGG 988
 909 ACGATCAGATACCGTCGTAGTCTTAACCATAAACTATGCCGACTAGGGATCGGACGGTGTTAGTATTTTGACCCGTTCGG 988
     ********************************************************************************

 191 CACCTTACGAGAAATCAAAGTTTTTGGGTTCTGGGGGGAGTATGGTCGCAAGGCTGAAACTTAAAGAAATTGACGGAAGG 270
1107 CACCTTACGAGAAATCAAAGTTTTTGGGTTCTGGGGGGAGTATGGTCGCAAGGCTGAAACTTAAAGAAATTGACGGAAGG 1186
 989 CACCTTACGAGAAATCAAAGTTTTTGGGTTCTGGGGGGAGTATGGTCGCAAGGCTGAAACTTAAAGAAATTGACGGAAGG 1068
 989 CACCTTACGAGAAATCAAAGTTTTTGGGTTCTGGGGGGAGTATGGTCGCAAGGCTGAAACTTAAAGAAATTGACGGAAGG 1068
     ********************************************************************************

 271 GCACCACCAGGCGTGGAGCCTGCGGCTTAATTTGACTCAACACGGGGAAA 320
1187 GCACCACCAGGCGTGGAGCCTGCGGCTTAATTTGACTCAACACGGGGAAA 1236
1069 GCACCACCAGGCGTGGAGCCTGCGGCTTAATTTGACTCAACACGGGGAAA 1118
1069 GCACCACCAGGCGTGGAGCCTGCGGCTTAATTTGACTCAACACGGGGAAA 1118
     **************************************************
```

Fig. 6　DNA sequence data of BADHU 0701 strain and similarities between its 18S rDNA sequence and DNA sequences in the GenBank, DDBJ, and EMBL.

It was found that there are *Cladosporidium* sp. and *Aspelgillus* sp. as living atmospheric bioaerosol at 820 m of altitude above the ground (1960 m of altitude above the sea level) over dryland, Dunhuang. Generally, the conidia of molds, *i. e. Cladosporidium* and *Aspelgillus*, have a tolerance for ultra-violet ray, dry, and low temperature. These characteristics seem to make these atmospheric bioaerosols able to live at high altitude. Yali *et al.* (2008) reported that fungi isolated from the milk vetch on the Loess Plateau contained *Cladospridium* sp. and *Aspergillus* sp. It was suggested that these atmospheric

```
BADHUN 0702 strain      (AB455105)      1 CGGGGGCGTCAGTATTCAGCTGTCAGAGGTGAAATTCTTGGATTTGCTGA 50
Aspergillus versicolor  (EU263603)    633 CGGGGGCGTCAGTATTCAGCTGTCAGAGGTGAAATTCTTGGATTTGCTGA 682
Aspergillus sp.         (DQ810193)    845 CGGGGGCGTCAGTATTCAGCTGTCAGAGGTGAAATTCTTGGATTTGCTGA 894
Aspergillus versicolor  (AF548069)    796 CGGGGGCGTCAGTATTCAGCTGTCAGAGGTGAAATTCTTGGATTTGCTGA 845
Aspergillus versicolor  (AF548068)    796 CGGGGGCGTCAGTATTCAGCTGTCAGAGGTGAAATTCTTGGATTTGCTGA 845
Aspergillus silvaticus  (AF548067)    796 CGGGGGCGTCAGTATTCAGCTGTCAGAGGTGAAATTCTTGGATTTGCTGA 845
Aspergillus versicolor  (AB008411)    834 CGGGGGCGTCAGTATTCAGCTGTCAGAGGTGAAATTCTTGGATTTGCTGA 883
Aspergillus ustus       (AB008410)    834 CGGGGGCGTCAGTATTCAGCTGTCAGAGGTGAAATTCTTGGATTTGCTGA 883
Aspergillus ustus       (AB002072)    853 CGGGGGCGTCAGTATTCAGCTGTCAGAGGTGAAATTCTTGGATTTGCTGA 902
Aspergillus sparsus     (AB002066)    853 CGGGGGCGTCAGTATTCAGCTGTCAGAGGTGAAATTCTTGGATTTGCTGA 902
                                          **************************************************

  51 AGACTAACTACTGCGAAAGCATTCGCCAAGGATGTTTTCATTAATCAGGGAACGAAAGTTAGGGGATCGAAGACGATCAGATACCGTCGT 140
 683 AGACTAACTACTGCGAAAGCATTCGCCAAGGATGTTTTCATTAATCAGGGAACGAAAGTTAGGGGATCGAAGACGATCAGATACCGTCGT 772
 895 AGACTAACTACTGCGAAAGCATTCGCCAAGGATGTTTTCATTAATCAGGGAACGAAAGTTAGGGGATCGAAGACGATCAGATACCGTCGT 984
 846 AGACTAACTACTGCGAAAGCATTCGCCAAGGATGTTTTCATTAATCAGGGAACGAAAGTTAGGGGATCGAAGACGATCAGATACCGTCGT 935
 846 AGACTAACTACTGCGAAAGCATTCGCCAAGGATGTTTTCATTAATCAGGGAACGAAAGTTAGGGGATCGAAGACGATCAGATACCGTCGT 935
 846 AGACTAACTACTGCGAAAGCATTCGCCAAGGATGTTTTCATTAATCAGGGAACGAAAGTTAGGGGATCGAAGACGATCAGATACCGTCGT 935
 884 AGACTAACTACTGCGAAAGCATTCGCCAAGGATGTTTTCATTAATCAGGGAACGAAAGTTAGGGGATCGAAGACGATCAGATACCGTCGT 973
 884 AGACTAACTACTGCGAAAGCATTCGCCAAGGATGTTTTCATTAATCAGGGAACGAAAGTTAGGGGATCGAAGACGATCAGATACCGTCGT 973
 903 AGACTAACTACTGCGAAAGCATTCGCCAAGGATGTTTTCATTAATCAGGGAACGAAAGTTAGGGGATCGAAGACGATCAGATACCGTCGT 992
 903 AGACTAACTACTGCGAAAGCATTCGCCAAGGATGTTTTCATTAATCAGGGAACGAAAGTTAGGGGATCGAAGACGATCAGATACCGTCGT 992
     ******************************************************************************************

 141 AGTCTTAACCATAAACTATGCCGACTA-AAATCGGGCGGCGTTTCTATGATGACCCGCTCGGCACCTTACGAGAAATCAAAGTTTTTGGGT 230
 773 AGTCTTAACCATAAACTATGCCGACTAGGGATCGGGCGGCGTTTCTATGATGACCCGCTCGGCACCTTACGAGAAATCAAAGTTTTTGGGT 863
 985 AGTCTTAACCATAAACTATGCCGACTAGGGATCGGGCGGCGTTTCTATGATGACCCGCTCGGCACCTTACGAGAAATCAAAGTTTTTGGGT 1075
 936 AGTCTTAACCATAAACTATGCCGACTAGGGATCGGGCGGCGTTTCTATGATGACCCGCTCGGCACCTTACGAGAAATCAAAGTTTTTGGGT 1026
 936 AGTCTTAACCATAAACTATGCCGACTAGGGATCGGGCGGCGTTTCTATGATGACCCGCTCGGCACCTTACGAGAAATCAAAGTTTTTGGGT 1026
 936 AGTCTTAACCATAAACTATGCCGACTAGGGATCGGGCGGCGTTTCTATGATGACCCGCTCGGCACCTTACGAGAAATCAAAGTTTTTGGGT 1026
 974 AGTCTTAACCATAAACTATGCCGACTAGGGATCGGGCGGCGTTTCTATGATGACCCGCTCGGCACCTTACGAGAAATCAAAGTTTTTGGGT 1064
 974 AGTCTTAACCATAAACTATGCCGACTAGGGATCGGGCGGCGTTTCTATGATGACCCGCTCGGCACCTTACGAGAAATCAAAGTTTTTGGGT 1064
 993 AGTCTTAACCATAAACTATGCCGACTAGGGATCGGGCGGCGTTTCTATGATGACCCGCTCGGCACCTTACGAGAAATCAAAGTTTTTGGGT 1083
 993 AGTCTTAACCATAAACTATGCCGACTAGGGATCGGGCGGCGTTTCTATGATGACCCGCTCGGCACCTTACGAGAAATCAAAGTTTTTGGGT 1083
     ***************************   ************************************************************

 231 TCTGGGGGGAGTATGGTCGCAAGGCTGAAACTTAAAGAAATTGACGGAAGGGCACCACAAGGCGTGGAGCCTGCGGCTTAATTTGACTCA 320
 864 TCTGGGGGGAGTATGGTCGCAAGGCTGAAACTTAAAGAAATTGACGGAAGGGCACCACAAGGCGTGGAGCCTGCGGCTTAATTTGACTCA 953
1076 TCTGGGGGGAGTATGGTCGCAAGGCTGAAACTTAAAGAAATTGACGGAAGGGCACCACAAGGCGTGGAGCCTGCGGCTTAATTTGACTCA 1165
1027 TCTGGGGGGAGTATGGTCGCAAGGCTGAAACTTAAAGAAATTGACGGAAGGGCACCACAAGGCGTGGAGCCTGCGGCTTAATTTGACTCA 1116
1027 TCTGGGGGGAGTATGGTCGCAAGGCTGAAACTTAAAGAAATTGACGGAAGGGCACCACAAGGCGTGGAGCCTGCGGCTTAATTTGACTCA 1116
1027 TCTGGGGGGAGTATGGTCGCAAGGCTGAAACTTAAAGAAATTGACGGAAGGGCACCACAAGGCGTGGAGCCTGCGGCTTAATTTGACTCA 1116
1065 TCTGGGGGGAGTATGGTCGCAAGGCTGAAACTTAAAGAAATTGACGGAAGGGCACCACAAGGCGTGGAGCCTGCGGCTTAATTTGACTCA 1154
1065 TCTGGGGGGAGTATGGTCGCAAGGCTGAAACTTAAAGAAATTGACGGAAGGGCACCACAAGGCGTGGAGCCTGCGGCTTAATTTGACTCA 1154
1084 TCTGGGGGGAGTATGGTCGCAAGGCTGAAACTTAAAGAAATTGACGGAAGGGCACCACAAGGCGTGGAGCCTGCGGCTTAATTTGACTCA 1173
1084 TCTGGGGGGAGTATGGTCGCAAGGCTGAAACTTAAAGAAATTGACGGAAGGGCACCACAAGGCGTGGAGCCTGCGGCTTAATTTGACTCA 1173
     ******************************************************************************************

 321 ACACGGGGAAAACTCACCAGGTCCAGACAAAATAAGGATTGACAGATTGAGAGCTCTTTCTTGATCTTTTGGATGG 396
 954 ACACGGGG-AAACTCACCAGGTCCAGACAAAATAAGGATTGACAGATTGAGAGCTCTTTCTTGATCTTTTGGATGG 1028
1166 ACACGGGG-AAACTCACCAGGTCCAGACAAAATAAGGATTGACAGATTGAGAGCTCTTTCTTGATCTTTTGGATGG 1240
1117 ACACGGGG-AAACTCACCAGGTCCAGACAAAATAAGGATTGACAGATTGAGAGCTCTTTCTTGATCTTTTGGATGG 1191
1117 ACACGGGG-AAACTCACCAGGTCCAGACAAAATAAGGATTGACAGATTGAGAGCTCTTTCTTGATCTTTTGGATGG 1191
1117 ACACGGGG-AAACTCACCAGGTCCAGACAAAATAAGGATTGACAGATTGAGAGCTCTTTCTTGATCTTTTGGATGG 1191
1155 ACACGGGG-AAACTCACCAGGTCCAGACAAAATAAGGATTGACAGATTGAGAGCTCTTTCTTGATCTTTTGGATGG 1229
1155 ACACGGGG-AAACTCACCAGGTCCAGACAAAATAAGGATTGACAGATTGAGAGCTCTTTCTTGATCTTTTGGATGG 1229
1174 ACACGGGG-AAACTCACCAGGTCCAGACAAAATAAGGATTGACAGATTGAGAGCTCTTTCTTGATCTTTTGGATGG 1248
1174 ACAC-GGGAAAACTCACCAGGTCCAGACAAAATAAGGATTGACAGATTGAGAGCTCTTTCTTGATCTTTTGGATGG 1248
     **** *** *******************************************************************
```

Fig. 7 DNA sequence data of BADHU 0702 strain and similarities between its 18S rDNA sequence and DNA sequences in the GenBank, DDBJ, and EMBL.

bioaerosols diffuse regionally from discussions of atmospheric conditions as shown in Figs. 2 and 3 since they are some of the most common molds. *Cladosporium* and *Aspergillus* are reported as atmospheric bioaerosols (airborne microorganisms) over African desert by Griffin and Kellog (Griffin *et al.*, 2001; 2003; 2006; 2007; Kellogg *et al.*, 2004; Griffin and Kellogg, 2004). In this study, *Cladosporidium* sp. and *Aspergillus* sp. were isolated from high altitude air over the dryland, Dunhuang in Northwest China. Microorganisms in this study agree with them over Affrican dryland. It was suggested that *Cladosporidium* and *Aspergillus* play an important role for going in and out of dryland ecosystem.

4 Conclusion

For the research of atomospheric bioaerosols over the dryland, Dunhuang in northwest China, we examined direct sampling of them using a tethered balloon on August 17, 2007 and the separated culture and identifications. The atomospheric bioaerosol sample collected would reflect the air quality in the boundary layer and two strains were isolated. From the observation by microscope, DNA sequence of 18S rDNA, and the similarities search, two isolates were identified as *Cladosporidium* sp. and *Aspergillus* sp. The atmospheric observation pointed out the local scale distribution of bioaerosols through the convective mixing of the air, and suggested a possibility of regional scale diffusion of bioaerosols. In order to clarify dryland ecosystem, it will be necessary to research soil microbial community in Taklamakan desert in detail and the future study will be focused on it.

Acknowledgments

The authors are deeply grateful to the staff of Dunhuang Meteorological Station. In addition, we wish to thank Prof. Teruya Maki, Prof. Makiko Kakikawa, Prof. Takeshi Naganuma, Dr. Yutaka Tobo, and Dr. Chun-Sang Hong for their technical support in this study. A part of this research was supported by the Global Environment Research Fund (B-0901 and RF-072) of the Ministry of the Environment, Japan, and the Grant-in-Aid for Scientific Research (A) (no. 20253005) from the Ministry of Education, Culture, Sports, Science and Technology of Japan, of which another part was supported by the project 2009DFA22650 of The Ministry of Science and Technology of China (MOST).

References

Altschul S F, Madden T F, Schaffer A A, *et al.* 1997. Gapped BLAST and PSI-BLAST: a new generation of protein database search programs, *Nucleic Acids Res.* **25**: 3389-3402.

Chen B, Yamada M, Shi G, *et al.* 2010a. Vertical changes in mixing state of aerosol particles in the boundary layer in Beijing, China: Balloon-borne measurements in summer and spring. *Journal of Ecotechnology*, in press.

Chen B, Shi G, Yamada M, *et al.* 2010b. Vertical change in extinction and atmospheric particle size in the boundary layers over Beijing: Balloon-borne measurement. *Asian Journal of Atmospheric Environment*, accepted.

Friedmann E I, Kibler A P. 1980. Nitrogen economy of endolithic microbial communities in hot and cold deserts. *Microbial Ecology* **6**: 95-108.

Friedmann, E. I. 1982. Endolithic microorganisms in the Antarctic cold desert. *Science*. **215**: 1045-1053.

Friedmann E I, Kappen L, Meyer M A, *et al*. 1993. Long-term productivity in the cryptoendolithic microbial community of the Ross Desert. Antarctica. *Microbial Ecology*. **25**: 51-69.

Griffin D W, Garrison V H , Herman J R, *et al*. 2001. African desert dust in the Caribbean atmosphere: Microbiology and public health. *Aerobiologia* **17**: 203-213.

Griffin D W, Kellogg C A, Garrison V H, *et al*. 2003. Atmospheric microbiology in the northern Caribbean during African dust events. *Aerobiologia*. **19**: 143-157.

Griffin D W, Kellogg C A. 2004. Dust storm and their impact on ocean and human health: Dust in earth's atmosphere. *EcoHealth*. **1**: 284-295.

Griffin D W, Westphal D L, Gray M A. 2006. Airbrne microorganisms in the African desert dust corridor over the mid-Atlantic ridge, ocean drilling program, Leg 209. *Aerobiologia*, **22**:211-226.

Griffin D W, Kubilay N, Kocak M, *et al*. 2007. Airborne desert dust and aeromicrobiology over the Turkish Mediterranean coastline. *Atmospheric Environment*. **41**: 4050-4062.

Iwasaka Y, Li J M, Shi G Y, *et al*. 2008. Mass transport of background Asian dust revealed by balloon-borne measurement: dust particles transported during calm periods by westerly from Taklamakan desert, *Advanced Environmental Monitoring*. 121-135.

Iwasaka Y, Shi G Y, Yamada M, *et al*. 2009. Mixture of Kosa (Asian dust) and bioaerosols detected in the atmosphere over the Kosa particles source regions with balloon-borne measurements, Possibility of long-range transport, *Air Qual. Atmos. Health*. **2**: 29-38.

Kellogg C A, Griffin D W, Garrison V H, *et al*. 2004. Characterization of aerosolized bacteria and fungi from desert dust events in Mali, West Africa. *Aerobiologia*. **20**: 99-110.

Kenzaka T, Sueyoshi A, Baba T, *et al*. 2010. Soil microbial community structure in an Asian dust source region (Loess Plateau). *Microbes Environments*. **25**: 53-57.

Pielou E C. 1979. *Biogeography*. Johon Wiley & Sons, NewYork, pp. 42-46.

Sankaran M, Hannan N P, Scholes R J , *et al*. 2005. Determinants of woody cover in African savannas. *Nature*. **438**: 846-849.

Scholes R J, Archer S R. 1997. Tree-grass interactions in savannas. *Annu. Rev. Ecol. Syst.* **28**: 517-544.

Trenton E F, Kelly K C, Jan M N, *et al*. 2010. An ecohydrological approach to predicting regional woody species distribution patterns in dryland ecosystems. *Advances in Water Resources*. **33**: 215-230.

Wang W, Ma Y, Ma X, *et al*. 2010. Seasonal variations of airborne bacteria in the Mogao Grottoes, Dunhuang, China. *International Biodeterioration & Biodegradation*, doi:10. 1016/j. ibiod. 2010. 03. 004.

Yin Y, Nan, Z B, Li C, Hou F. 2008. Root-invading fungi of milk vetch on the Loess Plateau, China. *Agriculture, Ecosystems and Environment*. **124**: 51-59.

SENSITIVITY STUDIES OF AEROSOL DATA ASSIMILATION AND DIRECT RADIATIVE FEEDBACKS IN MODELING DUST AEROSOLS①

WANG Hong and NIU Tao②

Chinese Academy of Meteorological Sciences(CAMS), Beijing 100081, China.

Abstract

In order to study dust aerosol assimilation and radiative forcing in modeling East Asian dust aerosols and their impacts on the regional atmosphere, a three-dimensional variational data assimilation (3DV) and an aerosol radiative feedback scheme (RAD) are online implemented into a mesoscale numerical weather prediction system GRAPES/CUACE _ Dust. Four modeling experiments are conducted: one control running (CTL) excluding 3DV and RAD as well as three sensitive running experiments respectively with 3DV, RAD, the integrated 3DV and RAD(3DV_RAD). The results indicate the 3DV-running shows a distinct improvement in the daily averaged dust concentrations, while the 3DV _ RAD performs the better modeling during strong dust storms. The comparisons of the model bias for air temperature, pressure and wind speed from the CTL, RAD, and 3DV _ RAD experiments present that the dust direct radiation leads to decreases in the lower tropospheric temperature and increases in the upper tropospheric temperature, which results in enhancing air pressure in the lower troposphere and declining air pressure in the upper troposphere. The 3DV_RAD modeling in the middle and upper troposphere is more reasonable than the RAD modeling, which suggests the importance of integration of aerosol assimilation and radiation forcing in modeling aerosols and meteorological fields.

Key words: dust radiative feedbacks, aerosol assimilation system, mesoscale dust forecast system

1 Introduction

The development of dust emission and transport models aiming on Asian dust episodes

① The paper published in *Atmospheric Environment*, **64**, 2013, 208-218.

② Corresponding author: NIU Tao, E-mail: niutao2001

began at the 1990s (Kotamarthi and Carmichael, 1993; Liu and Zhou 1997; Xiao *et al.*, 1997). Dust aerosol concentration and optical depth (AOD) over middle and east Asia are intensively modeled (Wang *et al.*, 2000; Shao *et al.*, 2001, 2004; Gong *et al.*, 2003; Uno *et al.*, 2001, 2004; Gong and Zhang, 2008; Zhou *et al.*, 2008; Wang *et al.*, 2009). However, large uncertainties still exist in modeling dust aerosols, and the modeled concentration levels could differ as high as 2-4 times and even up to more than two orders of magnitude (Uno *et al.*, 2006). Among the attempts reducing the modeling uncertainties and bias, the on-line calculation of dust radiative impacts and the introduction of data assimilation system are two eye-catching directions in the dust model development. They are also being regarded as the new steps to improve dust weather forecast (Baker *et al.*, 2004; Pe′rez *et al.*, 2006; Niu *et al.*, 2008; Wang *et al.*, 2010).

Dust aerosol has profound impacts on regional radiation balance, consequently on the atmospheric structures such as air temperature, air pressure, wind speeds and planetary boundary layer (PBL) circulation and cloud fields, etc, and the atmospheric changes may affect dust emissions, depositions and transport in turn (Haywood *et al.*, 2005; Ahn *et al.*, 2007; Heinold et al, 2008; Wang *et al.*, 2006). The dust forecast with the online integrated radiation forcing showed significant improvements in the modeling capability (Pe′rez *et al.*, 2006; Wang *et al.*, 2010), and this improvement highly depends on the accuracy of modeled dust concentrations or loadings. Data assimilation is an important way to decrease the bias of dust concentration simulations. A three-dimensional variational (3DV) data assimilation method was used for the sand and dust storm data assimilation system (SDS-DAS) in China (Niu *et al.*, 2008). The operational measurements of visibility and weather phenomena for dust aerosols from surface meteorological stations and the spatial coverage observed by the Chinese geostationary satellite FY-2C have been used in the SDS-DAS in China. The SDS-DAS has been applied in an operational dust forecast system MM5/CUACE-dust in East Asia providing the corrections to both under- and over-predictions of East Asian dust aerosols (Zhou *et al.*, 2008). However, the dust aerosol assimilation alone cannot improve modeling meteorological fields for lacking of the linking bridge, dust radiation, between the improved forecast dust concentrations and model dynamical core.

In this study, the SDS-DAS is introduced into the mesoscale two-way dust model GRAPES/CUACE_Dust and thereby enables the online interaction of dust forecast-assimilation-direct radiative effects. The assimilated dust concentration alters the meteorological fields via reforming radiation balance and finally results in improving of the model predictions of meteorological fields and dust concentrations themselves.

2 Study Method

By running the mesoscale dust aerosol model GRAPES/CUACE_Dust, four modeling experiments are conducted to evaluate the interactions of dust forecast-assimilation-radiation effects: one control test (CTL) excluding 3DV and RAD as well as three sensitive experiments respectively with 3DV, RAD, 3DV_RAD of the integrated 3DV and RAD.

2.1 Introduction of GRAPES/CUACE_Dust

The GRAPES/CUACE_Dust is a mesoscale operational dust forecasting system developed by the Chinese Academy of Meteorological Sciences (CAMS) (Wang *et al.*, 2009, 2010). The model top is set about 30 kilometers with 31 vertical layers, and the model domain covers the East Asian region (15°—60°N, 70°—145°E) with the horizontal resolution of 0.25° × 0.25° in all studies. The dust particles are divided into 12 size bins respectively with the diameter ranges of 0.01—40.96 μm: 0.01—0.02 μm, 0.02—0.04 μm, 0.04—0.08 μm, 0.08—0.16 μm, 0.16—0.32 μm, 0.32—0.64 μm, 0.64—1.28 μm, 1.28—2.56 μm, 2.56—5.12 μm, 5.12—10.24 μm, 10.24—20.48 μm and 20.48—40.96 μm (Gong *et al.*, 2003) according to the measurements of soil dust size distribution in Chinese desert regions during 1994—2001 (Zhang *et al.*, 2003). The model initial time is at 00 UTC, and the assimilation time is at 03 UTC of the start days in April 2006.

2.2 Radiative Transfer Model and Aerosol Parameterization

The short wave and long wave radiative transfer model (CLIRAD-SW and CLIRAD-LW) developed by the Climate and Radiation Branch, NASA/Goddard Space Flight Center (Chou *et al.*, 1999), for a wide variety of weather and climate applications, is introduced into GRAPES/CUACE_Dust. The CLIRAD-SW and -LW include the absorption due to water vapor, O_3, O_2, CO_2, clouds and aerosols. The interactions among absorption and scattering by clouds, aerosols as well as Rayleigh scattering are considered. The entire solar spectrum is divided into 11 bands, from 0.175 to 10 μm and thermal infrared spectrum into 9 bands from 3.333 to 40 μm. The absorption and scattering of solar and thermal radiations due to dust are considered with this radiative transfer model in this paper. The dust refractive index data collected in Chinese deserts during the international project "Studies on the Origin and Transport of Aeolian Dust and Its Effects on Climate (ADEC)" have been proved to be the most representative optical data for East Asia dust (Wang *et al.*, 2006, 2010). It is used to calculate dust aerosol optical depth (AOD), single-scattering albedo (SSA), and asymmetry factor (ASY) as inputs for the CLIRAD-SW and CLIRAD-LW.

2.3 Dust Data Assimilation System

The SDS-DAS was developed for CUACE/Dust model. It is based on a three dimensional variational method (Zhang *et al.*, 2004; Zhuang *et al.*, 2005) with the measurements of surface visibility, dust loading derived from the Chinese geostationary satellite FY-2C (Hu *et al.*, 2008) and the modeled dust concentrations for the initial date.

The background error covariance matrix is important to the analysis system, which controls how the information from the observation influences the value of model grids nearby the observational position. A statistical harmonious correctionis given via the background error covariance matrix to the model grid nearby observational position in order to make sure the dynamical harmony of model variables. However, the background error covariance cannot be calculated accurately because the true situation of the atmosphere cannot be known. The observation method or Hollingsworth-Lonnberg (1986) method, NMC method (Parrish and Derber, 1992) and analysis ensemble method (Fisher, 2001) are used to solve this problem, which has been described in detail in the previous paper of Niu *et al.* (2008). As the SDS-DAS had been found to provide the reasonable corrections to both under- and over estimates of dust concentration forecast in MM5/CUACE-Dust (Zhou *et al.*, 2008), it is introduced in GRAPES/CUACE_Dust model to study the improvements of the modeling performance through dust radiation feedbacks.

3 The Impacts of Dust Radiation and Assimilation on Dust Modeling

In order to evaluate the impacts of dust radiation, assimilation as well as their composite impacts on model capability, the PM_{10} (particles with diameter less than 10 μm) forecast biases of CTL, 3DV and 3DV_RAD experiments are analyzed. The model biases of air temperature, wind speed, air pressure and geopotential height (GPH) of CTL, RAD and 3DV_RAD experiments are contrastively studied in the following sections. The spring 2006 is observed as one of the most frequent SDS season in the recent 10 years with the SDS-peak in April in Northeast Asia (Yang *et al.*, 2008). In the sensitivity modeling studies, the typical severe dust storms occurred in April 2006 over East Asia are discussed in the following sections.

3.1 The Impacts on Surface Dust Concentration Predictions

Using the PM_{10} observation data of six stations in CMA SDS monitoring network (Wang *et al.*, 2008), the daily averaged PM_{10} biases of CTL, 3DV and 3DV_RAD experiments from April 1 to 30, 2006 are calculated as displayed in Figure 1. The station identification number STID, latitude, longitude of the six stations are listed in Table 1. At Tazhong Station (Fig. 1a), two dust storms occurred on April 9—12 and April 23—25. It can be seen that the modeled PM_{10} bias of 3DV_RAD is smallest among the three experiments for the two dust events, which indicates effective reduction of model bias by the interaction of dust assimilation and radiation over Taklimakan desert. The modeled PM_{10} bias of 3DV experiment is also obviously smaller than that of CTL experiment. Four dust events were observed in Minqin on April 5, 9—11, 14—17 and 23—25 (Fig. 1b). The

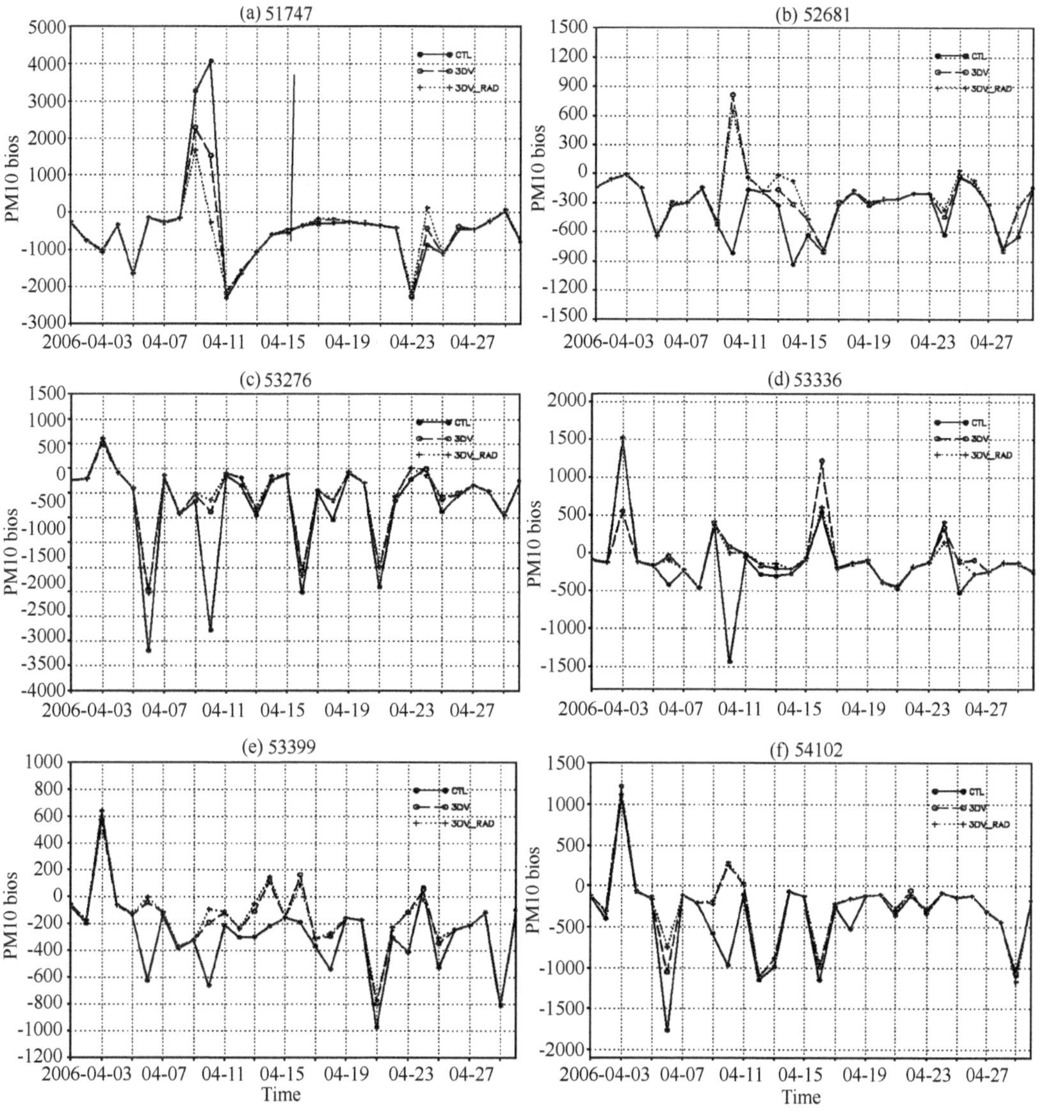

Fig. 1　Modeled daily averaged PM_{10} bias of CTL, 3DV and 3DV_RAD experiments ($\mu g \cdot m^{-3}$) April 2006 for (a) Tazhong, (b) Minqin, (c) Zhurihe, (d) Wulatezhongqi, (e) Zhangbei, and (f) Xilinhaote.

3DV_RAD experiment shows the best performance among the three experiments and exhibits better capability than 3DV experiment by correcting the positive bias of SDS concentrations occurred during April 9—11 and the negative bias during 14—17 and 23—25. For Wulatezhongqi Station (Fig. 1d), the 3DV experiment shows a better performance than the CTL by reducing its positive and negative PM_{10} bias and the 3DV_RAD experiment seems to reduce the PM_{10} bias of 3DV experiment slightly. At Zhurihe (Fig. 1c) and Zhangbei (Fig. 1e) Stations, The 3DV experiment shows the corrections of CTL experiment, but the 3DV_RAD experiment doesn't show obvious improvements further. For Xilinhaote Station (Fig. 1f), the 3DV experiment displays less PM_{10} bias than

CTL experiment generally and 3DV_RAD experiment shows a better performance for the strong dust storms occurred on April 5—7 than the 3DV experiment by decreasing the negative PM_{10} bias.

Table 1 Observation stations for the dust concentration comparisons

Name	STID	Lat(N)	Lon(E)	Representing deserts and regions
Tazhong	51747	39.00	83.67	Taklimakan desert
Minqin	52681	38.63	103.08	Tengger and Badanjinlin desert
Zhurihe	53276	42.40	112.90	middle, east of Inner Mongolia and Mongolia deserts
Wulatezhongqi	53336	41.57	108.52	sandy deserts over Sino-Mongolia border
Zhangbei	53339	41.25	114.70	Sourth of Hunshandake desert
Xilinhaote	54102	43.95	116.07	Hunshandake desert

Generally, it can be seen form the above discussion, the 3DV experiment shows obvious corrections of modeled 0—24 h averaged PM_{10} bias for the six sites, especially for the strong and severe dust storms. Comparing to the 3DV experiment, the 3DV_RAD experiment achieves more reasonable dust concentration predictions during strong dust storms at four sites. For weak dust events, the 3DV_RAD experiment doesn't show better capability of PM_{10} forecasting than the 3DV experiment. 3DV and 3DV_RAD experiments show the desirable revision on the dust concentrations during the severe dust storm of April 9—11 at most sites. This dust storm is selected as the case to study the impacts on meteorological fields due to dust data assimilation and radiation in the following sections.

3.2 The Impacts on Meteorological Fields

The 3DV alone has no influence on the model meteorological field predictions because there is no linkage between the improved dust concentrations and meteorological fields. The simulation results of CTL, RAD and 3DV_RAD experiments are performed to study the impacts of dust radiation and assimilation on meteorological fields.

Considering the geography locations of East Asian deserts and the main downwind regions affected by dust storms, nine surface meteorological stations (Table 2) representing Taklimakan, Mongolia, western Inner Mongolia, middle Inner Mongolia, east Inner Mongolia, east China, northeast China, Korea, and Japan are selected to study the impacts of dust radiation and assimilation on meteorological fields. The model bias of temperature at 2 meters above surface (T2m in Fig. 2), sea level pressure (Psl, Fig. 3) and surface wind speed (SWS, Fig. 4) of CTL, RAD and 3DV_RAD experiments of the nine sites are analyzed. It can be seen from Figure 2 that the RAD experiment presents a remarkable reduction of T2m errors compared with the CTL experiment due to the inclusion of the dust absorption and scattering on solar and infrared radiation interactively in the model. The positive T2m errors at nine sites (Fig. 2a—i) are almost all corrected to

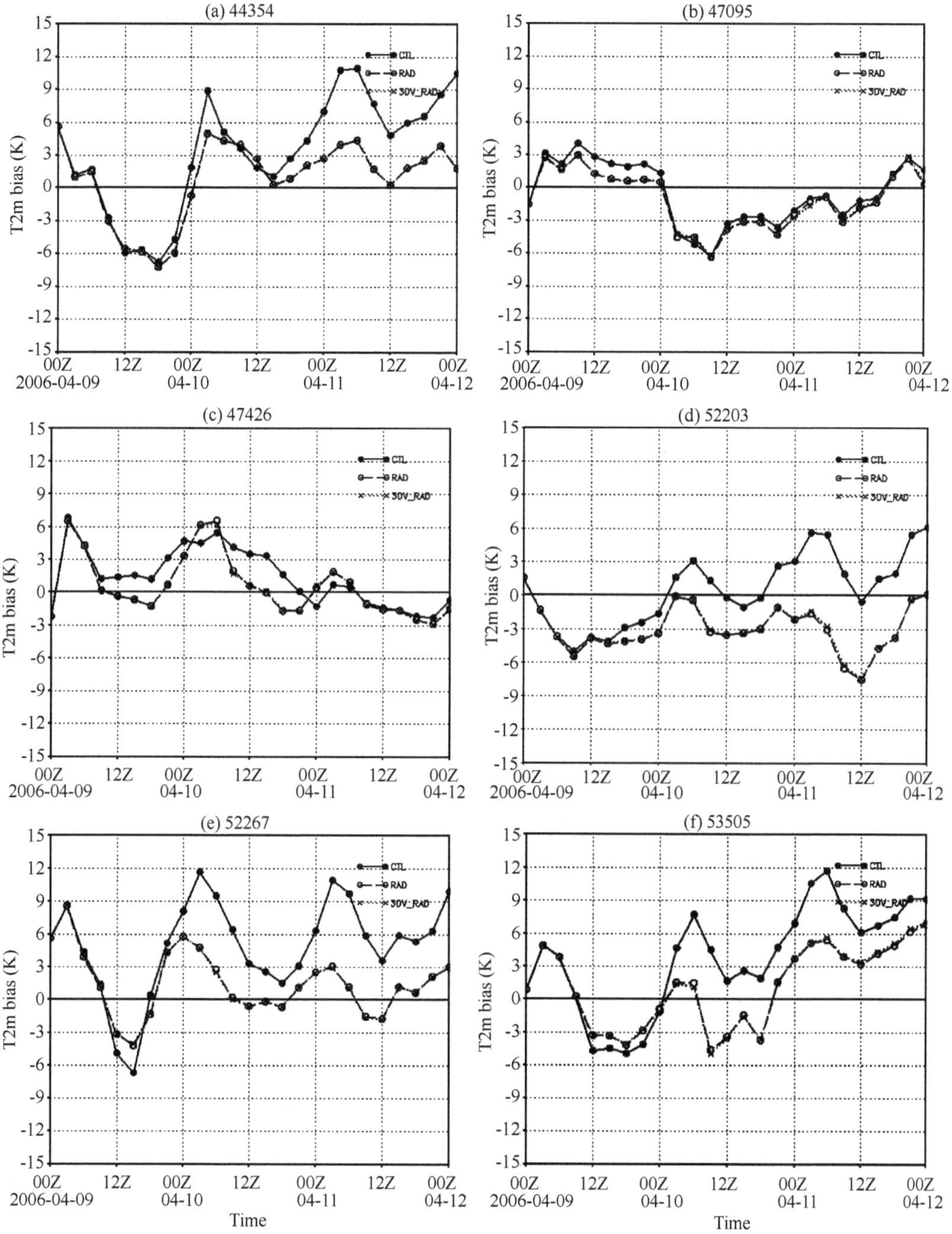
(a) 44354
(b) 47095
(c) 47426
(d) 52203
(e) 52267
(f) 53505
CTL
RAD
3DV_RAD
T2m bias (K)
15
12
9
6
3
0
-3
-6
-9
-12
-15
00Z
12Z
2006-04-09
04-10
04-11
04-12
Time

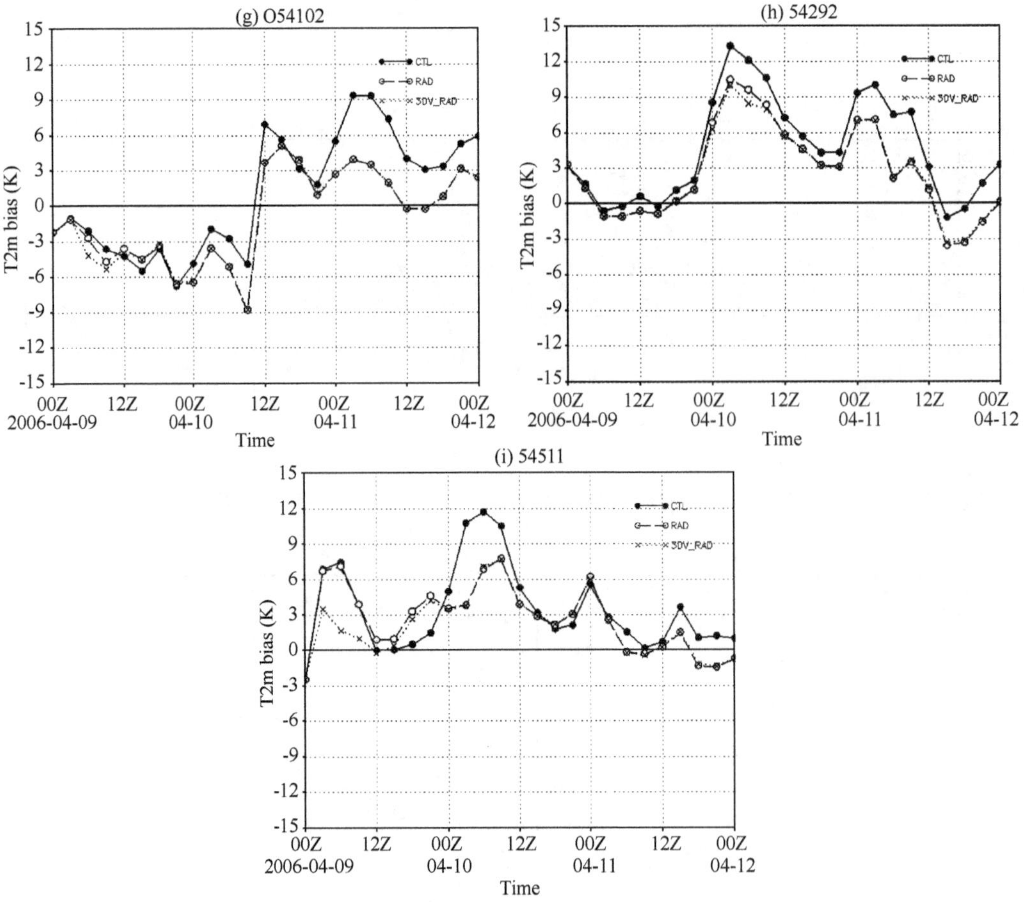

Fig. 2 T2m bias within 72 hours modeled by CTL, RAD, and RAD_3DV experiments of the 9 stations (a—i)

a certain degree during the dust storm events. At the sites located in the source and near downwind areas, such as for the stations with STID 44354 (a), 52313 (b), 52267 (c), 53505 (d), 54102 (e), 54511 (f) and 54292 (g) in Figure 2, the T2m error corrections reach as high as 5K or even more. The results are consistent with the previous study on the Asian dust storm of April 16-18, 2006 (Wang *et al.*, 2010). Even for the stations located as far as in Korea (STID: 47095)and Japan(STID: 47426), the corrections of T2m can be observed clearly for certain period of time. Furthermore, the RAD experiment also reduces the negative T2m errors from the CTL experiment during local night time (12 UTC April 9 to 00 UTC April 10)at sites 52267(c), 53305(d) and 54102(e) due to the online calculation of dust infrared radiative effects in this study, which was not be found in the previous study (Wang *et al.*, 2010). It should be attributed to the long wave radiation emitting by the dust layers near the surface, which may heat the earth surface and thereby lead to the increase of T2m during night time. This result indicates although the dust infrared radiative effect is smaller than the solar radiative effects during daytime, it is important during night time and may result in opposite impacts on modeling T2m. When the dust infrared radiation is calculated online, the negative bias of modeled T2m during night time can be corrected.

Table 2　Surface (S) and radiosonde (H) meteorology comparison stations

Name	STID	Lat(N)	Lon(E)	S/H	Representing deserts and regions
1a	52203	42.82	93.52	S,H	Taklimakan
1b	52313	41.53	94.66	S	Taklimakan
2	44354	44.90	110.12	S	Mongolia
3	52267	41.95	101.07	S	west Inner Mongolia
4	53505	39.08	105.38	S	middle Inner Mongolia
5	54102	43.95	116.07	S,H	east Inner Mongolia
6	54511	39.80	116.47	S,H	east China
7	54292	42.88	129.47	S,H	north China
8	47095	38.08	127.82	S	Korea
9	47426	42.17	142.78	S	Japan
10	44292	47.93	106.98	H	Mongolia
11	52652	38.93	100.43	H	west Inner Mongolia
12	53336	41.57	108.52	H	middle Inner Mongolia
13	47090	38.20	128.60	H	Korea
14	47590	38.27	140.90	H	Korea

Compared with the RAD experiment, the T2m predicted by the 3DV_RAD experiment doesn't exhibit the obvious improvement during the dust storms at the sites except for the site 54511(f) and 54292 (g). The assimilation scheme used in this study is done only once a day at 03 UTC in the model running. For the sites located in or near desert sources such as sites 44354 (a), 52313 (b), 52267 (c), 53505 (d), and 54102 (e), the modeled dust concentrations are mainly affected by the dust emission sources. It is easy for the data assimilation to revise the dust concentrations at 03 UTC, but it is hard to continuously correct the dust concentrations over the 72h forecasting time. Consequently, the 3DV_RAD does not further improve the T2m forecast at these stations. At the stations located in downwind area in Korea and Japan (h and i), the 3DV_RAD experiment also fails to forecast T2m more reasonably than the RAD experiment. T2m by the 3DV_RAD experiment shows better results than RAD experiment during 0—36 h forecasts at the stations located in downwind area as east and northeast China (f and g), especially within 0—12 h at 54511 station (f).

Figure 3 shows that the RAD experiment revises the Psl negative bias of CTL experiment evidently, which suggests that, corresponding with T2m decreasing, the RAD experiment increased Psl at all the nine sites (a—i) during almost the whole dust storm process with the more reasonable simulations. The RAD_3DV experiment doesn't show any further improvement over the RAD experiment. This may be because Psl as a relative stable physical parameter can not be changed so easily and rapidly with the instant changes of dust concentrations and air temperature by the SDS-DAS.

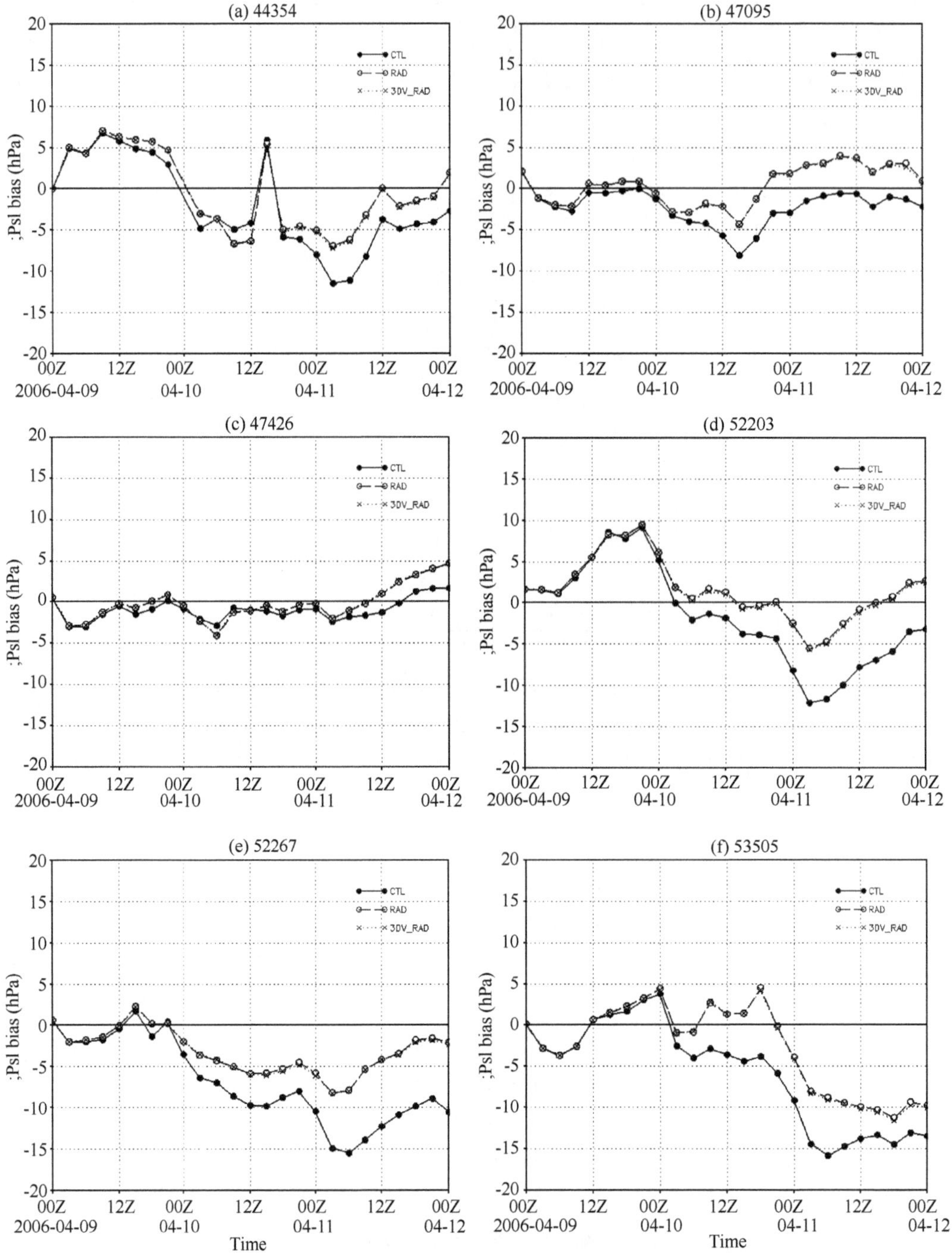

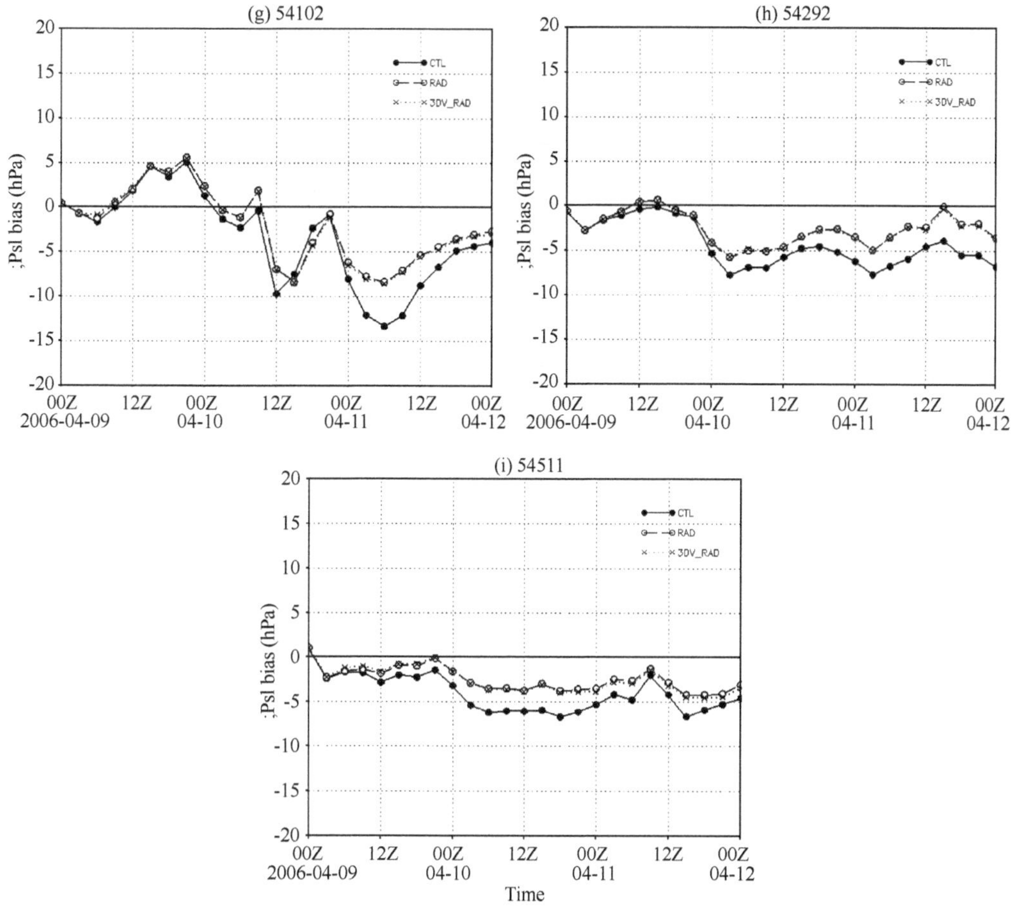

Fig. 3 Psl bias within 72 hours modeled by CTL, RAD, and RAD_3DV experiments of the 9 stations (a—i)

An accurate and quantitative prediction of SWS is difficult for most mesoscale models and this is also a big challenge for dust models. The changes of the SWS due to dust radiative effects eventually determine whether the radiative feedbacks on dust storm are positive or negative (Wang *et al*, 2010). Figure 4 displays the SWS errors for the CTL, RAD, and 3DV_RAD experiments. It is found that the RAD experiment obtains better results at almost all sites except for site 47095 (h). Comparing study of SWS errors of the CTL and RAD experiments during daytime (from 00 to 12 UTC) with that during night time (from 12 to 00 UTC), it is found that both the positive SWS errors during daytime and the negative SWS bias during night are corrected to some extent by the RAD experiment. This means that the dust solar radiation effects can decrease the SWS during daytime, suggesting negative radiative feedbacks on dust emission, while infrared effects can increase the SWS during night time, suggesting positive radiative feedbacks. The dust's infrared effects are usually concealed by its solar radiation effects. This can lead to a misunderstanding of dust radiative feedback on dust emission during night.

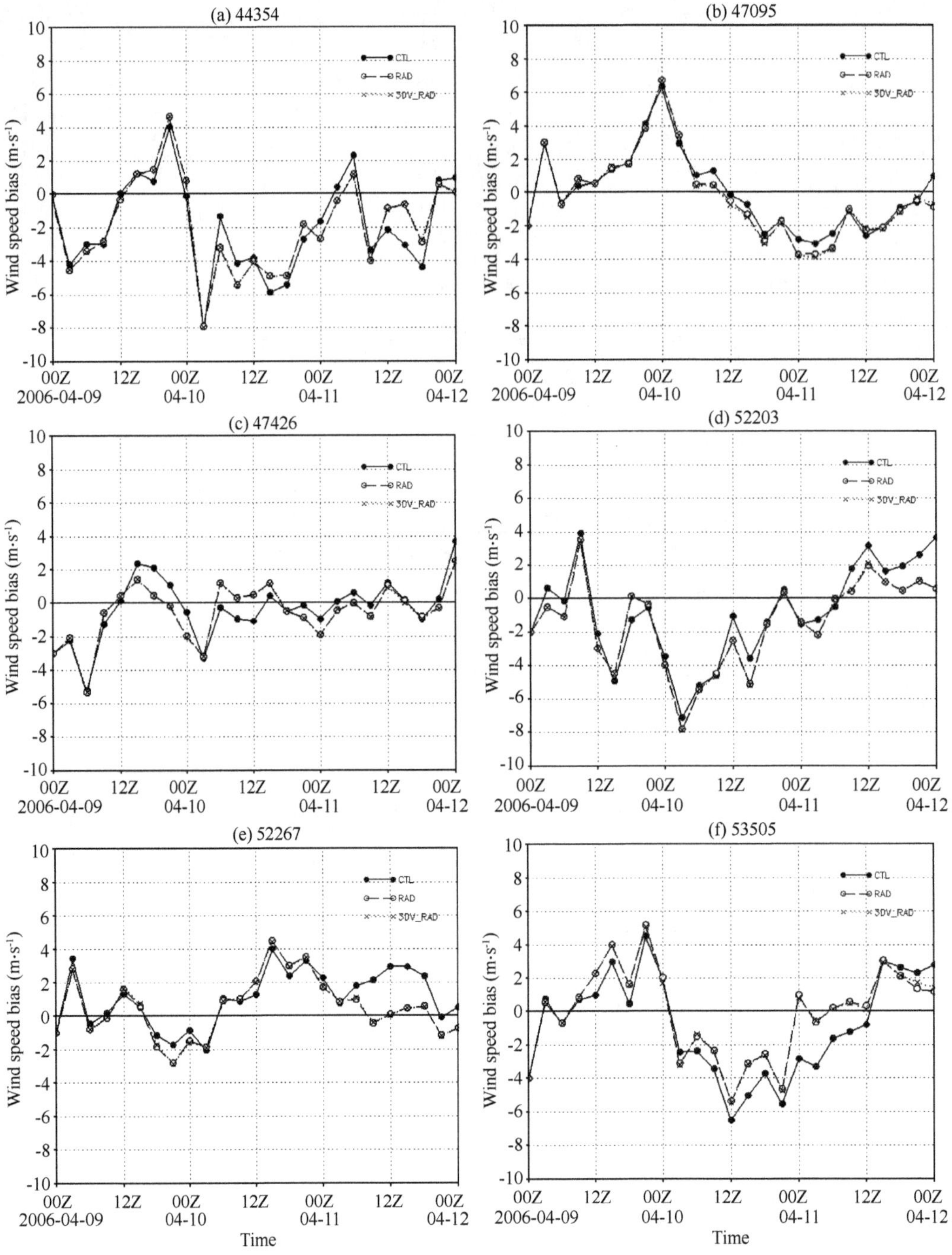

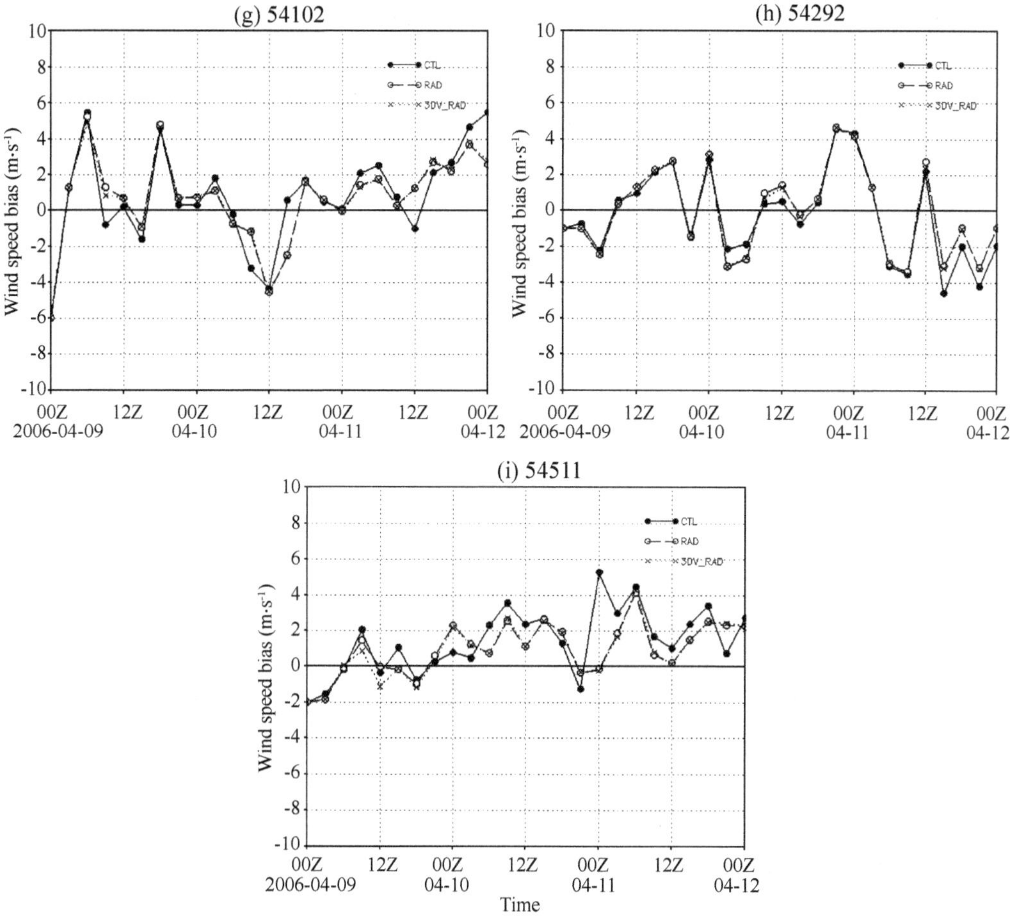

Fig. 4　Surface wind speed bias within 72 hours modeled by CTL, RAD, and RAD_3DV experiments of the 9 stations (a—i)

Generally, the bias analysis on T2m, Psl, and SWS from the CTL, RAD, and 3DV_RAD experiments shows that the dust radiation could continuously improve surface meteorological prediction with the more accurate simulations, but its improvement on surface meteorological fields are also influenced by the dynamical characteristics of these parameters themselves and the error structure. It should be also noted that the dust radiative feedbacks may be different or opposite during day and night time over East Asia. The data assimilation is not as effective as expected on the improving of surface meteorological field simulation, probably because it is difficult for the instant dust concentration corrections to revise surface meteorological forecast distinctly in the model. On the other hand, SDS-DAS's improvement on dust concentration is also restricted by both the dust emission processes of the model and the observation data used in the SDS-DAS.

3.3 The Impacts on Vertical Structures of Meteorological Fields

Dust radiation and SDS-DAS can modify the meteorological fields at surface and also their vertical structures. Previous studies (Pe′rez *et al.*, 2006; Wang *et al.*, 2010) indicated that the impacts of dust on the atmosphere following vertical altitudes are different or even opposite. As mentioned in Section 3.2., nine radiosonde meteorological stations (Table 2) are selected representing Taklimakan, Mongolia, western Inner Mongolia, middle Inner Mongolia, east Inner Mongolia, East China, Northeast China, Korea, and Japan to study the impacts of SDS-DAS and dust radiation on the air temperature, pressure and wind speed fields from surface to 100hPa. Figures 5, 6 and 7 present the vertical profiles of air temperature, geopotential height (GPH), and wind speed bias of CTL, RAD and 3DV_RAD experiments at the nine stations (a—i). At present, the regular meteorological radiosonde stations can only offer observation data at 00 and 12 UTC (08 am in the morning and 08 pm in the evening in local time) and the results of Figures 5—7 are all at these two moments of April 9—11. It can be seen from Figure 5 that the positive bias in upper air temperature and negative errors in middle and low air temperature of RAD experiment are obviously less than those of the CTL experiment. This means that the dust radiation cools the low and middle air but heats the upper air at the same time, which leads to better modeling of temperature from surface to 100 hPa. The altitudes where the dust radiation effects change from heating to cooling atmosphere alter from 300 hPa to 500 hPa at different sites. A further improvement of air temperature by 3DV_RAD experiment can not be seen as expected in Fig. 5. SDS-DAS introduced at 03 UTC weakens its effect on temperature after 12 hours (Fig. 2a). Therefore, neither 00 nor 12 UTC is the best time to show the impacts by 3DV. If the air temperature bias at 06 UTC or 09 UTC is available, possibly better revision on temperature by the 3DV_RAD than that by RAD experiment might be seen in Figure 5. GPH is usually used to describe the atmosphere pressure patterns at the vertical constant-pressure layers above ground for weather-prognostics and it is an important parameter describing atmospheric circulation. Figure 6 indicates that both GPH negative errors below 600—700 hPa and the positive errors above 600—700 hPa in the RAD experiment are less than that of CTL experiment. This suggests the low level air pressure might be increased by the dust radiation, as the upper air pressure might be decreased. The more reasonable GPH and air pressure prediction from ground to 100 hPa result from the online calculating of dust radiation. It is notable that 3DV_RAD experiment shows further reduction of GPH negative bias of middle and upper air (Figs. 6a—i). The changes are maintained in GPH during the modeling period because GPH is more stable compared to air temperature. The impact of dust radiation on wind is a little more complex than that on temperature and GPH. However, Figure 7 shows the positive biases of wind speed of upper air in the RAD are less than those in the CTL experiment at sites 44292(Fig. 7a), 52652(Fig. 7e), 54511(Fig. 7i), 54292

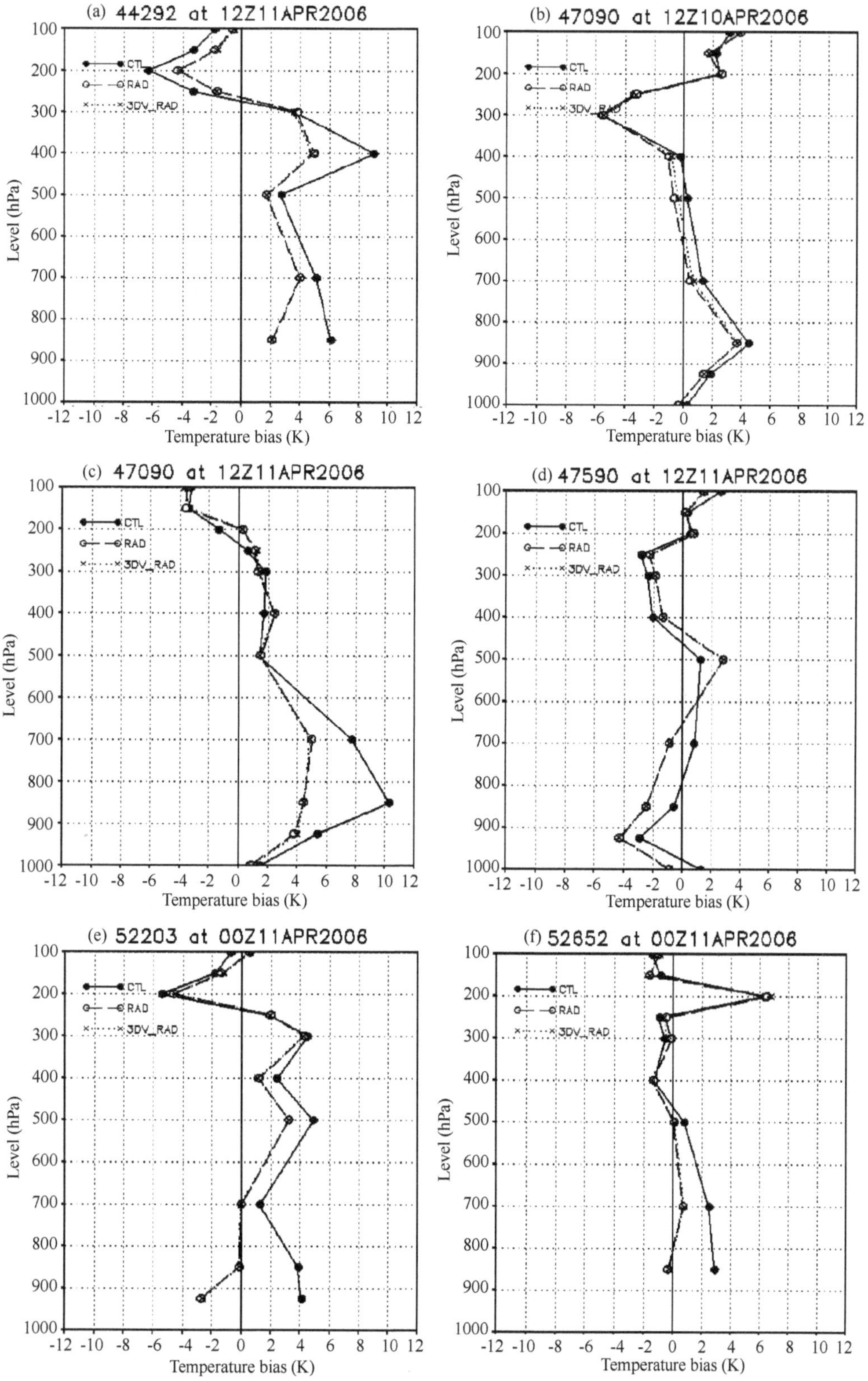
(a) 44292 at 12Z11APR2006
(b) 47090 at 12Z10APR2006
(c) 47090 at 12Z11APR2006
(d) 47590 at 12Z11APR2006
(e) 52203 at 00Z11APR2006
(f) 52652 at 00Z11APR2006
CTL
RAD
3DV_RAD
Level (hPa)
Temperature bias (K)

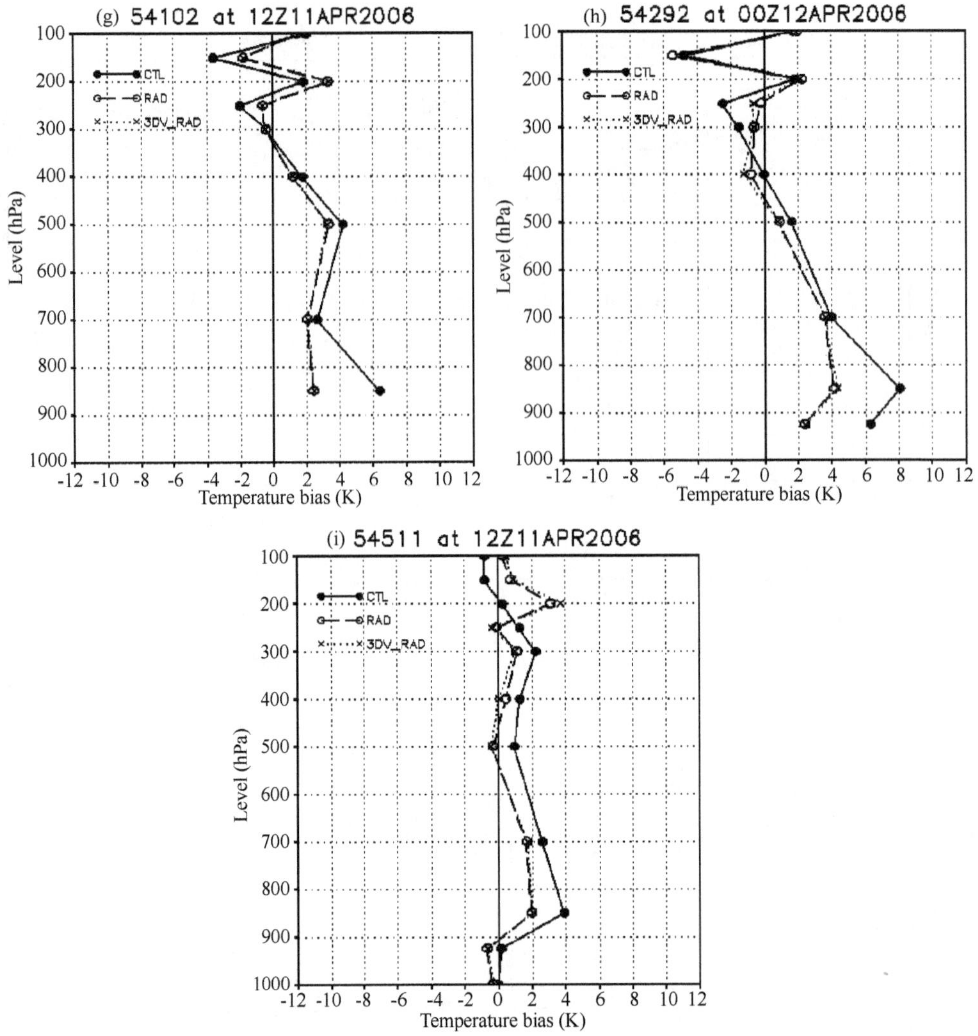

Fig. 5 Vertical profiles of temperature bias modeled by CTL, RAD and RAD_3DV experiment from surface to 100 hPa (a—i)

(Fig. 7h) and 47090(Fig. 7b), indicating a decrease of wind speed by dust radiation. The negative bias of wind speed in upper air is also decreased by dust radiation at other sites such as 53336 (Fig. 7f) and 47590 (Fig. 7c). The RAD experiment provides an improvement on the wind speed forecast in upper air. The 3DV_RAD experiment shows a better performance on reducing the wind speed bias in upper air than RAD experiment at sites 44292 (Fig. 7a), 52652 (Fig. 7e), 54511 (Fig. 7i), 54292 (Fig. 7h), and 47090 (Fig. 7b).

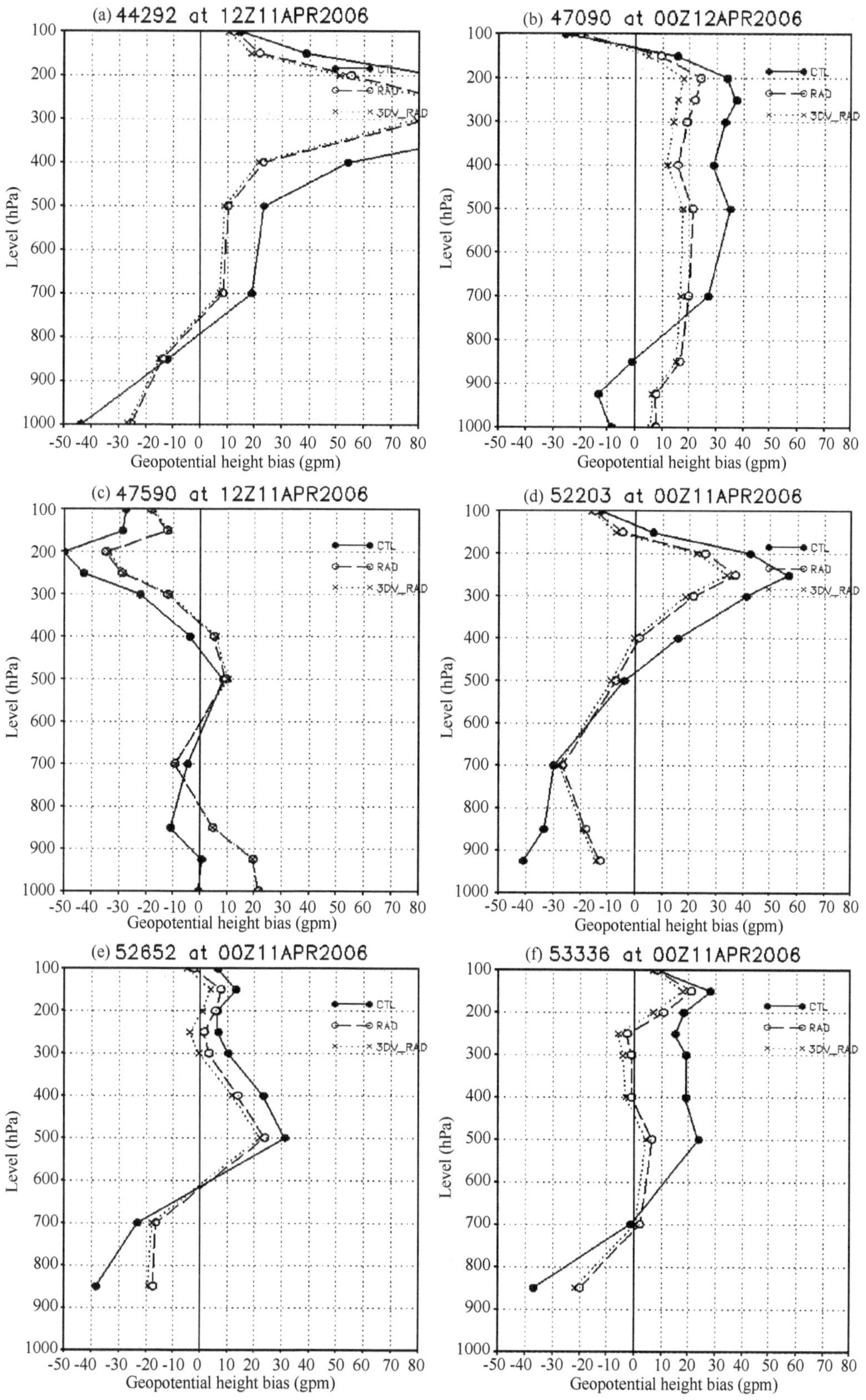
(a) 44292 at 12Z11APR2006
(b) 47090 at 00Z12APR2006
(c) 47590 at 12Z11APR2006
(d) 52203 at 00Z11APR2006
(e) 52652 at 00Z11APR2006
(f) 53336 at 00Z11APR2006
CTL
RAD
3DV_RAD
Level (hPa)
Geopotential height bias (gpm)
100
200
300
400
500
600
700
800
900
1000
-50 -40 -30 -20 -10 0 10 20 30 40 50 60 70 80

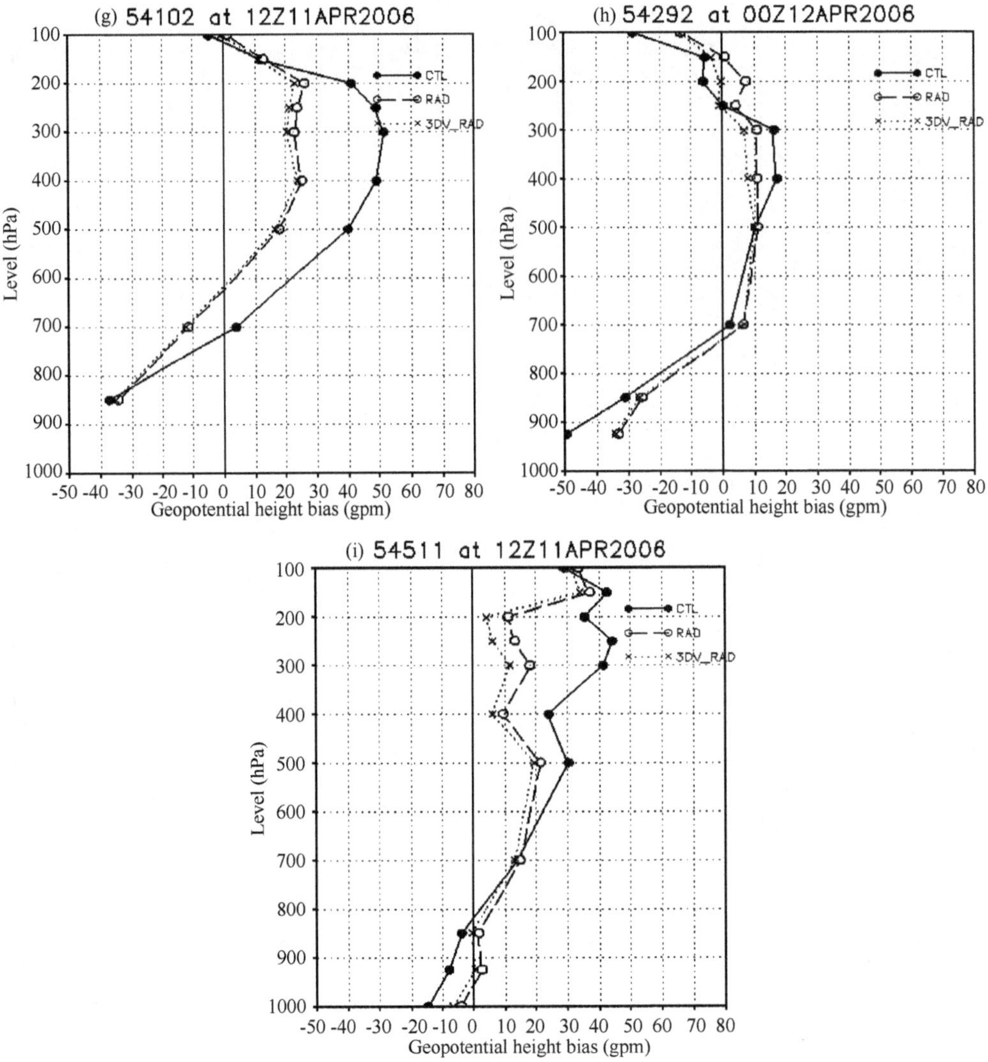

Fig. 6 Vertical profiles of geopotential height bias modeled by CTL, RAD and RAD_3DV experiments from surface to 100 hPa (a—i)

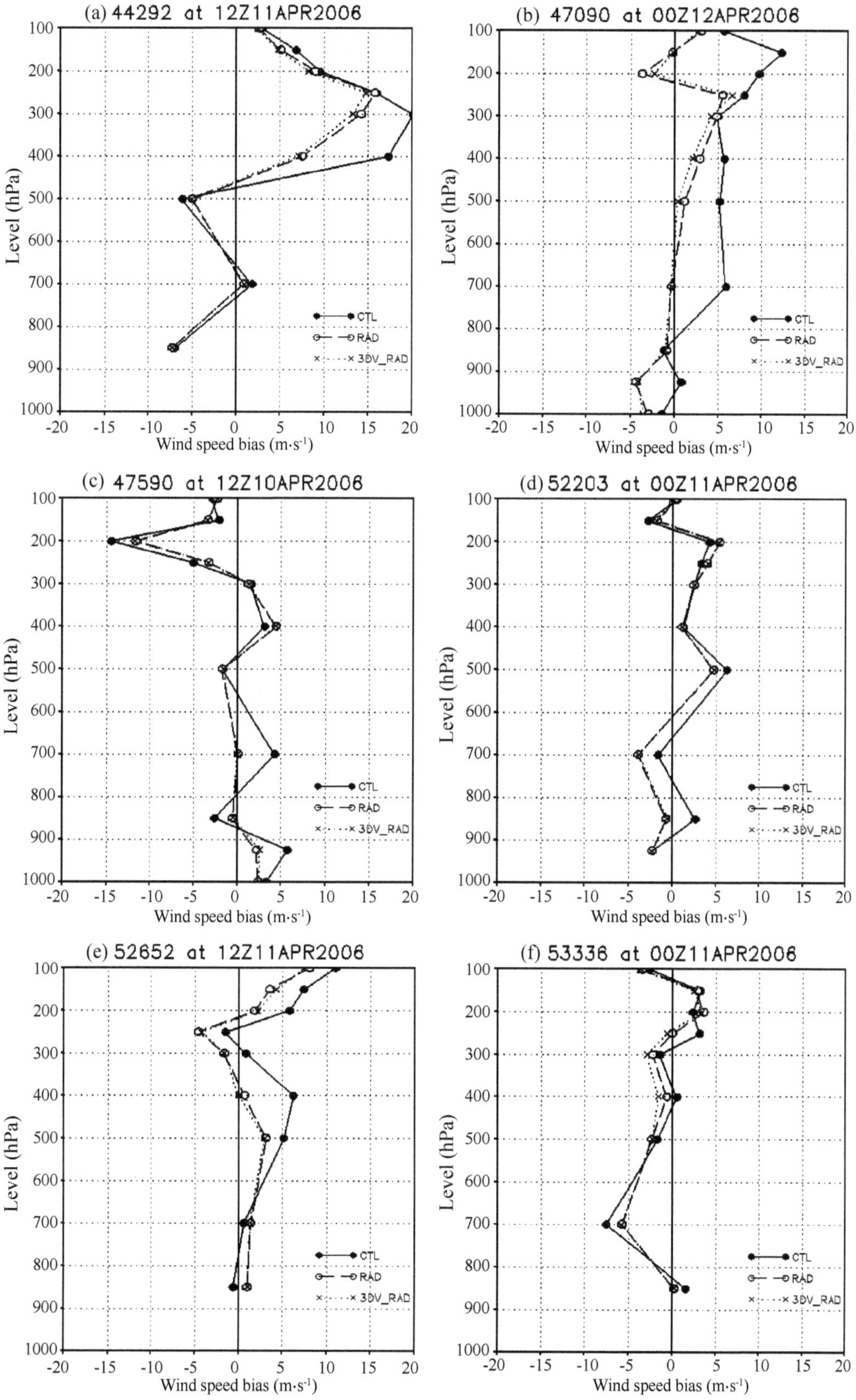
(a) 44292 at 12Z11APR2006
(b) 47090 at 00Z12APR2006
(c) 47590 at 12Z10APR2006
(d) 52203 at 00Z11APR2006
(e) 52652 at 12Z11APR2006
(f) 53336 at 00Z11APR2006
Level (hPa)
Wind speed bias (m·s-1)
CTL
RAD
3DV_RAD

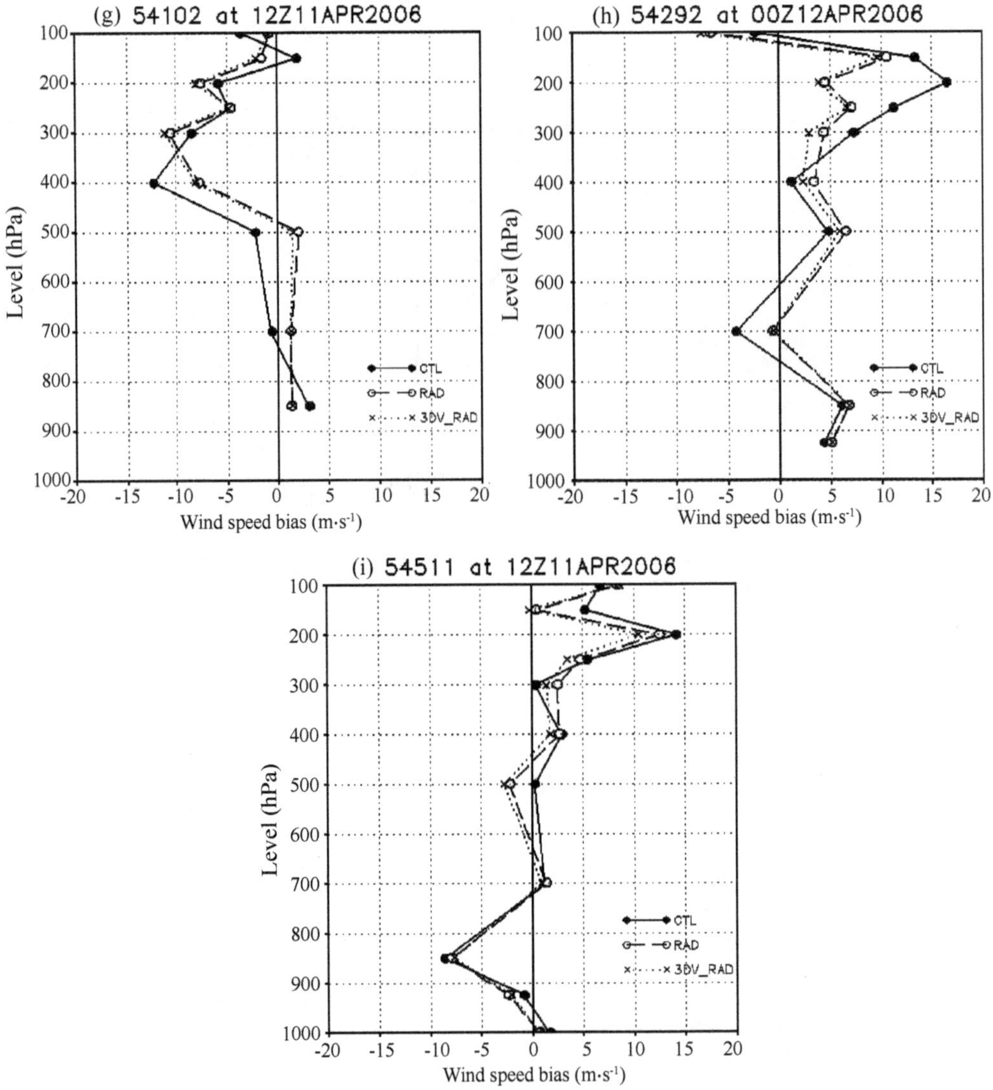

Fig. 7 Vertical profiles of wind speed bias modeled by CTL, RAD and RAD_3DV experiments from surface to 100 hPa (a—i)

4 Concluding Remarks

Dust direct radiation and data assimilation are online integrated into the mesoscale dust model GRAPES/CUACE_Dust to study their impacts on the regional circulation in East Asia during spring time dust storms with four sensitive experiments, i. e. CTL, RAD, 3DV and 3DV_RAD. Comparing study of modeled daily averaged PM_{10} bias of CTL, 3DV and 3DV_RAD experiments in April 2006 shows that the 3DV experiment can improve modeling dust concentration distinctly and the 3DV_RAD experiment may further correct the modeling capacity when strong dust storms occurred. A case study on the dust storm during April 9－11 2006 shows dust radiation cools the lower atmosphere and heats the upper atmosphere, which leads to an increase of air pressure in lower atmosphere and decrease of air pressure in upper atmosphere. The impacts of dust radiation also leads to the continuous improvement in the surface air temperature and pressure prediction during the 72 h modeling interval, and the surface wind speed is also corrected to some extent. It is noteworthy that dust radiation may decrease the surface temperature during daytime while increase it during night time at some stations, which is associated with the surface wind speed drop at daytime and rise at night time. This indicates a negative feedback during daytime and positive feedback during night on dust emission. Based on the RAD experiment, the 3DV_RAD experiment further improves the air temperature and pressure simulations over 12 hours after the dust data assimilation at certain stations. 3DV_RAD experiment seems to be more efficiency in enhancing model prediction ability in upper air meteorological fields than these in lower and surface layers.

More detailed study and cases are needed to further illustrate the different radiative feedbacks of dust aerosols between day and night, which requires observational data at a specific time, i. e. at noon, for the comparisons. This kind of data could be also used to further evaluate the influence of 3DV_RAD on the model forecasting ability. The responses of different meteorological fields to the interaction of dust radiation and assimilation should be studied based on their dynamic and physical characteristics in future work.

Acknowledgement

The authors wish to thank for the financial supports from the National Basic Research Program (973) (2011CB403404), the National Natural Science Foundation of China (Grant Nos. 41130104), Ministry of Science and Technology of China(2010DFA22770), the open project of Key Laboratory of Semi-Arid Climate Change, Lanzhou University, Ministry of Education, China and CAMS key project (2009z001).

References

Ahn H J, Park S U, Chang L S, 2007. Effect of direct radiative forcing of Asian dust on the meteorological fields in East Asia during an Asian dust event period. *J. Appl. Meteorol. Climatol.* **46**:1655-1681.

Barker D M, Huang W, Guo Y R, *et al*. 2004. A three-dimensional variational data assimilation system for MM5: Implementation and initial results. *Mon. Weather Rev.* **132**: 197-914.

Chou M D, Suarez M J. 1999. A solar radiation parameterization for atmospheric studies. NASA Tech. Memo. 10460, 15, http:// climate. gsfc. nasa. gov/chou.

Fisher M. 2001. Assimilation techniques (1), 3D-Var, ECMWF meteorological training course lecture series, 220 pp.

Gong S L, Zhang X Y, Zhao T L, *et al*. 2003. Characterization of soil dust aerosol in China and its transport and distribution during 2001 ACE-Asia: 2. Model simulation and validation. *J. Geophys Res.* **108**(D9), 4262, doi: 10.1029/2002JD002633.

Gong S L, Zhang X Y. 2008. CUACE/Dust—an integrated system of observation and modeling systems for operational dust forecasting in Asia. *Atmos. Chem. Phys.* **8**: 2,333-82,340.

Haywood J M, Allan R P, Culverwell I, *et al*. 2005. Can desert dust explain the outgoing longwave radiation anomaly over the Sahara during July 2003?. *J. Geophys. Res.* **110**(D05105), doi:10.1029/2004JD005232.

Heinold B, Tegen I, Schepanski K, *et al*. 2008. Dust radiative feedback on Saharan boundary layer dynamics and dust mobilization. *Geophys. Res. Lett.* **35**: L20817, doi:10.1029/2008GL035319.

Hollingsworth A, Lonnberg P. 1986. The statistical structure of short-range forecast errors as determined from radio sonde data. Part I: the wind field. *Tellus*. 111-136.

Hu X Q, Lu N M, Niu T, *et al*. 2008. Operational Retrievalof Asian Dust Storm from FY-2C Geostationary Meteorological Satellite and its Application to real time Forecast in Asia. *Atmos. Chem. Phys.* **8**: 1649-1659.

Kotamarthi V R, Carmichael G R. 1993. A modeling study of the long range transport of Kosa using particles trajectory methods. *Tellus*. Ser. B, **45**: 426-441.

Liu Y, Zhou M Y. 1997. A numerical simulatioin of dust storm and its long range transport process. *J. Nanjing Inst. of Meteorol*. **20**: 511-517.

Niu T, Gong S L, Zhu G F, *et al*. 2008. Data assimilation of dust aerosol observations for CUACE/Dust forecasting system. *Atmos. Chem. Phys.* **8**: 3473-3482.

Parrish D F, Derber J D. 1992. The national meteorological center spectral statistical interpolation analysis system. *Mon. Weather. Rev.* **120**: 1747-1763.

Perez C, Nickovic S, Pejanovic G, *et al*. 2006. Interactive dust-radiation modeling: A step to improve weather forecast. *J. Geophy. Res.* **111**:(D16206), doi:10.1029/2005JD006717.

Shao Y. 2001. A model for mineral dust emission. *J. Geophys. Res.* **106**: 20239-20254.

Shao Y. 2004. Simplification of a dust emission scheme and comparison with data. *J. Geophy. Res.* **109**: D10202, doi:10.1029/2003JD004372.

Uno I, Hiroyasu A, Emori S, *et al*. 2001. Trans-Pacific yellow sand transport observed in April 1998: A numerical simulation. *J. Geophys Res*. **106**: D22, 18331-18344.

UnoI, Satake S, Gregory R C, *et al*. 2004. Numerical study of Asian dust transport during the spring time of 2001 simulated with the Chemical Weather Forecasting System (CFORS) model. *J. Geophys. Res.* **109**: D19S24, doi:10.1029/2003JD004222.

Uno I, Wang Z, Chiba M, *et al*. 2006. Dust model intercomparison (dmip) study over Asia: Overview. *J. Geophys. Res.*, **111**: D12213, doi:12210.11029/12005JD006575.

Wang H, Shi G Y, Li W, Wang B. 2006. The impacts of optical properties on radiative forcing due to dust aerosol. *Adv. Atmos. Sci.* **23**(2), 431-441.

Wang H, Gong S L, Zhang H L, *et al*. 2009. A new-generation sand and dust storm forecasting system GRAPES/CUACE_Dust: Model development, verification and numerical simulation. *Chin. Sci. Bull*, doi. **10**:1007/s11434-009-0481-z.

Wang H, Zhang X Y, Gong S L, *et al*. 2010. Radiative feedback of dust aerosols on the East Asian dust storms. *J. Geophys. Res.* **115**: D23214, doi:10.1029/2009JD013430.

Wang Y Q, Zhang X Y, Gong S L, *et al*. 2008. Surface observation of sandand dust storm in east Asia and its application in CUACE/Dust forecasting system. *Atmos. Chem. Phys.* **8**: 545-553.

Wang Z F, Hiromasa U, Huang M Y. 2000. A deflatio in module for use in modeling long-rang transport of yellow sand over East Asia. *J. Geophys. Res.* **105**: D22, 26947-26959.

Xiao H, Carmichael G R, Durchenwald J. 1997. Long-range transport of Sox and dust in East Asia during the PEMB experiment, *J. Geophys. Res.* **102**: 28589-28612.

Yang Y Q, Hou Q, Zhou C H, *et al*. 2008. A study on sand/dust storms over northeast Asia and associated large-scale circulations in spring 2006. *Atmos. Chem. Phys.* **8**: 25-33.

Zhang H, Xue J S, Zhuang S Y, *et al*. 2004. Idea experiments of GRAPES three dimensional variational data assimilation system. *Acta. Meteor. Sinica.* **62**: 31-41(in Chinese).

Zhang X Y, Gong S L, Zhao T L, *et al*., 2003. Sources of Asian dust and role of climate change versus desertification Asian dust emission, *J. Geophys. Res.* **30**(24),2272, doi:10.1029/2003 GL018206, 20.

Zhou C H, Gong S L, Zhang X Y, *et al*. 2008. Development and Evaluation of an Operational SDS Forecasting System for EastAsia: CUACE/Dust, *Atmos. Chem. Phys.* **8**: 787-798.

Zhuang S Y, Xue J S, Zhu G F, *et al*. 2005. GRAPES global 3D-Var system-Basic scheme design and single observation test. *Chinese J. Atmo. Sci.* **29**: 872-884.

VARIATION OF AEROSOL OPTICAL PROPERTIES OVER TAKLIMAKAN DESERT OF CHINA①

CHE Huizheng[1②], WANG Yaqiang[1], SUN Junying[1], ZHANG Xiaochun[2], ZHANG Xiaoye[1] and GUO Jianping[1]

[1)] Key Laboratory of Atmospheric Chemistry (LAC), Chinese Academy of Meteorological Sciences (CAMS), CMA, Beijing 100081, China

[2)] National Meteorological Center, CMA, Beijing 100081, China

Abstract

Aerosol optical properties at the center of Taklimakan Desert in Northwest China are investigated based on the measurements of aerosol optical depth (AOD) and Angstrom exponent from 2004 to 2008. A seasonal variation is found with high AOD and low Angstrom exponent values in spring and summer due to the effect of dust storm events, and low AOD in autumn and winter. The maximum and minimum AOD occur in April (0.83±0.41) and November (0.19±0.10), respectively, with the maximum and minimum Angstrom exponent in January (0.70±0.25) and May (0.09±0.06), respectively. The diurnal variation of AOD (Angstrom exponent) shows the characteristic of high (low) values about 0.50—0.60 in the morning and evening and constant around 0.40 during daytime. Relationship between AOD and Angstrom exponent can be fitted by a power equation with a R^2 of 0.55. The frequency distributions of AOD and Angstrom exponent occurrence probability show a single peak distribution characteristic and can be well fitted by a two-mode distribution.

Key words: aerosol optical depth (AOD); Angstrom exponent; Taklimakan Desert; China

1 Introduction

Aerosol particles can influence the radiative energy and the conversion of water vapor into cloud droplets through direct and indirect effects (Hansen *et al.*, 2000). Many studies addressed that aerosol optical property is one of the largest sources of uncertainty in the current estimating climate forcing (Ramanathan *et al.*, 2001).

Dust aerosols from arid and semi-arid regions can be transported thousands of kilometers far from their original resource regions (Wang *et al.*, 2001; Gong *et al.*, 2003; Zhang *et al.*, 2003a). It was estimated that the dust emission is of the order of 1500 Tg/yr

① The paper published in *Aerosol and Air Quality Research*, **13**: 777-785. 2013.

② Corresponding author. Tel.: 86-10-5899-3116; Fax: 86-10-6217-6414; E-mail: chehz@cams.cma.gov.cn globally (Tegen and Fung, 1995). The emission from East Asia is about 800 Tg/yr with half of them deposited back to the source and adjacent regions (Zhang *et al.*, 1997). The dust aerosol particles have great effect on global and regional climate change (Li *et al.*, 1996; Mikami *et al.* 2006).

Despite many dust aerosol studies, the optical properties are still far from being sufficient (Sokolik and Toon, 1999). In recent years, there were many studies on the optical properties of dust aerosols in Sahara desert (Zakey *et al.*, 2004), West Asia (Nakajima *et al.*, 1996; Smirnov *et al.*, 2002), India (Dey *et al.*, 2004), Australia (Kalashnikova *et al.*, 2007), and East Asia (Kim *et al.*, 2004, Eck *et al.*, 2005). As far as East Asia was concerned, arid and semi-arid regions in Western and Northern China is one of the major source regions of dust aerosols (Zhang *et al.*, 2003b, Zhang *et al.*, 2012). Some scientists have begun to investigate the optical properties (e. g., Alfaro *et al.*, 2003; Xia *et al.*, 2005; Cheng *et al.*, 2006, Gai *et al.*, 2006, Che *et al.*, 2009; Wu *et al.*, 2012) and radiative forcing (Huang *et al.*, 2009; Xia and Zong, 2009) of dust aerosols in this area. These studies are very important to understand the essential properties and variations of the dust aerosols in East Asia.

The aim of this work is to study the dust climatological aerosol optical properties in the center of Taklimakan Desert of Western China during 2004 to 2008, which will benefit the estimation of the effect of East Asian dust aerosols on global and regional climate change in future.

2 Measurement and Data

The research site of Tazhong (39°00′, 83°40′, 1099. 3 m) is located in the center of the Taklimakan Desert, which is known as one of the largest sandy deserts in the world. Taklimakan Desert covers an area of 270000 km^2 with 1000 km long and 400 km wide which is regarded as one of the largest resources of Asian aeolian dust aerosol particles (Mikami *et al.*, 2006; Huang *et al.*, 2009). Tazhong site is the only meteorological observatory in the world located in 229 km deep of desert hinterland (Lu *et al.*, 2010). There are few anthropogenic activities surrounding the observation site. Annual precipitation at Tazhong is just 25. 9 mm. Dust events happen more than 500 times annually (Li *et al.*, 2006). The aerosol measurements at Tazhong site could represent the characteristic of Taklimakan Desert.

A Cimel 318 sunphotometer was installed at Tazhong from 2004 and has been running at this site continuously. The sunphotometer makes direct spectral solar radiation measurements within a 1. 2° full field-of-view around 15 minutes at 4 normal bands (440, 675, 870, and 1020 nm), 3 polarization bands at 870nm and 1 water vapor band at 940 nm. Measurements at 440, 675, 870, and 1020 nm are used to calculate the aerosol optical depth (AOD) (Holben *et al.*, 1998; Eck *et al.* 2005). The signals are measured by the instrument three times at one scenario. The error of these three measurements is about 1%—2% at different channels, which cause the error of retrieved AOD is about 0.01—0.02. Thus the total uncertainty is about 0.01 to 0.02 (Eck *et al*. 1999).

The sunphotometer is calibrated by using CARSNET (CMA Aerosol Remote Sensing NETwork) reference instrument annually to make sure the accuracy and reliability of the measurement data. The reference instrument has been calibrated in Izana, Spain (the WMO-GAW station) by using Langley calibration method, which follows the AERONET protocol. The inter-comparison calibration protocol has been given by Che *et al*. (2009b). During the inter-calibration process, measurements from 2:00 AM to 6:00 AM (UTC) on the clear days with AOD at 500 nm less than 0.20 were used. The interval of the measurements between the reference instrument and the instrument to be calibrated was defined as less than 10 seconds. The AOD difference between the reference instrument and the recalibrated instrument should be less than 0.01.

The AOD data are calculated by using the ASTP win software (Cimel Ltd. Co.) for Level 1.0 AOD (raw result without cloud-screening), Level 1.5 AOD (cloud-screened AOD based on Smirnov *et al*., 2000) and Angstrom Exponent between 440 to 870 nm. To make sure the data quality more accurately, all the data were checked manually site by site and unreasonable data were deleted. e.g., some exceptional large AOD points were usually caused by the cloud accumulation after checking the MODIS Level-1B granule (MOD02_1km) images (http://modis-atmos.gsfc.nasa.gov/IMAGES/index_mod021km.html). Furthermore, daily averaged AOD were computed and those data with measurements less than 10 times in a day were eliminated. Daily and monthly mean values of AOD and Angstrom exponent were investigated by statistical analysis to characterize the aerosol columnar properties.

3 Result and Analysis

Frequency Distribution of AOD and Angstrom Exponent

Frequency histogram of AOD at Tazhong is shown in Fig. 1 for all the instantaneous data. There is one peak distribution composed for the AOD occurrence frequency. The accumulated frequency in the range of 0.20 to 0.50 is about 57%. The frequency distribution of AOD can be well fitted ($r^2 = 0.95$) by a bi-mode normal distribution centered about 0.23 and 0.50 with the standard deviation of 0.006 and 0.06, respectively (Fig. 1). The equation could be expressed as following:

$$Y = \left(\frac{1.96}{0.16} \times \sqrt{\frac{\pi}{2}}\right) \times \exp\left(-2 \times \left(\frac{x-0.23}{0.16}\right)^2\right) + \left(\frac{2.79}{0.50} \times \sqrt{\frac{\pi}{2}}\right) \times \exp\left(-2 \times \left(\frac{x-0.50}{0.50}\right)^2\right) \tag{1}$$

O'neill *et al*. (2000) suggested that multiple peaks could usually reveal the presence of

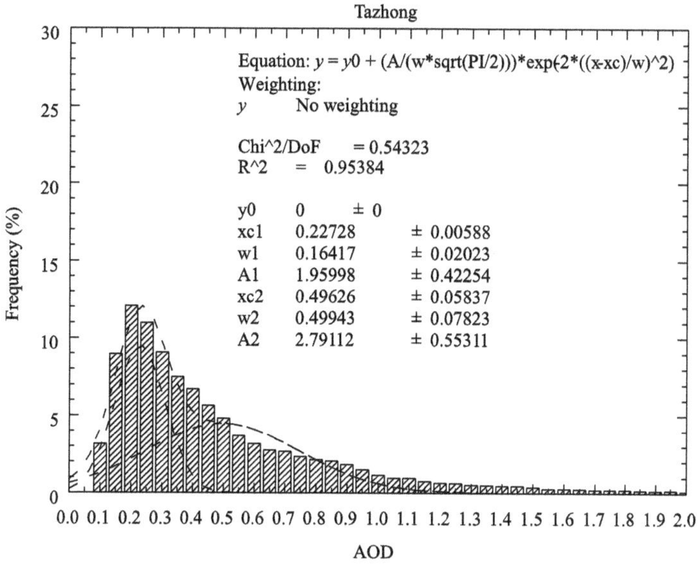

Fig. 1　Frequency of occurrences of AOD at 440nm at Tazhong. The dash lines mean the Gauss fitting curves.

different aerosol populations and types. The mode centered −0.23 probably corresponds to the non-dust atmospheric conditions at Tazhong and the mode centered −0.50 to the high mineral dust burden in atmosphere, such as Asian dust from deserts in spring and early summer.

Frequency histogram of Angstrom exponent at Tazhong is shown in Fig. 2 for all the instantaneous data. The probability distribution of Angstrom exponent is similar to that of

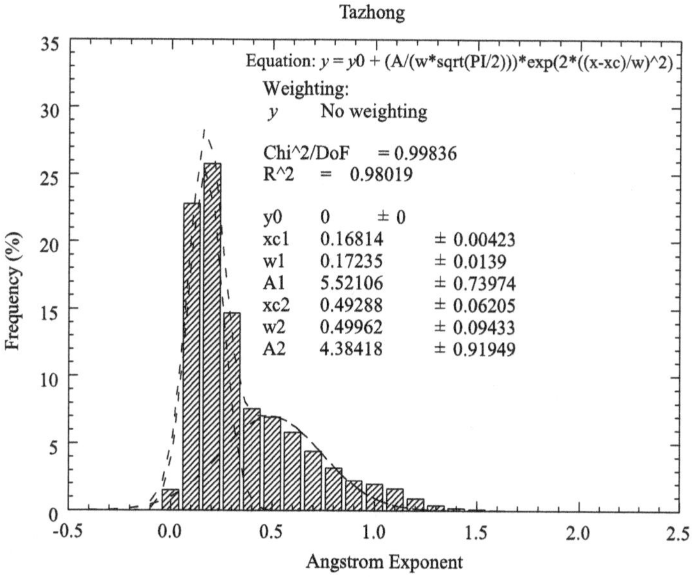

Fig. 2　Frequency of occurrences of Angstrom exponent between 440nm and 870nm at Tazhong. The dash lines mean the Gauss fitting curves.

AOD. There is one peak distribution for the Angstrom exponent. The Angstrom exponent frequency distribution can be well fitted ($r^2 = 0.98$) by a bi-modal normal distribution centered about 0.17 and 0.50 with standard deviations of 0.004 and 0.06, respectively (Fig. 2). The equation could be expressed as following:

$$Y = \left(\frac{5.52}{0.17} \times \sqrt{\frac{\pi}{2}}\right) \times \exp\left(-2 \times \left(\frac{x - 0.17}{0.17}\right)^2\right) + \left(\frac{4.38}{0.50} \times \sqrt{\frac{\pi}{2}}\right) \times \exp\left(-2 \times \left(\frac{x - 0.49}{0.50}\right)^2\right) \tag{2}$$

The first mode includes more than half of the data and corresponds to coarse particles which are usually associated with sand storm events. Since Tazhong is at the middle of Taklimakan Desert, plenty of coarse aerosol particles could be emitted in atmosphere during the sand storm events with strong wind at surface land. The second mode corresponds to aerosols whose size distribution is also dominated by coarse particles, reflecting the effect of floating dust or dust blowing events occurring at Taklimakan Desert. One can also found that there are a few cases of Angstrom exponent larger than 1.0 in the frequency distribution. This may reflect the effect of anthropogenic activities. It has been proved that the fine particles (such as black carbon) could contribute to the composition of aerosol in Taklimakan desert. These anthropogenic particles emitted specially by coal combustion could be transported from north and south part of Xinjiang region to the middle of Taklimakan desert (Li *et al.*, 2010).

Seasonal Variation of AOD and Angstrom Exponent

Fig. 3 illustrates the seasonal variation of AOD at Tazhong. In general, the mean AOD values in spring and summer are larger than those in autumn and winter. The AODs in spring and summer are about 0.75 and 0.65, respectively. In contrast, the mean AODs in autumn and winter are lower than 0.30. The 75th percentile AOD values are about 0.93, 0.84, 0.30, and 0.32 in spring, summer, autumn, and winter, respectively. High AODs in spring and summer reflect the contribution of dust events. Dust events are very frequent during spring and early summer, which causes large aerosol loading in atmosphere over Taklimakan Desert (Xue *et al.*, 2009). During autumn and winter period, there were few dust events. Although some anthropogenic activities could have effect on aerosol particles of Tazhong (Li *et al.*, 2010), the anthropogenic aerosol particles are mainly transported from outside resources. Comparing with mineral dust during spring and summer, the anthropogenic aerosol burden in autumn and winter is much less.

Eck *et al.* (2005) addressed that Angstrom exponent less than 0.80 mean coarse mode dominated aerosol cases. The 75th percentile Angstrom exponent value at Tazhong is less than 0.80 all the year (Fig. 4), which suggests coarse mode aerosol strongly dominated the AOD in Taklimakan Desert regions of China. The seasonal variation of Angstrom exponent at Tazhong shows the characteristic of small values (−0.15) in spring and summer but larger values in autumn (−0.36) and winter (−0.55), which suggests the aerosol

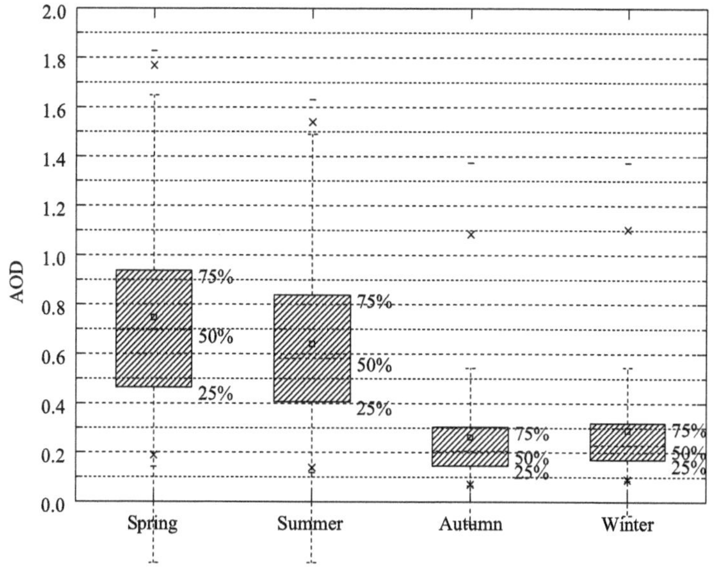

Fig. 3　Seasonal mean and standard deviation values of AOD at Tazhong
(The extreme "—" means the maximum and minimum value; the "×" means 99% and 1% percentile value; the "□"means the mean value).

particles are larger in spring and summer than in autumn and winter. There are some cases with Angstrom exponent larger than 0.80 in autumn and winter, which probably reflects the anthropogenic effect on aerosol particles in Taklimakan Deserts (Li *et al.*, 2010).

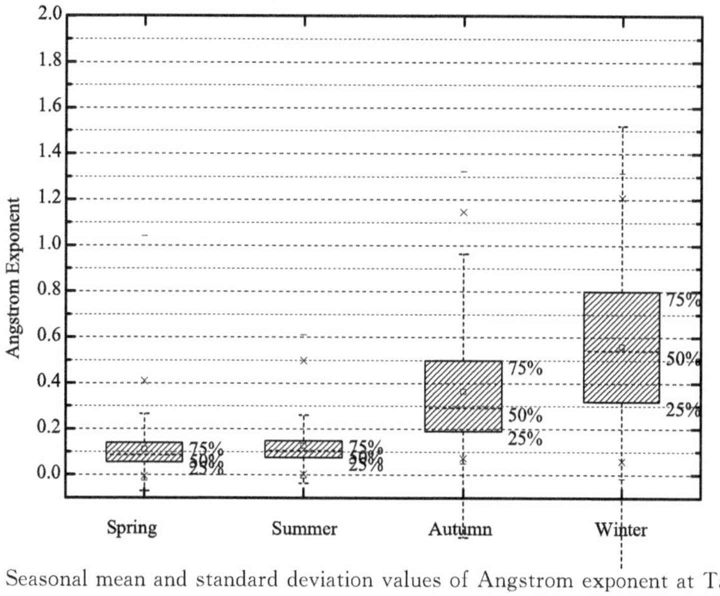

Fig. 4　Seasonal mean and standard deviation values of Angstrom exponent at Tazhong
(The extreme "—" means the maximum and minimum value;
the "×" means 99% and 1% percentile value; the "□"means the mean value).

Monthly Variation of AOD and Angstrom Exponent

The monthly variation of AOD and Angstrom exponent is shown in Table 1. The intra-annual variation of AOD and Angstrom exponent is very obvious. The AOD increases during January to April and decreases till December while the Angstrom exponent varies contrary to AOD. The AOD variation is similar to that of Total Suspended Particles (TSP). Liu *et al.* (2011) investigated the TSP variation at Tazhong. The TSP concentration has large value range between April to August. In this article, high AOD and low Angstrom exponent occur during March to July with values larger than 0.60 for AOD and less than 0.15 for Angstrom exponent. The maximum AOD occurs in April with value of 0.83±0.41 and the minimum Angstrom exponent occurs in May with a value of 0.09±0.06. The AODs from November to January are less than 0.30 and the Angstrom exponent varies around 0.58 to 0.70, which indicates the aerosol loading during this period at Taklimakan Desert is low because of infrequent dust events and the coarse particles are less comparing with the period during March to July.

Table 1 The monthly means of AOD and Angstrom exponent at Tazhong.

	AOD	Std(Δ)	Angstrom exponent	Std(Δ)	No. (d)
Jan	0.22	0.08	0.70	0.25	74
Feb	0.41	0.26	0.42	0.32	86
Mar	0.65	0.34	0.13	0.09	79
Apr	0.83	0.41	0.11	0.14	70
May	0.77	0.36	0.09	0.06	77
Jun	0.76	0.33	0.10	0.07	83
Jul	0.60	0.30	0.12	0.06	88
Aug	0.58	0.29	0.15	0.11	106
Sep	0.46	0.22	0.20	0.09	116
Oct	0.30	0.15	0.31	0.14	134
Nov	0.19	0.10	0.59	0.26	111
Dec	0.24	0.17	0.58	0.28	107

Diurnal Variation of AOD and Angstrom Exponent

The diurnal variation of AOD and Angstrom exponent is shown in Figs. 5 and 6, respectively. The mean values are based on the statistic of all the instantaneous data. The diurnal variation shows that AOD is higher in the morning and evening. AOD is −0.50 before 9:00 (local time) and decrease to −0.40 until 18:00 and then increases to over 0.50 again from 19:00. The variation of Angstrom exponent is contrary to that of AOD. Angstrom exponent increases from −0.15 to −0.35 from 8:00 to 10:00 and varies very little until on 18:00 at afternoon. From 18:00, the Angstrom exponent decreases till to the minimum about 0.13 at 20:00.

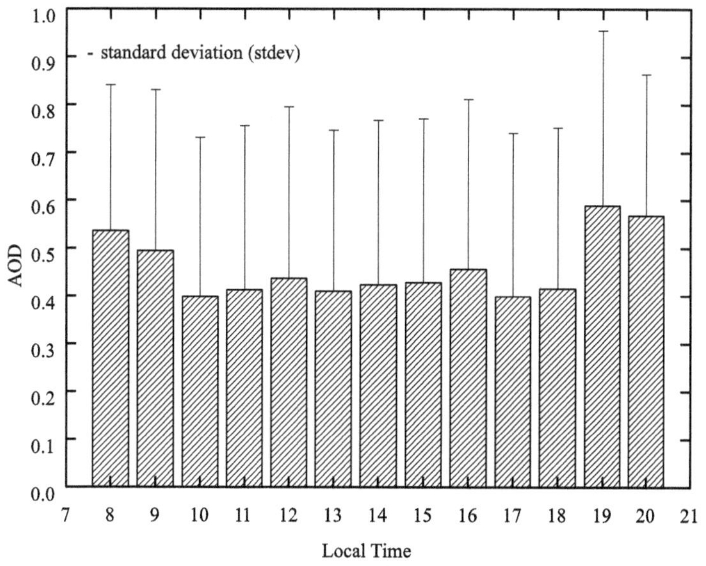

Fig. 5 Diurnal variation of AOD and standard deviation (stdev) at Tazhong.

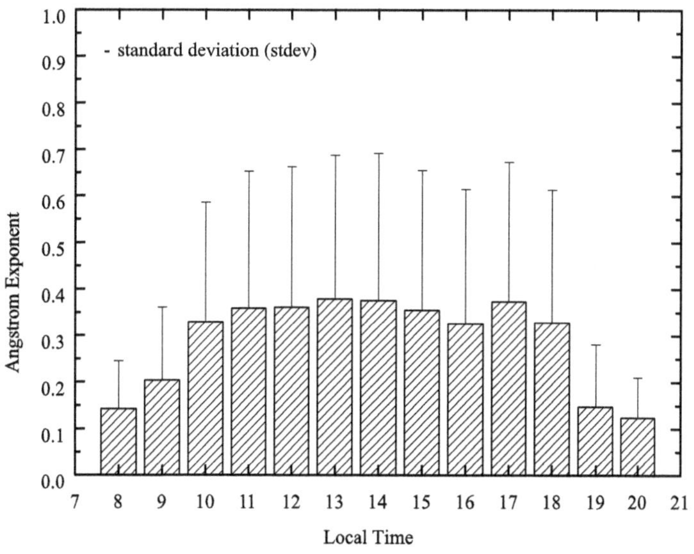

Fig. 6 Diurnal variation of Angstrom exponent between 440nm and 870nm and standard deviation (stdev) at Tazhong.

The diurnal variation in this study is different from that of Gai *et al.* (2006), which could be due to the different measurement period and processing method. The diurnal variation of AOD in this study is similar to that of scattering coefficient. Lu *et al.* (2010) measured the scattering coefficient (σ) by using M9003 nephenometer and also found diurnal variation of σ with the large values in the morning and evening. Tazhong is located in the middle of Taklimakan Desert with large temperature difference and affected mainly by mineral particles (Xue *et al.*, 2009). From Fig. 7, one can see that the diurnal temperature and wind speed variations at Tazhong are very similar. Temperature and wind speed decrease continuously during middle night (00:00 local time)to early morning (08:00 local time) and increase rapidly from early morning about 08:00 to noon about 12:00. Inversion layer is easily formed in the morning, which is not in favor of the diffuse of aerosol particles. From evening about 18:00, both the temperature and wind speed begin to decrease rapidly. The near-surface atmosphere becomes stable because of the rapid temperature and wind speed decrease, which could block the diffuse of aerosol particles. However, during the daytime high temperature and wind speed could cause the turbulence exchange and convection very actively, and the aerosol particles are easily emitted into the atmosphere and diffused (Zhang *et al.*, 2008). This classic meteorological condition probably results in large AOD in the morning and evening but stable during daytime at Tazhong station.

The Angstrom exponent varies differently to that of AOD, which shows lower values in the morning and evening than daytime. This could probably because aerosol loading includes more coarse particles in the morning and evening. However, some coarse particles could be transported away with the strong turbulence exchange and convection during the daytime.

Relationship between AOD and Angstrom Exponent

The scattergram of AOD versus Angstrom exponent is shown in Fig. 8. This representation often allows the physical definition of interpretable cluster regions for different types of aerosols (Smirnov *et al.*, 2002). One can see that obviously there is a trend of increasing AOD with decreasing Angstrom exponent, which stands for large particles. Most likely, the origin of this type of aerosol is local or regional dust events. There are also some cases with AOD around 0.50 and Angstrom exponent around 1.0, which probably reflects the presence of some fine particles in the aerosol size distribution, such as sulfate and black carbon (Li *et al.*, 2010). In general, the relationship between AOD and Angstrom exponent can be well fitted by a power curve with the equation of $Y=0.1055X^{-0.8568}$ with a R^2 about 0.55.

Summary

It was found that seasonal variation of the optical properties was significant over Taklimakan Desert. The AOD increases from January to April and decreases until December while the Angstrom exponent shows an opposite trend to the AOD, with a minimum Angstrom exponent of (0.09±0.06) in May. The maximum AOD occurs in April with a value of (0.83±0.41). The dust aerosols in spring and summer are the major contributors to AOD and Angstrom exponent variations. AOD is lower in autumn and winter seasons than in spring and summer.

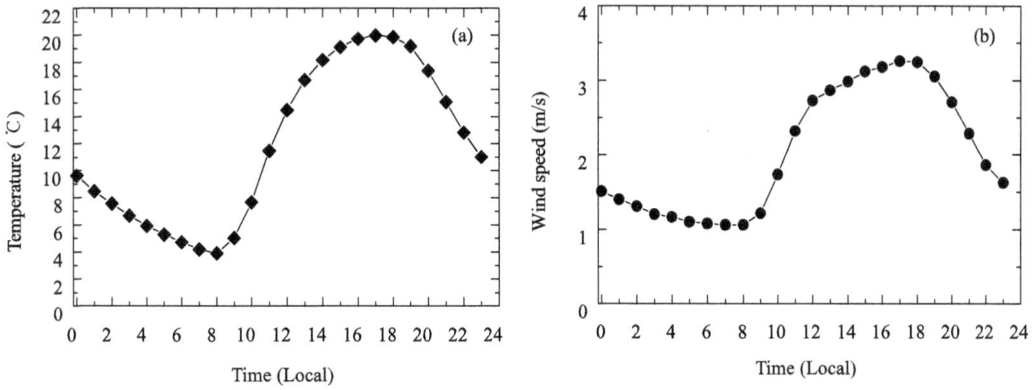

Fig. 7　Diurnal variation of temperature (a) and wind speed (b) at Tazhong.

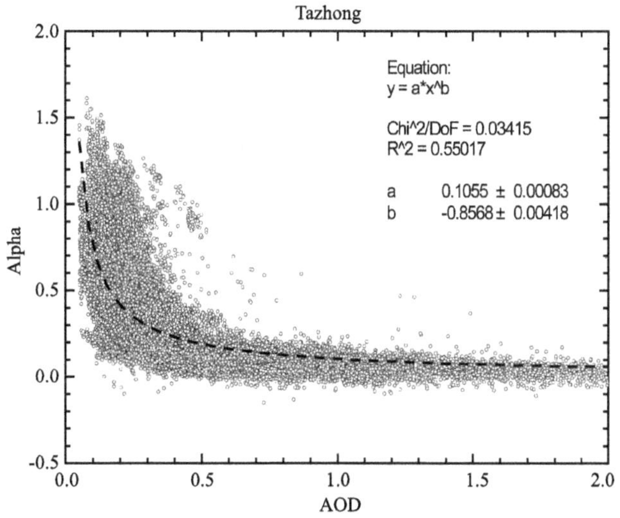

Fig. 8　Relationship between AOD and Angstrom exponent.

The diurnal variation shows that AOD (Angstrom exponent) is high (low) in the morning and evening and rather constant during the day. This pattern is linked to the meteorological conditions of Taklimakan Desert.

Relationship between AOD and Angstrom exponent shows that coarse particles are the major parts of aerosol in Taklimakan Desert. There is one peak probability distribution for the AOD and the Angstrom exponent. Both the AOD and Angstrom exponent distributions could be well fitted by a bi-mode normal distribution.

Acknowledgments

This work is financially supported by grants from the National Key Project of Basic Research (2011CB403401), the Project (41005086 &41130104) supported by NSFC, CAMS Basis Research Project (2012Y02 &2010Z002), the Meteorological Special Project of China (GYHY-200906038&201206037), and the Project supported by Ministry of Science and Technology of China (2010DFA22770).

References

Alfaro S C, Gomes L, Rajot J L, *et al*. 2003. Chemical and Optical Characterization of Aerosols Measured in Spring 2002 at the ACE-Asia Supersite, Zhenbeitai, China. *J. Geophys. Res.* **108**: 8641, doi: 10.1029/2002JD003214.

Che H, Zhang, X, Alfraro S, *et al*. 2009a. Aerosol Optical Properties and Its Radiative Forcing over Yulin, China in 2001 and 2002. *Adv. Atmos. Sci.* **26**: 564-576.

Che H, Zhang X, Chen H, *et al*. 2009b. Instrument Calibration and Aerosol Optical Depth Validation of the China Aerosol Remote Sensing Network. *J. Geophys. Res.* **114**: D03206, doi: 10.1029/2008JD011030.

Cheng T T, Liu Y, Lu D R, *et al*. 2006. Aerosol Properties and Radiative Forcing in Hunshan Dake Desert, Northern China. *Atmos. Environ.* **40**: 2169-2179.

Dey S, Tripathi S N, Singh R P, *et al*. 2004. Influence of Dust Storms on the Aerosol Optical Properties over the Indo-Gangetic Basin. *J. Geophys. Res.* **109**: D20211, doi:10.1029/2004JD004924.

Eck T F, Holben B N, Reid J S, *et al*. 1999. Wavelength Dependence of the Optical Depth Of Biomass Burning, Urban and Desert Dust Aerosols. *J. Geophys. Res.* **104**: 31333-31350.

Eck T F, Holben B N, Dubovik O, *et al*. 2005. Columnar Aerosol Optical Properties at AERONET Sites in Central Eastern Asia and Aerosol Transport to the Tropical Mid-Pacific. *J. Geophys. Res.* **110**: D06202, doi:10.1029/2004JD005274.

Gai C S, Li X Q, Zhao, F S. 2006. Mineral Aerosol Properties Observed in the Northwest Region of China. *Global Planet. Change.* **52**: 173-181.

Gong S L, Zhang X Y, Zhao T L, *et al*. 2003. Characterization of Soil Dust Aerosol in China and Its Transport/Distribution during 2001 ACE-Asia, 2. Model Simulation and Validation. *J. Geophys. Res.* **108**: 4262, doi:10.1029/2002JD002633.

Hansen J, Sato M, Ruedy R, *et al*. 2000. Global Warming in the Twenty-first Century: An Alternative Scenario. *Proc. Nat. Acad. Sci. U. S. A.* **97**: 9875-9880.

Holben B N, Eck T F, Slutsker I, *et al*. 1998. AERONET - A Federated Instrument Network and Data Archive for Aerosol Characterization. *Remote Sens. Environ.* **66**: 1-16.

Huang J, Fu Q, Su J, *et al*. 2009. Taklimakan Dust Aerosol Radiative Heating Derived from CALIPSO Observations Using the Fu-Liou Radiation Model with CERES Constraints. *Atmos. Chem. Phys.* **9**: 4011-4021.

Kalashnikova O V, Franklin P M, Eldering A, *et al*. 2007. Application of Satellite and Ground-based Data to Investigate the UV Radiative Effects of Australian Aerosols. *Remote Sens. Environ.* **107**: 65-80.

Kim D H, Sohn B J, Nakajima T, *et al*. 2004. Aerosol Optical Properties over East Asia Determined from Ground-based Sky Radiation Measurements. *J. Geophys. Res.* **109**: D02209, doi: 10.1029/2003JD003387.

Li J, Huang K, Wang Q, *et al*. 2010. Characteristics and source of black carbon aerosol over Taklimakan Desert. *Sci. China Chem.* **53**: 1202-1209.

Li S, Lei J, Xu X, *et al*. 2006. Features of Sandstorms in Hinterland of Taklimakan Desert: A Case of Tazhong Area. *J. Nat. Disasters.* **15**: 14-19 (in Chinese).

Li X, Maring H, Savoie D, *et al*. 1996. Dominance of Mineral Dust in Aerosol Light-scattering in the North Atlantic Trade Winds. *Nature.* **380**: 416-419.

Liu X, Zhong Y, He Q, *et al*. 2011. Observation Study on Mass Concentration of Dust Aerosols in the

Taklimakan Desert Hinterland. *China Environ. Sci.* **31**:1609-1617.

Lu H, Wei W, Liu M, *et al.* 2010. Aerosol Scattering Properties in the Hinterland of Taklimakan Desert. *J. Desert Res.* **30**: 660-667 (in Chinese).

Mikami M, Shi G Y, Uno I, *et al.* 2006. The Impact of Aeolian Dust on Climate: Sino-Japanese Cooperative Project ADEC. *Global Planet. Change.* **52**: 142-172.

Nakajima T, Hayasaka T, Higurashi A, *et al.* 1996. Aerosol Optical Properties in the Iranian Region Obtained by Ground-based Solar Measurements in the Summer of 1991. *J. Appl. Meteorol.* **35**: 1265-1278.

O'neill N T, Ignatov A, Holben B N, *et al.* 2000. The Lognormal Distribution as a Reference for Reporting Aerosol Optical Depth Statistics; Empirical Tests Using Multi-year, Multi-site AERONET Sunphotometer Data. *Geophys. Res. Lett.* **27**: 3333-3336.

Ramanathan V, Crutzen P J, Kiehl J T, *et al.* 2001. Aerosols, Climate, and the Hydrological Cycle. *Science.* **294**: 2119-2124.

Smirnov A, Holben B N, Eck T F, *et al.* 2000. Cloud Screening and Quality Control Algorithms for the AERONET Database. *Remote Sens. Environ.* **73**: 337-349.

Smirnov A, Holben B N, Dubovik O, *et al.* 2002. Atmospheric Aerosol Optical Properties in the Persian Gulf. *J. Atmos. Sci.* **59**: 620-634.

Sokolik I N, Toon O B. 1999. Incorporation of Mineralogical Composition into Models of the Radiative Properties of Mineral Aerosol from UV to IR Wavelengths. *J. Geophys. Res.* **104**: 9423-9444.

Tegen I, Fung I. 1995. Contribution to the Atmospheric Mineral Aerosol Load from Land Source Modification. *J. Geophys. Res.* **100**: 18707-18726.

Wang M, Zhang R, Pu Y. 2001. Recent Researches on Aerosol in China. *Adv. Atmos. Sci.* **18**: 576-586.

Wu Y, Zhang R, Pu Y, *et al.* 2012. Aerosol Optical Properties Observed at a Semi-arid Rural Site in Northeastern China. *Aerosol Air Qual. Res.* **12**: 503-514.

Xia X, Zong X. 2009. Shortwave versus Longwave Direct Radiative Forcing by Taklimakan Dust Aerosols. *Geophys. Res. Lett.* **36**: L07803, doi:10.1029/2009GL037237.

Xia X A, Chen H B, Wang P C, *et al.* 2005. Aerosol Properties and Their Spatial and Temporal Variations over North China in Spring 2001. *Tellus* Ser. B. **57**: 28-39.

Xue F, Liu X, Ma Y, *et al.* 2009. Variation Characteristics of Dust Weather in the Hinterland of Taklimakan Desert during 1997-2007. *Desert Oasis Meteorol.* **3**: 31-34 (in Chinese).

Zakey A S, Abdelwahab M M, Makar P A. 2004. Atmospheric Turbidity over Egypt. *Atmos. Environ.* **38**: 1579-1591.

Zhang Q, Niu S, Shen J, *et al.* 2008. Observational Study on Aerosol Scattering Properties in Semiarid Area. *J. Desert Res.* **28**: 755-761 (in Chinese).

Zhang R, Wang M, Zhang X, *et al.* 2003a. Analysis on the Chemical and Physical Properties of Particles in a Dust Storm in Spring in Beijing. *Powder Technol.* **137**:77-82.

Zhang R, Tao J, Ho K F, *et al.* 2012. Characterization of Atmospheric Organic Carbon and Elemental Carbon of $PM_{2.5}$ in a Typical Semi-arid Area of Northeastern China. *Aerosol Air Qual. Res.* **12**: 792-802.

Zhang X, Arimoto R, An Z. 1997. Dust Emission from Chinese Desert Sources Linked to Variations in Atmospheric Circulation. *J. Geophys. Res.* **102**: 28041-28047.

Zhang X Y, Gong S L, Zhao T L, *et al.* 2003b. Sources of Asian Dust and Role of Climate Change versus Desertification in Asian Dust Emission. *Geophys. Res. Lett.* **30**: 2272 10.1029/2003GL018206.

TRANSPORT OF A SEVERE DUST STORM IN MARCH 2007 AND IMPACTS ON CHLOROPHYLL-*a* CONCENTRATION IN THE YELLOW SEA[①]

TAN Saichun and SHI Guangyu

State Key Laboratory of Numerical Modeling for Atmospheric Sciences and Geophysical Fluid Dynamics, Institute of Atmospheric Physics, Chinese Academy of Sciences, Beijing, China

Abstract

The transport process of a severe Asian dust storm event generated in the Gobi Desert during 30—31 March 2007 was examined by several datasets. Results clearly showed that this storm was transported eastward to the northwest Pacific and southeastward to the China seas. Dust particles were deposited in the Yellow Sea accompanied by precipitation on 31 March 2007 and the average deposition flux at an offshore region in the Yellow Sea was 0.56 $g \cdot m^{-2} \cdot d^{-1}$. After 4 days of dust passage over this offshore region, a phytoplankton bloom event appeared on 3 April 2007 and continued until 4—6 April 2007. When compared to non-dust year in 2005, the initial timing of the bloom in 2007 was about 18 days earlier than that in 2005 and peak chlorophyll-*a* concentration in 2007 was 45% higher than that in 2005. Similar results were found in the dust storm cases in 2004 and 2006. Results indicated that besides increased SST, PAR, and nutrients accumulated in winter from strong mixing, dust input may play important roles in spring bloom in the dust years. That provided evidence of biotic response to natural fertilization caused by dust deposition.

Key words: transport; severe dust storm; dust particle; impact

1 Introduction

The dust aerosols carried by dust storm events in East Asia can be transported to downwind areas through long-range transport(Mikami *et al*. 2006). Many studies reported that the long-range transport of Asian dust storms to Hong Kong, Taiwan, and the China seas, including the Bohai Sea, the Yellow Sea, the East China Sea, and the South China Sea (Fang *et al*. 1999; Lin *et al*. 2007; Yang *et al*. 2010; Lin *et al*. 2012), Korea (Chun *et al*. 2001), Japan (Matsuki *et al*. 2003), and even the North Pacific and the West Coast of North America (Husar *et al*. 2001). The marine plankton can be affected by the

① The paper published in *Scientific Online Letters on the Atmosphere*, **8**: 85-89. 2012.

nutrient-rich dust particles deposited into the oceans, and then it would impact the climate through complex biogeochemical processes (Martin and Fitzwater 1988; Jickells *et al.* 2005). Many studies proposed that a correlation exited between dust deposition and biological activity and bloom events in the western North Pacific, the Sea of Japan (also known as the East Sea), and the coastal seas of China (Yuan and Zhang 2006; Jo *et al.* 2007; Han *et al.* 2011; Tan *et al.* 2011).

However, few studies have investigated that the supportive evidence of natural eolian iron fertilization on phytoplankton growth directly (Bishop *et al.* 2002). It is very difficult to measure a complete process of dust generation, transport, deposition and the effects on ocean ecosystem during one dust storm event owing to sporadic property. Meteorological station observation can record the occurring time of dust storm at each station. Conventional satellite can obtain dust aerosol properties (such as aerosol index, aerosol optical depth, etc) throughout the atmosphere, while space-borne lidar measurements (Cloud Aerosol Lidar with Orthogonal Polarization, CALIOP) can provide vertical profiles of dust aerosols.

Combination of meteorological stations and satellite observations and air parcel trajectories, this study analyzed the transport of a severe dust storm occurring in the Gobi Desert (across Mongolia-Inner Mongolia Autonomous Region, China) during 30－31 March 2007 from source to the sea and examined the relationship between dust deposition and phytoplankton bloom in the Yellow Sea; aiming to better understand how phytoplankton growth in response to dust depositions.

2 Data and Methods

Several datasets have been employed in this study to track the transport process of the dust storm generated in the Gobi Desert during 30－31 March 2007. This dust storm was observed by some meteorological stations in Inner Mongolia and the data were obtained from the National Meteorological Information Center of the China Meteorological Administration. Aerosol index product of Ozone Monitoring Instrument (OMI) onboard Aura satellite (resolution $1° \times 1°$) was used to identify the transport of dust since it provided useful information on the daily spatial distribution of the dust cloud (Husar *et al.* 2001). It is reported that aerosol index >2 can be used to identify the outbreak of dust storms over the Yellow Sea according to comparison with historical record of decades of dust storm episodes (Tan *et al.* 2011). The CALIOP on board the Cloud-Aerosol Lidar and Infrared Pathfinder Satellite Observations (CALIPSO) satellite Level 2 Version 3 aerosol profiles products (Winker *et al.* 2009) were also used. They include both 532 and 1064-nm aerosol extinction coefficient and particulate depolarization ratio at 532 nm. The vertical resolution was 60 m below and 180 m above 20 km, respectively. Dust aerosols could be identified from CALIPSO measurements because dust aerosols have a large depolarization ratio due to their nonsphericity, while the depolarization ratio is near zero for water clouds and other types of aerosols (Huang *et al.* 2009).

A forward trajectory analysis was performed using the Hybrid Single Particle Lagrangian Integrated Trajectory (HYSPLIT) model (Draxler and Rolph 2010) to examine the transport path of dust storms. Input meteorological data was the National Centers for Environmental Prediction/the National Center for Atmospheric Research (NCEP/NCAR) global reanalysis data with a $2.5°\times2.5°$ spatial resolution. An air parcel appearing in the middle of a dust storm event at the source stations was traced forward. In addition, NCEP reanalysis geopotential height, mean sea level pressure and wind speed data were applied to analyze the synoptic pattern during the dust storm.

Chlorophyll-*a* concentrations were used as an indicator for phytoplankton bloom. Sea Surface Temperature (SST) and Photosynthetically Available Radiation (PAR) were used for temperature and light condition for phytoplankton growth, respectively. Both chlorophyll-*a* concentration and PAR data are SeaWiFS Level 3 Standard Mapped Image products with a 9×9 km spatial resolution. The SST data were Advanced Very High Resolution Radiometer (AVHRR) products and have 4×4 km spatial resolution. Moreover, Global Precipitation Climatology Project merged daily precipitation data with spatial resolution of $1°\times1°$ and monthly averaged ocean surface winds at 10 m height with $0.25°\times0.25°$ spatial resolution from microwave scatterometer QuikSCAT were used to examine the environmental condition. These two dataset were provided by Asia-Pacific Data Research Center at the University of Hawaii (http://apdrc.soest.hawaii.edu).

3 Results and Discussion

3.1 Transport of Dust from Source Regions to the Seas

Asian dust is usually associated with frontal systems and/or cyclones (Tsai *et al.* 2008). A deep 500 hPa trough extended from about 45°N to 25°N on 30 March 2007 over China (Fig. 1a), indicating the strong intensity of the synoptic system. At the surface, a cyclone ($L=996$ hPa) was located across the border of the eastern Mongolia and Inner Mongolia and another one ($L=1004$ hPa) arrived at the southern China (Fig. 1b). Behind the surface cyclone, there was a high pressure system ($H=1032$ hPa). Tsai *et al.* (2008) indicated that the synoptic pattern with 500 hPa trough and strong surface low center are suitable for dust generation and transport. Thus, such synoptic conditions appeared to be favorable for dust transport southeastward to the China seas and eastward to the northwest Pacific, as supported by the CALIPSO lidar data and forward trajectories (Fig. 2 and Fig. 3).

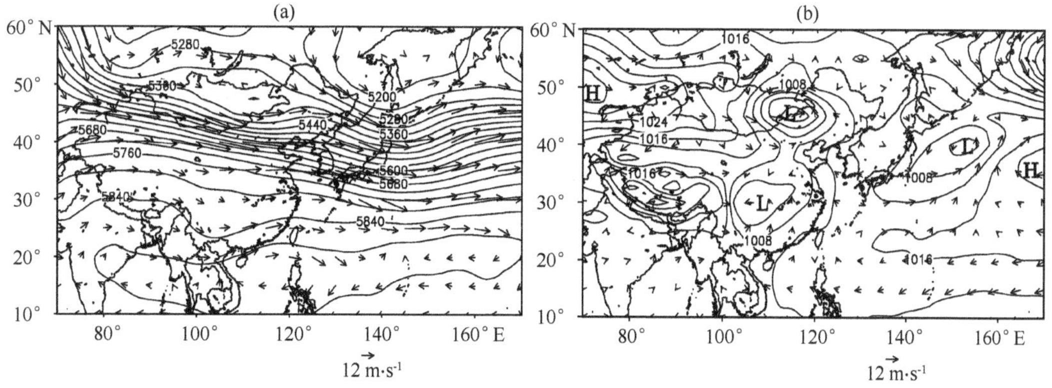

Fig. 1　Synoptic maps at (a) 500 hPa and (b) surface on 30 March 2007. Solid contour lines show geopotential heights in meters at 500 hPa and mean sea level pressures in hPa at surface.

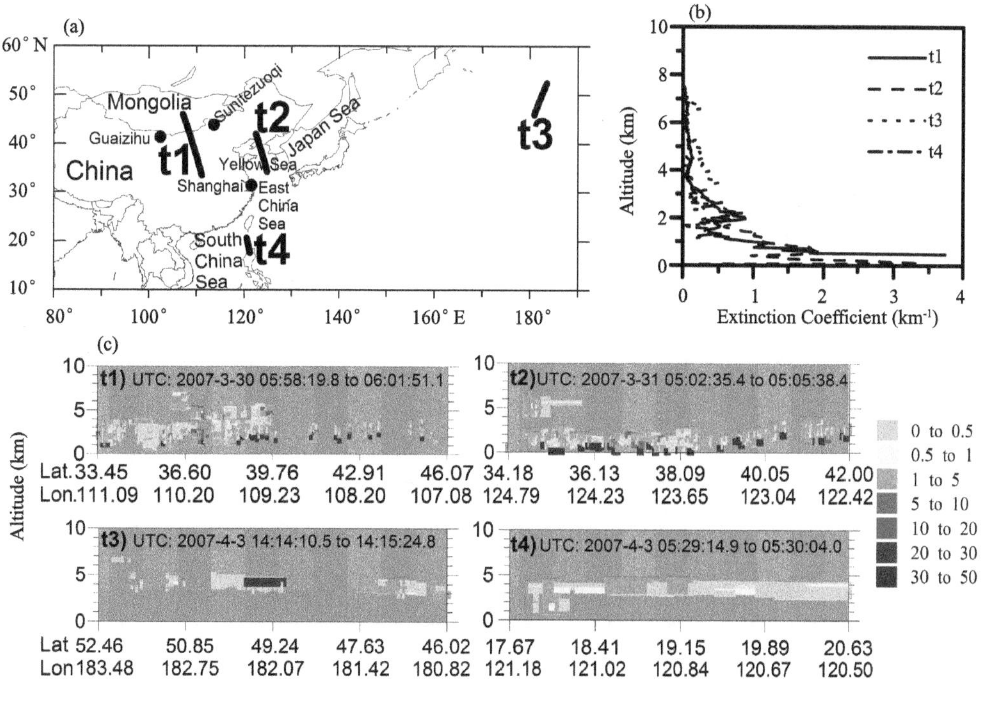

Fig. 2　(a) The location of satellite tracks. (b) The orbit averaged vertical profile of aerosol extinction coefficient for tracks shown in (a), the value for track t3 was multiplied by ten for being recognized clearly. (c) The altitude-orbit cross-section of depolarization ratio for tracks shown in (a). The altitude refers to above local mean sea level.

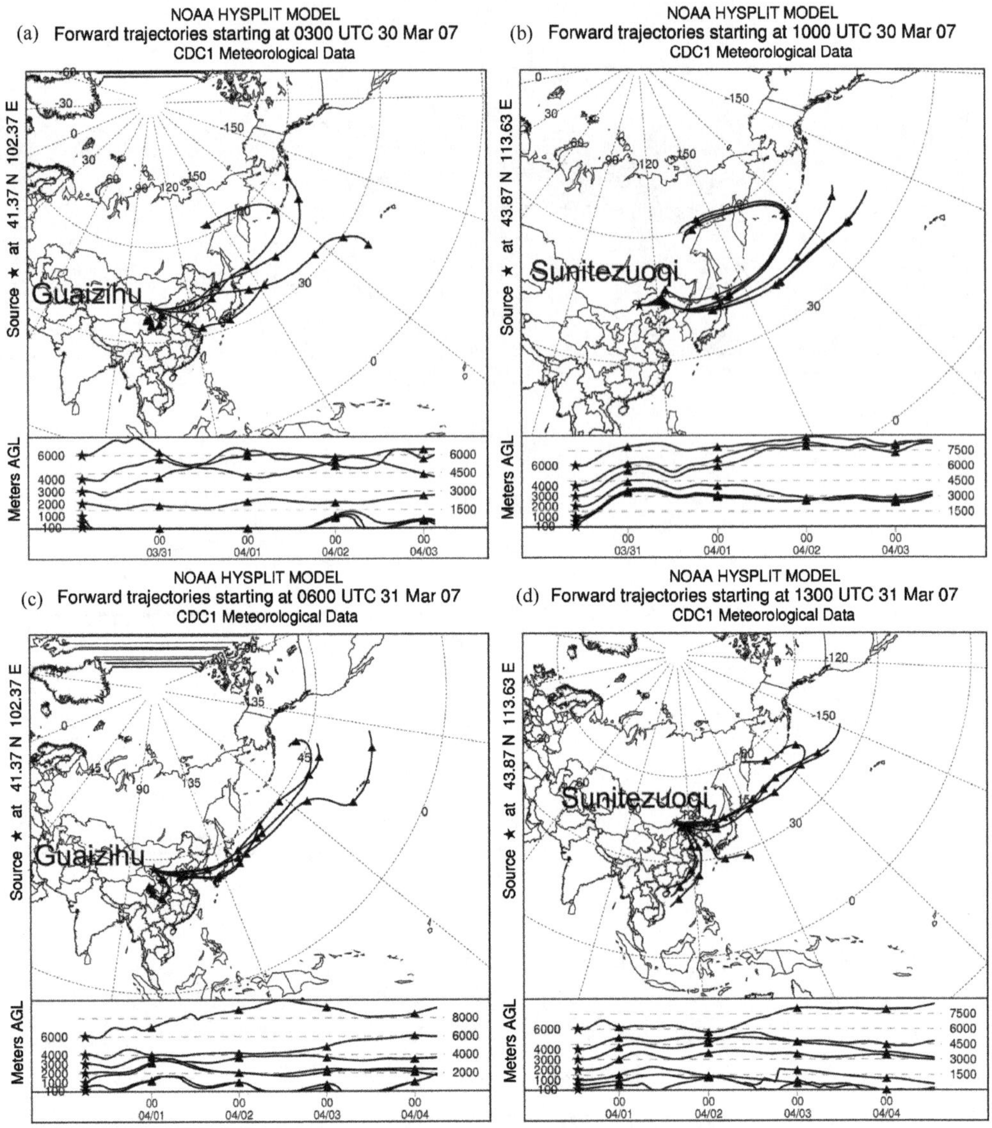

Fig. 3 The 96-hour forward trajectories of air parcels at heights of 100 m, 500 m, 1000 m, 2000 m, 3000 m, 4000 m, and 6000 m over Guaizihu and Sunitezuoqi stations (pentacle in the figure) during the dust event on 30 March 2007 (a and b) and 31 March 2007 (c and d). The top and bottom panels show horizontal and vertical motion, respectively. AGL means above ground level.

Meteorological station observations indicated that the dust storm occurred at Guaizihu station in Inner Mongolia at UTC 01:26—04:15 and Sunitezuoqi station in Inner Mongolia at UTC 07:35—12:00 on 30 March 2007, and occurred at Guaizihu at UTC 04:47—06:45 and Sunitezuoqi at UTC 12:00—15:26 on 31 March 2007. Figure 2 shows the average vertical aerosol extinction profiles and particulate depolarization ratio of CALIPSO. The data showed that the dust layer over the source regions, the Mongolia-Inner Mongolia region (satellite track t1 in Fig. 2a), was mainly located up to 6 km above mean sea level

(Fig. 2b and c). The dust layer peaked at around 0.5 km and 2 km with extinction coefficient about 3.7 km^{-1} and 0.9 km^{-1}, respectively. The dust storm arrived over the Yellow Sea on 31 March (satellite track t2), and the dust particles were located in the altitude from surface to 4 km. The maximum extinction coefficient was 3.4 km^{-1} at about 0.1 km. The dust storm arrived over the northwest Pacific (satellite track t3) and the South China Sea (satellite track t4) on 3 April. The dust layer was located in 3—5 km and 1—4 km over the northwest Pacific and the South China Sea, respectively. This result was consistent with the forward trajectories from dust source regions at Guaizihu and Sunitezuoqi stations in Inner Mongolia (Fig. 3). The trajectories showed that the dust storm was transported to the Yellow Sea, the Sea of Japan, the northwest Pacific, the East China Sea, and the South China Sea. Yan *et al.* (2011) also found this dust storm arrived at Shanghai which is close to the East China Sea on 2 and 3 April.

This typical dust storm processes were also tracked by aerosol index images clearly (Fig. 4). On 30 March 2007, the dust storm was generated in the Gobi Desert. The dust plume was transported to the North and Northeast China and the Yellow Sea after 1 day on 31 March 2007. On 2 April 2007, the dust cloud has passed the Yellow Sea and reached Japan. The dust cloud appeared over the South China Sea and the northwest Pacific on 3 April.

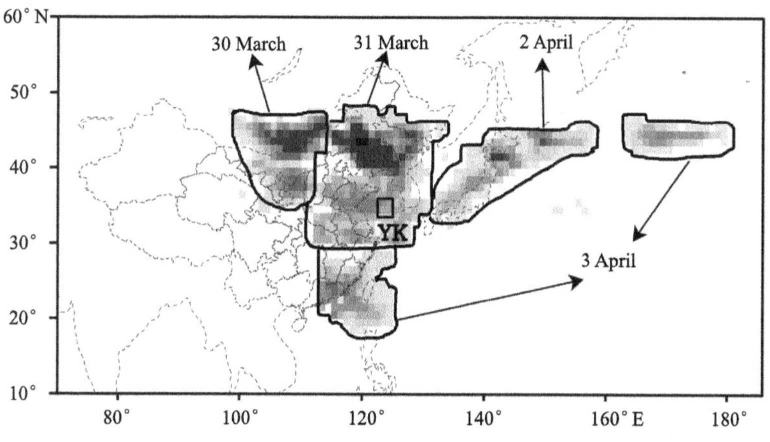

Fig. 4　Approximate location of dust cloud between 30 March and 3 April derived from OMI aerosol index data. YK is a small offshore sampling region (122.5°—125°E, 33.5°—36°N) located in the Yellow Sea.

3.2 Dust Deposition

Tan *et al.* (2012) reported a close correlation between dust deposition and aerosol index (AI) over the China seas in spring 2006, with a relationship of deposition (g/m^2/spring) $=22.292-50.370\times AI+28.127\times AI^2$ ($R^2=0.55$, significance <0.0001). Dust deposition (including dry and wet depositions) was obtained from the Global/Regional Assimilation and Prediction System/the Chinese Unified Atmospheric Chemistry Environment for Dust Atmospheric Chemistry Module (GRAPES/CUACE-Dust) numerical modeling. The model domain covers 70°—140°E and 15°—60°N, including the source regions of dust storms in East Asia. Details of the model have been described by Wang *et al*. (2010). This relationship was then applied to estimate the dust deposition flux on 31 March 2007. A small offshore sampling region (122. 5°—125°E, 33. 5°—36°N) labeled YK (Fig. 4) located in the Yellow Sea was selected in order to examine the response of phytoplankton to dust deposition in the following section. The average deposition flux in YK region was 0. 56 $g \cdot m^{-2} \cdot d^{-1}$, which included dry and wet deposition because of 8. 9 mm rainfall amount appearing on that day. The mean dry deposition flux over the Yellow Sea on 24 April 2006 estimated by a two-layer model was about 0. 44 $g \cdot m^{-2} \cdot d^{-1}$ (Yang *et al*. 2010), which is comparable to our result.

3.3 Chlorophyll-*a* Concentration in Response to Dust Deposition

Daily chlorophyll-*a* concentrations in YK region before, during and after the dust storm are shown in Fig. 5. SST, PAR, aerosol index during the same period are also shown. The initiation of spring bloom was defined as chlorophyll-*a* concentration > 3 $mg \cdot m^{-3}$, which is twice the average in the winter of 2004—2007.

It was clearly shown in Fig. 5a that aerosol index on 31 March 2007 was higher than the other days. Before dust deposition, chlorophyll-*a* concentration was lower than bloom level and the largest value was only 1. 9 $mg \cdot m^{-3}$ (Fig. 5b). After 4 days of dust deposition, a phytoplankton bloom event appeared during 4—6 April 2007. Chlorophyll-*a* concentration increased to 5. 8 $mg \cdot m^{-3}$ on 6 April 2007, and then decreased to lower than 1. 6 $mg \cdot m^{-3}$ after 7 April 2007. In comparison with 2007, no dust events occurred during the same period in 2005. Aerosol index was lower than 1. 8 in 2005 (Fig. 5c). Chlorophyll-*a* concentration showed that a bloom event occurred on 22 April with a peak value of 4. 0 $mg \cdot m^{-3}$ (Fig. 5d). The bloom event in 2007 developed about 18 days earlier than non-dust event year in 2005 and peak chlorophyll-*a* concentration in 2007 was 45% higher than 2005.

In addition, the dust storm cases in 2004 (Fig. 5e) and 2006 (Fig. 5f) were also taken into account as an additional support. Similarly, the peak chlorophyll-*a* concentration in 2004 was 38% higher than 2005 and the timing of bloom in 2004 was 17 days earlier. The peak chlorophyll-*a* concentration in 2006 was about 2. 6 times that in 2005 and the timing of bloom was 11 days earlier than 2005. The lag time of bloom relative to dust passage was 2—3 days in 2004 and 2006. A lag of several days was reasonable as aerosol iron dissolution could take place in short-term (minutes to hours) or long-term (days to weeks) within the euphotic zone (Boyd *et al*. 2010), which was comparable to 5—10 days' lag in the Sea of Japan (Jo *et al*. 2007).

During the research period in the dust years and non-dust year, PAR was higher than

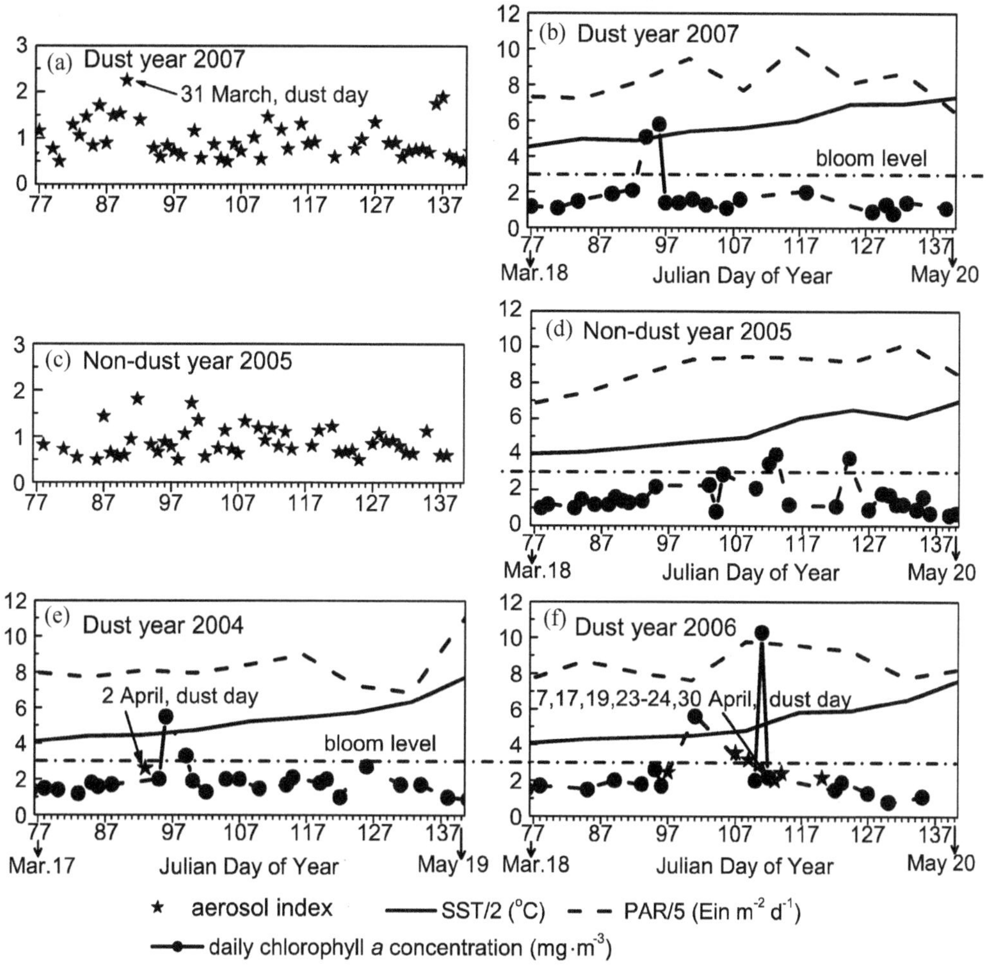

Fig. 5　Variations in daily aerosol index, daily chlorophyll-*a* concentration, and 8-day surface PAR and SST in the YK region from 18 March through 20 May. Dash dot line shows the bloom level defined as chlorophyll-*a* concentration is 3 $mg \cdot m^{-3}$ (twice as large as winter average).

34 E in $m^{-2} \cdot d^{-1}$ and SST was over 8℃. First, high SST and PAR are very important for spring bloom in the dust years owing to the lowest SST in the dust years was 1%—12% higher than that in 2005 and PAR was 6% — 13% higher than that in 2005. Second, nutrients are another important limiting factor for phytoplankton growth. Region YK was far from the impact region of the Yangtze River diluted water in spring according to Wang *et al*. (2003). From spring to late fall, the existence of thermocline and pycnocline constrained the transport of nutrients from bottom to surface (Wang 2000). Model simulations indicated that strong winter mixing affecting nutrients supply from deep waters to surface waters is prerequisite to spring bloom in the central southern Yellow Sea (Tian *et al*. 2005). The mean wind speed from December in last year to next year March in 2005 (8.3 $m \cdot s^{-1}$) was higher than that in 2004 (7.5 $m \cdot s^{-1}$) and 2007 (7.3 $m \cdot s^{-1}$) and comparable to 8.4 $m \cdot s^{-1}$ in 2006, which indicated that mixing due to winter monsoon in the dust years was not stronger than non-dust year. That suggests besides increased SST,

PAR, and nutrients accumulated in winter from strong winter mixing, dust deposition may also play important roles in spring bloom in the dust years.

The southern central Yellow Sea could be limited by macronutrients (nitrogen and phosphorus) and micronutrients (such as iron) according to previous studies(Zou *et al*. 2000; Wang *et al*. 2003). Dust storm could provide more nitrogen in the Yellow Sea in spring because the concentrations of organic nitrogen in dust aerosols significantly increased during dust storm events (Shi *et al*. 2010). In addition, dust deposition associated with precipitation may deliver more iron in the dust years due to higher solubitlity (Spokes *et al*. 1994), especially in 2007. YK region has low rainfall (0.02 mm) on the dust day in 2004 and only 0.2 mm rainfall on 17 April 2006 among the six dust deposition days (Fig. 5f) and 8.9 mm rainfall appeared on the dust day in 2007. Jo *et al*. (2007) also indicated that wet dust deposition played an important role in the spring bloom in the Sea of Japan.

It was needed to be noted that not all dust inputs would induce bloom. There was no bloom occurring after the passage of the dust storms on 23—24 April 2006 and 30 April 2006. It was likely because the PAR decreased after passage of these dust events (Fig. 5f) and dust particles transported at higher altitude deprived of nutrients accruement(Tan *et al*. 2011).

4 Conclusions

We report the transport processes of a severe dust storm event generated in the Gobi Desert during 30 — 31 March 2007 based on meteorological stations and satellite observations and trajectory analysis. Results indentified that dust particles were transported from source regions to the downwind seas, including the Yellow Sea and the northwest Pacific Ocean. The deposition flux was estimated to 0.56 $g \cdot m^{-2} \cdot d^{-1}$ in the Yellow Sea on 31 March 2007 according to an empirical formula.

Based on dust deposition, the relationship of dust with phytoplankton bloom in the Yellow Sea was investigated. Results show that bloom could be initiated after several days' passage of dust events in the Yellow Sea, which strongly implicates the possibility that aeolian dust sometimes associated with precipitation providing bio-available nutrients such as nitrogen and iron to the ocean. When compared to the situation in non-dust year, the timing of bloom in dust years were about two weeks earlier than non-dust year and peak chlorophyll-*a* concentrations in dust years were also higher than that in non-dust year. In agreement with previous studies (e.g., Bishop *et al*., 2002) the results of this work provide a direct support to the natural fertilization owing to dust deposition.

Acknowledgements

The authors would like to thank the Ministry of Science and Technology of China (Grant Nos. 2014CB953703 and 2010DFA22770) and the National Natural Science Foundation of China (Grant Nos. 41005080 and 41130104) for funding this study.

References

Bishop J K B, Davis R E, Sherman J T. 2002. Robotic observations of dust storm enhancement of carbon biomass in the North Pacific. *Science*, **298**: 817-821.

Boyd P W, Mackie D S, Hunter K A. 2010. Aerosol iron deposition to the surface ocean - Modes of iron supply and biological responses. *Marine Chemistry*, **120**:128-143.

Chun Y, Boo K O, Kim J, *et al*. 2001. Synopsis, transport, and physical characteristics of Asian dust in Korea. *J. Geophys. Res.*, **106**:18461-18469.

Draxler R R, Rolph G D. 2010. HYSPLIT (HYbrid Single-Particle Lagrangian Integrated Trajectory) Model access via NOAA ARL READY Website (http://ready.arl.noaa.gov/HYSPLIT.php). NOAA Air Resources Laboratory, Silver Spring, MD.

Fang M, Zheng M, Wang F, *et al*. 1999. The long-range transport of aerosols from northern China to Hong Kong—a multi-technique study. *Atmos. Environ.*, **33**: 1803-1817.

Han Y, Zhao T, Song L, Fang X, *et al*. 2011. A linkage between Asian dust, dissolved iron and marine export production in the deep ocean. *Atmos. Environ.*, **45**:4291-4298.

Huang J, Fu Q, Su J, *et al*. 2009. Taklimakan dust aerosol radiative heating derived from CALIPSO observations using the Fu-Liou radiation model with CERES constraints. *Atmos. Chem. Phys.*, **9**: 4011-4021.

Husar R B, Tratt D M, Schichtel B A, *et al*. 2001. Asian dust events of April 1998. *J. Geophys. Res.*, **106**:18317-18330.

Jickells T D, An Z S, Andersen K K, *et al*. 2005. Global iron connections between desert dust, ocean biogeochemistry, and climate. *Science*, **308**:67-71.

Jo C O, Lee J Y, Park K A, *et al*. 2007. Asian dust initiated early spring bloom in the northern East/Japan Sea. *Geophys. Res. Lett.*, **34**: L05602, doi:05610.01029/02006GL027395.

Lin C Y, Sheng Y F, Chen W N, *et al*. 2012. The impact of channel effect on Asian dust transport dynamics: a case in southeastern Asia. *Atmos. Chem. Phys.*, **12**: 271-285.

Lin I I, Chen J P, Wong G T F, *et al*. 2007. Aerosol input to the South China Sea: Results from the MODerate Resolution Imaging Spectro-radiometer, the Quick Scatterometer, and the Measurements of Pollution in the Troposphere Sensor. *Deep Sea Res.*, *Part II*, **54**: 1589-1601. doi: 1510.1016/j.dsr1582.2007.1505.1013.

Martin J H, Fitzwater S E. 1988. Iron deficiency limits phytoplankton growth in the north-east Pacific subarctic. *Nature*, **331**:341-343.

Matsuki A, Iwasaka Y, Osada K, *et al*. 2003. Seasonal dependence of the long-range transport and vertical distribution of free tropospheric aerosols over east Asia: On the basis of aircraft and lidar measurements and isentropic trajectory analysis. *J. Geophys. Res.*, **108**:8663-8676, doi:8610.1029/2002JD003266.

Mikami M, Shi G Y, Uno I, *et al*. 2006. Aeolian dust experiment on climate impact: An overview of Japan-China joint project ADEC. *Global and Planetary Change*, **52**:142-172.

Shi J, Gao H, Qi J, *et al*. 2010. Sources, compositions, and distributions of water-soluble organic nitrogen in aerosols over the China Sea. *J. Geophys. Res.*, **115**:D17303, doi:17310.11029/12009JD013238.

Spokes L J, Jickells T D, Lim B. 1994. Solubilisation of aerosol trace metals by cloud processing: A laboratory study. *Geochimica et Cosmochimica Acta*, **58**:3281-3287.

Tan S C, Shi G Y, Shi J H, *et al*. 2011. Correlation of Asian dust with chlorophyll and primary productivity in the coastal seas of China during the period from 1998 to 2008. *J. Geophys. Res.*, **116**:

G02029, doi:02010.01029/02010JG001456.

Tan S C, Shi G Y, Wang H. 2012. Long-range transport of spring dust storms in Inner Mongolia and impact on the China seas. *Atmos. Environ.*, **46**:299-308.

Tian T, Wei H, Su J, *et al.* 2005. Simulations of Annual Cycle of Phytoplankton Production and the Utilization of Nitrogen in the Yellow Sea. *Journal of Oceanography*, **61**:343-357.

Tsai F, Chen G T J, Liu T H, *et al.* 2008. Characterizing the transport pathways of Asian dust. *J. Geophys. Res.*, **113**:D17311, doi:17310.11029/12007JD009674.

Wang B D, Wang X L, Zhan R. 2003. Nutrient conditions in the Yellow Sea and the East China Sea. *Estuarine, Coastal and Shelf Science*, **58**:127-136.

Wang B D. 2000. Characteristics of variations and interrelations of biogenie elements in the Yellow Sea Cold Water Mass (in Chinese). *Acta Oceanologica Sinica*, **22**:47-54.

Wang H, Gong S L, Zhang H L, *et al.* 2010. A new-generation sand and dust storm forecasting system GRAPES_ CUACE/Dust: model development, verification and numerical simulation. *Chinese Sci. Bull.*, **55**:635-649.

Winker D M, Vaughan M A, Omar A, *et al.* 2009. Overview of the CALIPSO Mission and CALIOP Data Processing Algorithms. *J. Atmos. Oceanic Technol.*, **26**:2310-2323.

Yan P, Wang Z, Wang X, *et al.* 2011. Impact of Pollutant Transport on the Air Quality of Shanghai in 2007. *SOLA*, **7**:85-88.

Yang D X, Liu Y, Chen W Z. 2010. Estimation of the Total Dust Column and Dry Deposition Flux over the Yellow Sea, China Based on Shipboard Sunphotometer Measurements: Case Study. *Atmospheric and Oceanic Science Letters*, **3**:64-69.

Yuan W, Zhang J. 2006. High correlations between Asian dust events and biological productivity in the western North Pacific. *Geophys. Res. Lett.*, **33**: L07603, doi:07610.01029/02005GL025174.

Zou L, Chen H T, Zhang J. 2000. Experimental examination of the effects of atmospheric wet deposition on primary production in the Yellow Sea. *Journal of Experimental Marine Biology and Ecology*, **249**: 111-121.

SIMULATED AEROSOL KEY OPTICAL PROPERTIES OVER GLOBAL SCALE USING AN AEROSOL TRANSPORT MODEL COUPLED WITH A NEW TYPE OF DYNAMIC CORE①

DAI Tie[1,2]②, GOTO D[3], SCHUTGENS N A J[4], DONG Xiquan[5], SHI Guangyu[1] and NAKAJIMA T[2]

1) Institute of Atmospheric Physics, Chinese Academy of Sciences, Beijing, China
2) Atmosphere and Ocean Research Institute, University of Tokyo, Kashiwa, Japan
3) National Institute for Environmental Studies, Tsukuba, Japan
4) Atmosphere, Ocean and Planetary Physics, University of Oxford, UK
5) Department of Atmospheric Sciences, University of North Dakota, USA

Abstract

Aerosol optical depth (AOD), Ångström Exponent (AE), and single scattering albedo (SSA) simulated by a new aerosol-coupled version of Nonhydrostatic ICosahedral Atmospheric Model (NICAM) have been compared with corresponding AERONET retrievals over a total of 196 sites during the 2006—2008 period. The temporal and spatial distributions of the modeled AODs and AEs match those of the AERONET retrievals reasonably well. For the 3-year mean AODs and AEs for all sites show the correlations between model and AERONET of 0.753 and 0.735, respectively, and 82.1% of the modeled AODs agree within a factor of two with the retrieved AODs. The primary model deficiency is an underestimation of fine mode aerosol AOD and a corresponding underestimation of AE over pollution region. Compared to the retrievals, the model underestimates the global 3-year mean AOD and AE by 0.022 (10.5%) and 0.329 (31.2%), respectively. The probability distribution function (PDF) of the modeled AODs is comparable to that of the retrieved ones, however, the model overestimates the occurrence frequencies of small AEs and SSAs.

Key words: global aerosol model, aerosol optical depth (AOD), Ångström Exponent (AE), single scattering albedo (SSA), evaluation, AERONET

---

① The paper published in *Atmospheric Environment*, **82**: 71-82. 2014.

② Corresponding author. Tel.: +86 10 82995452; fax: +86 10 8299 5097.

E-mail address: daitie@mail.iap.ac.cn (T. Dai).

Postal address: Institute of Atmospheric Physics, Chinese Academy of Sciences, Beijing 100029.

1 Introduction

Atmospheric aerosols greatly impact the Earth's climate in many ways, and to date, not all of them are well known. Aerosols are considered to be one of the factors inducing climate change primarily through two effects: (a) a direct effect in which aerosol particles scatter and absorb the solar and thermal radiation (Coakley *et al.*, 1983), and (b) an indirect effect in which they change the microphysical and optical properties of cloud droplets acting as cloud condensation nuclei (Albrecht, 1989).

To evaluate aerosol effects on the climate system, we need to accurately estimate aerosol optical properties, such as aerosol optical depth (AOD), Ångström Exponent (AE), and single scattering albedo (SSA). Aerosol optical properties are determined not only by aerosol amount but also by physical and optical parameters such as size distribution of particles, mixing state of particles, and refractive index (especially for absorbing particles, e.g., soot and dust). These parameters are usually described differently within global aerosol models, and there are large model diversities in aerosol dispersal and consequently optical properties (Textor *et al.*, 2006; Textor *et al.*, 2007). It has become evident that aerosol modeling suffers from both poorly known emission inventories and aerosol physical and optical parameters. Thus, the modeled aerosol properties have to be validated by observations to ensure high confidence in the modeled results.

AErosol RObotic NETwork (AERONET) is to date the most dedicated effort in establishing a global surface network with the purpose of observing aerosol behavior, and its data have been commonly used for model validations. Monthly mean AODs, AEs and SSAs simulated by Spectral Radiation Transport Model for Aerosol Species (SPRINTARS) coupled with an atmospheric general circulation model, MIROC (Model for Interdisciplinary Research on Climate), were compared with the observations collected at dozens of AERONET sites (Takemura *et al.*, 2002). In an attempt to provide an absolute measure for model skill, AODs simulated with aerosol modules of seven global models were compared to the observations from 20 AERONET sites (Kinne *et al.*, 2003). To investigate the ability of the Community Multiscale Air Quality (CMAQ) model to simulate the aerosol distribution in Europe, the modeled results were compared with surface-measured PM10 values and AERONET AODs (Matthias, 2008). Compared to AERONET retrievals, the simulated AODs with a global chemical and transport model (GEOS-Chem) were systematically overestimated over northern Africa and southern Europe (Generoso *et al.*, 2008). Chin *et al.* (2009) evaluated the Goddard Chemistry Aerosol Radiation and Transport (GOCART) model simulated key aerosol optical parameters against AERONET retrievals at seven different regions worldwide and concluded the model underestimated AODs for biomass burning aerosols by 30%—40%.

The SPRINTARS module hasalso been implemented into the Nonhydrostatic ICosahedral Atmospheric Model (NICAM). The simulated results with a spatial resolution of 7 km were compared with satellite observations, but the period was very limited only during July 1—8, 2006 (Suzuki *et al.*, 2008). In addition, although NICAM model with a coarse spatial resolution of 224 km was also applied for a passive tracer model to simulate CO_2 distribution (Niwa *et al.*, 2011), NICAM+SPRINTARS with a coarse resolution has not been evaluated using ground-based remote sensing measurements. Therefore, we simulate the global temporal and spatial distributions of aerosol characteristics using this new aerosol-coupled version of NICAM and evaluate the simulated aerosol key optical properties with the AERONET retrievals over a total of 196 sites during the period 2006—2008 in this study.

2 Model Description, Data, and Methodology

2.1 Model Description

The Nonhydrostatic ICosahedral Atmospheric Model (NICAM) (Tomita and Satoh, 2004; Satoh *et al.*, 2008) is designed to perform cloud-resolving simulations by directly calculating deep convection and meso-scale circulations. It has been used for several types of global cloud-resolving experiments with a horizontal resolution of 3.5 km (Satoh *et al.*, 2008, and references therein), including a realistic simulation of the Madden-Julian Oscillation (Miura *et al.*, 2007). The aerosol module called Spectral Radiation Transport Model for Aerosol Species (SPRINTARS), a global three-dimensional aerosol transport-radiation model (Takemura *et al.*, 2000; Takemura *et al.*, 2002; Takemura *et al.*, 2009), has been implemented into NICAM (Suzuki *et al.*, 2008). In this aerosol-coupled version of NICAM, the aerosol effects are incorporated into cloud microphysical and radiative transfer processes so that the direct and indirect effects of aerosols are represented (Suzuki *et al.*, 2008; Suzuki *et al.*, 2011).

We use this new aerosol-coupled model to perform the global simulation with a horizontal resolution of 224 km (a total of 10242 grid points). The Lorenz grid is used for the vertical grid configuration (Satoh *et al.*, 2008), and there are a total of 40 vertical layers with the top of model located at 40 km. The SPRINTARS uses a single-moment scheme to track only aerosol mass by considering the transport processes including emission, advection, diffusion, and deposition. Bulk aerosol mass of sulfate and carbonaceous aerosols are predicted, whereas sea salt mass is tracked in 4 bins and dust mass is tracked in 10 bins (Takemura *et al.*, 2002). The carbonaceous aerosols include one pure black carbon (BC), one pure organic carbon (OC), and five different internal mixtures of OC and BC according to the carbonaceous aerosols sources (Takemura *et al.*, 2000). Therefore, 22 tracers are used in SPRINTARS to predict the mass of the four aerosol species. The combinations of the aerosol mass and the pre-calculated optical parameters with prescribed size distributions provide the modeled aerosol optical properties. AOD is derived as $\tau = \beta m$, where β is the mass extinction coefficient and m is the aerosol mass. AE represents the spectral change in AOD and is calculated as $-\log(\tau_1/\tau_2)/\log(\lambda_1/\lambda_2)$, where τ_1 and τ_2 represent AODs at wavelengths λ_1 and λ_2, respectively. SSA measures the contribution of scattering to total extinction and is calculated as τ_s/τ, where τ_s is the scattering optical depth. The main model physics used are the *k*-distribution radiation scheme (Nakajima *et al.*, 2000; Sekiguchi and Nakajima, 2008), the prognostic Arakawa-Schubert type cumulus convection scheme (Arakawa and Schubert, 1974; Pan and Randall, 1998), and the MATSIRO land surface scheme (Takata *et al.*, 2003). NCEP Final (FNL) Operational Global Analysis data are used for the initial and boundary conditions (e. g., SST and sea ice). For proper simulations, the modeled wind, water vapor, pressure and temperature fields are also nudged to the NCEP FNL analysis data with a time-scale of six hours.

The emission inventories of aerosols (primary OC and BC) and aerosol precursors emitted from anthropogenic sources, including fossil fuel combustion and biomass burning, come from the AeroCom Phase-II dataset (Diehl *et al.*, 2012). Other sources such as volcanic SO_2 and DMS are same as those used in Takemura (2012). The emission flux of dust aerosol depends on the near-surface wind speed, vegetation, leaf area index (LAI), soil moisture, and amount of snow (Takemura *et al.*, 2009). Sea salt emission depends on the near-surface wind speed, and is typically not possible over area covered by sea ice (Takemura *et al.*, 2009). In order to calculate sulfate chemistry (Takemura *et al.*, 2000), the oxidant concentrations (ozone, hydroxyl radicals, and H_2O_2) are given off-line by CHASER, a global chemical transport model (Sudo *et al.*, 2002).

Theoriginal pre-calculated optical parameters of simulated aerosols used in the standard SPRINTARS (Takemura *et al.*, 2002) are modified as proposed by Schutgens *et al.*, (2010). The main parameters for dry aerosols used in this study are listed in Table 1. It is observed that smaller dry particle is generally much more efficient in light extinction at 550 nm and its specific extinction is much more sensitive to the increase of wavelength. The absorbing aerosols are dust and carbonaceous aerosols, and the larger the dust aerosol particle size is, the more absorbing it has. Hygroscopic growth is considered for sulfate, carbonaceous aerosols except the BC, and sea salt as a function of relative humidity, which is prescribed at eight relative humidity values as in Takemura *et al.* (2002). The aerosol optical properties for any humidity value are calculated online through linear interpolation.

As shown in Fig. 1, the higher the humidity is, the more efficient total extinction but the smaller AE and the lower absorption the aerosol has.

Table 1　Particle density, particle radius, refractive index, mass extinction coefficient (β), AE derived from β at 440 and 870 nm, and SSA for dry aerosol as used in the model

Species	density ($g \cdot cm^{-3}$)	dry particle radius(μm)	refractive index at 550 nm	β at 550 nm ($m^2 \cdot g^{-1}$)	AE	SSA
Sulfate	1.769	0.0695	$1.43-10^{-8}i$	1.775	2.740	1.000
BC	1.25	0.0118	1.75−0.44i	8.185	1.119	0.323
OC	1.5	0.02	1.53−0.006i	1.062	2.683	0.915
OC/BC (1)	1.473	0.1	1.558−0.063i	5.455	1.520	0.766
OC/BC (2)	1.468	0.1	1.563−0.072i	5.547	1.478	0.741
OC/BC (3)	1.442	0.1	1.588−0.126i	6.036	1.278	0.638
OC/BC (4)	1.462	0.1	1.569−0.085i	5.666	1.426	0.712
OC/BC (5)	1.468	0.1	1.563−0.072i	5.547	1.478	0.741
Dust	2.5	0.13	1.53−0.0055i	2.239	2.811	0.965
	2.5	0.20	1.53−0.0055i	3.809	1.838	0.976
	2.5	0.33	1.53−0.0055i	3.652	0.608	0.972
	2.5	0.52	1.53−0.0055i	2.059	−0.405	0.950
	2.5	0.82	1.53−0.0055i	1.013	−0.348	0.903
	2.5	1.27	1.53−0.0055i	0.687	−0.081	0.876
	2.5	2.02	1.53−0.0055i	0.397	−0.010	0.831
	2.5	3.20	1.53−0.0055i	0.248	−0.047	0.778
	2.5	5.06	1.53−0.0055i	0.155	−0.0530	0.719
	2.5	8.02	1.53−0.0055i	0.089	−0.041	0.652
Sea salt	2.2	0.178	$1.50-10^{-8}i$	4.257	1.850	1.000
	2.2	0.562	$1.50-10^{-8}i$	1.715	−0.352	1.000
	2.2	1.78	$1.50-10^{-8}i$	0.493	0.011	1.000
	2.2	5.62	$1.50-10^{-8}i$	0.168	−0.078	1.000

Note: OC/BC (1−5) represent the internal mixture of OC and BC for tropical forest fire, other forest fire, fossil fuel, fuel wood, and agriculture source, respectively. The β and SSA for sulfate, BC, OC, OC/BC, and sea salt are calculated from the Mie−scattering theory using a mono-modally lognormal size distribution. The modal radii (r_m) for sulfate, BC, OC, OC/BC, and sea salt are 0.0695, 0.0118, 0.02, 0.1, and 0.114 μm, while the geometric standard deviations (σ_g) are 1.526, 2.3, 1.8, 1.562, and 2.305, respectively. The size distribution for dust aerosol is assumed a bi−modally lognormal distribution with r_m=0.202, 0.994 μm and σ_g=2.397, 1.110.

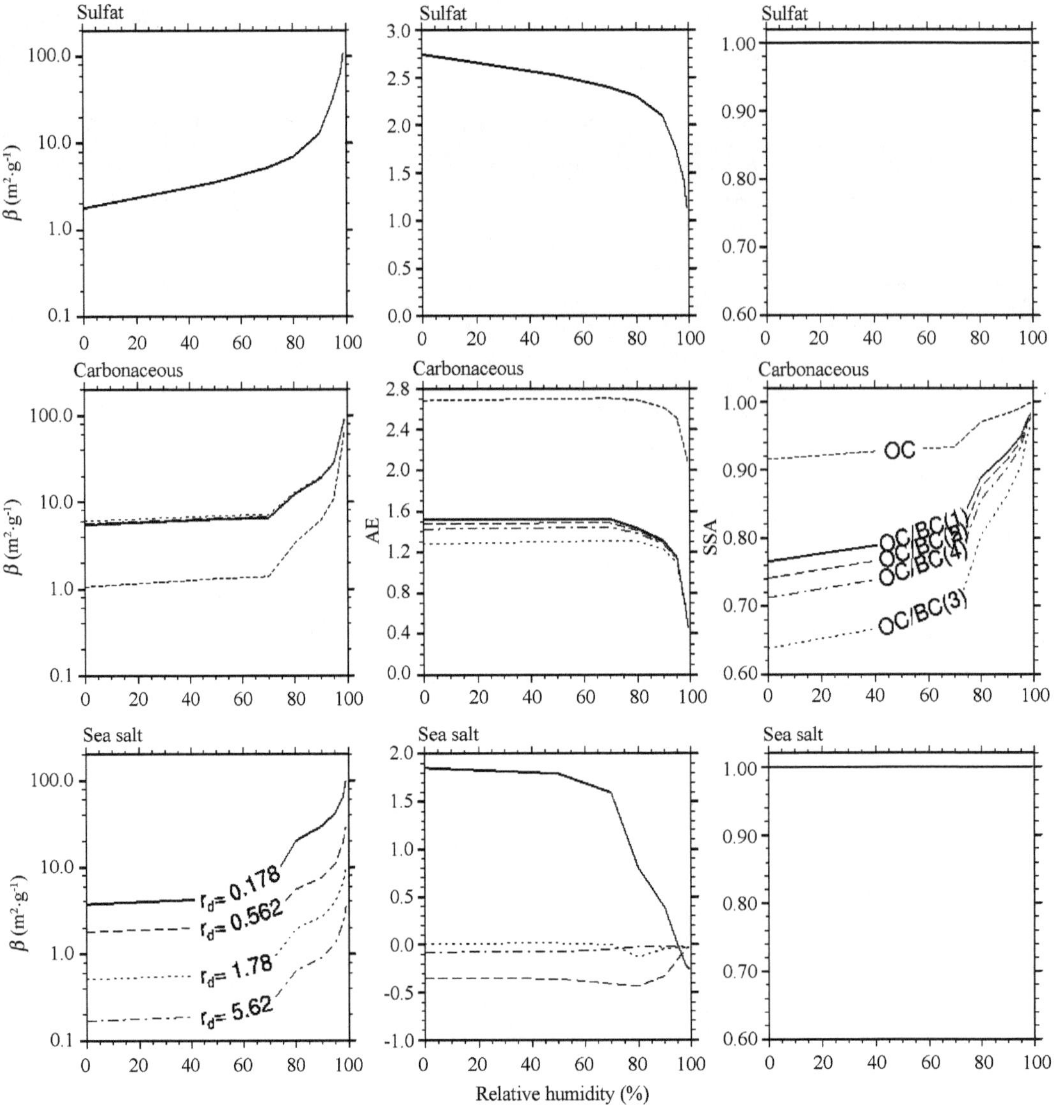

Fig. 1 Mass extinction coefficient (β) at 550 nm (left column), AE derived from β at 440 and 870 nm (middle column), and SSA at 550 nm (right column) for sulfate, carbonaceous aerosols, and sea salt as a function of relative humidity. Here, OC/BC (1—5) represent the internal mixture of OC and BC for tropical forest fire, other forest fire, fossil fuel, fuel wood, and agriculture source, respectively. Four different sizes of sea salt are shown with respective dry radius (r_d).

2.2 AERONET Dataset and Methodology

AERONET derives aerosol optical properties from sun/sky-radiometers, ground-based remote sensing instruments (Holben *et al.*, 1998). The inversion retrievals represent a wide number of parameters and characteristics that are important for comprehensive interpretation of the optical aerosol regime (Dubovik and King, 2000; Dubovik *et al.*, 2000). In this study, the AERONET level 2.0 AODs, AEs, and SSAs are used. The model numerical experiment is done for four years (2005—2008) with the first year used for spin up, and the simulation is performed year by year. The modeled and observed instantaneous AODs, AEs and SSAs from 2006 to 2008 have been compared at every three hours. The difference in sampling times between model and AERONET is within 15 minutes because AERONET's automatic-tracking Sun- and sky-scanning radiometers make direct Sun measurements at least every 15 minutes (Holben *et al.*, 2001), see also Goto *et al.*, (2012). Because AERONET AOD at 500nm is not available at some sites and AERONET retrieves SSAs at 440 and 675nm, the AERONET AODs and SSAs at both 440 and 675 nm are interpolated to compare with the modeled results at 550nm under the assumption that the AODs are proportional to wavelength on a logarithmic scale. The AE used for comparison is determined from the AODs at 440 and 870nm. Model performances against the observations are measured by three statistical parameters: correlation coefficient (R), mean bias (B) and the skill score (S) (Taylor, 2001). B is defined as the average difference between all model-observed pairs (Boylan and Russell, 2006) and can be written as

$$B = \frac{1}{N}\sum_{i=1}^{N}(C_m - C_o) \tag{1}$$

where C_m is the model-estimated value, C_o is the observed one and N equals the number of estimate - observation pairs for the comparison. S is a relatively comprehensive statistical measurement for model evaluation, because it uses both the R and the standard deviations of the observed and modeled results in the following equation

$$S = \frac{4(1+R)}{(\sigma_f + 1/\sigma_f)^2(1+R_0)} \tag{2}$$

where σ_f is the ratio of the standard deviation of model to that of observation and R_0 is the maximum attainable R which is set to 1 (Taylor, 2001; Chin *et al.*, 2009). We use only AERONET sites that yield at least 20 model-collocated observations in any one year from 2006 to and including 2008 in the comparison. Fig. 2 shows the location of the selected AERONET stations (a total of 196 sites). We also select six stations which are located in different regions of the world representing distinct aerosol sources and have retrievals of AOD, AE and SSA in all of the three years for a detailed comparison. An list of these six stations is given in Table 2.

Table 2 Location and altitude of the six selected AERONET stations for a detailed comparison

Station name	Country	Lat. /°N	Lon. /°E	Alt. /m
IER_Cinzana	Mali	13.28	−5.93	285.0
Capo Verde	Cape Verde	16.73	−22.94	60.0
Rome Tor Vergata	Italy	41.84	12.65	130.0
Taihu	China	31.42	120.22	20.0
GSFC*	America	38.99	−76.84	87.0
Sao Paulo	Brazil	−23.56	−46.73	865.0

* The full station name of GSFC is Goddard Space Flight Center.

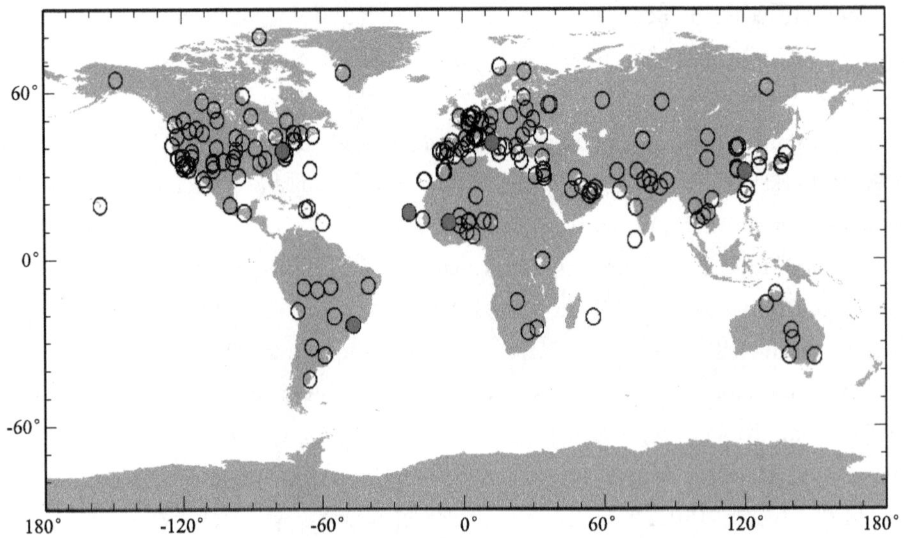

Fig. 2 Location of all AERONET sites used in this study (a total of 196 sites). Six sites in red circles are re-selected for detailed analysis.

3 Results

3.1 Evaluation of the Model Using 6 Specific sites

In this section, the modeled and retrieved instantaneous AODs, AEs, and SSAs are compared over the six sites mentioned in Table 2. The average annual cycle (monthly mean) is derived from the available instantaneous model-observed pairs.

Over the site IER_Cinzana, located within the Sahel (one of the most active dust sources in North Africa), aerosol type is dominated by the natural dust source (Ridley *et al.*, 2012). As shown in Figs. 3d, 3e and 3f, model and AERONET reveal similar seasonal variations of AOD, AE and SSA. Larger AODs and smaller AEs are found during spring and early summer periods, which are mainly caused by frequent outbreaks of North African dust events during these seasons as illustrated in the modeled aerosol components. Modeled results further indicate that a clear seasonal cycle of carbonaceous aerosols cause the maximum AE during winter season, when biomass burning occurs in the Sahel region (Crutzen and Andreae, 1990). As shown in Figs. 3a and 3b, the correlations between modeled and AERONET retrieved AODs and AEs are 0.532 and 0.762, respectively. For the skill score, we find the model also has a high skill in reproducing the retrieved AODs and AEs with scores of 0.732 and 0.789, respectively. On average, the model overestimates the AOD by 0.038 (8.7%) and underestimates the AE by 0.006 (1.6%). These results suggest that the parameterizations of the processes determining dust availability over the source region are generally acceptable. We also find that the model severely overestimates the mean AOD in January. A further investigation of the model results suggests that this is caused by an incorrect modeled dust event between 21 and 24 January 2008, when model severely overestimates the dust AOD because of the large emission caused by a fast 10 m height wind speed. This indicates more constraints are needed to improve the parameterization of dust emission, because the emission flux is assumed to be proportional to the square of the 10 m height wind speed in the model. With respect to SSA (Fig. 3c), both the correlation and skill score are low. Observed SSAs have a wider spread than modeled ones, and the modeled values are mostly confined within a narrow range 0.88 to 0.92 except the winter season (Fig. 3c). This may be caused by the less variation of the simulated aerosol components. The model generally underestimates the SSAs especially during the Sahara desert dust outbreak seasons, presumably due to too much absorption in the assumed dust optical parameters (Table 1) because the dust mainly contributes both the total and absorption AODs during the dust outbreak seasons.

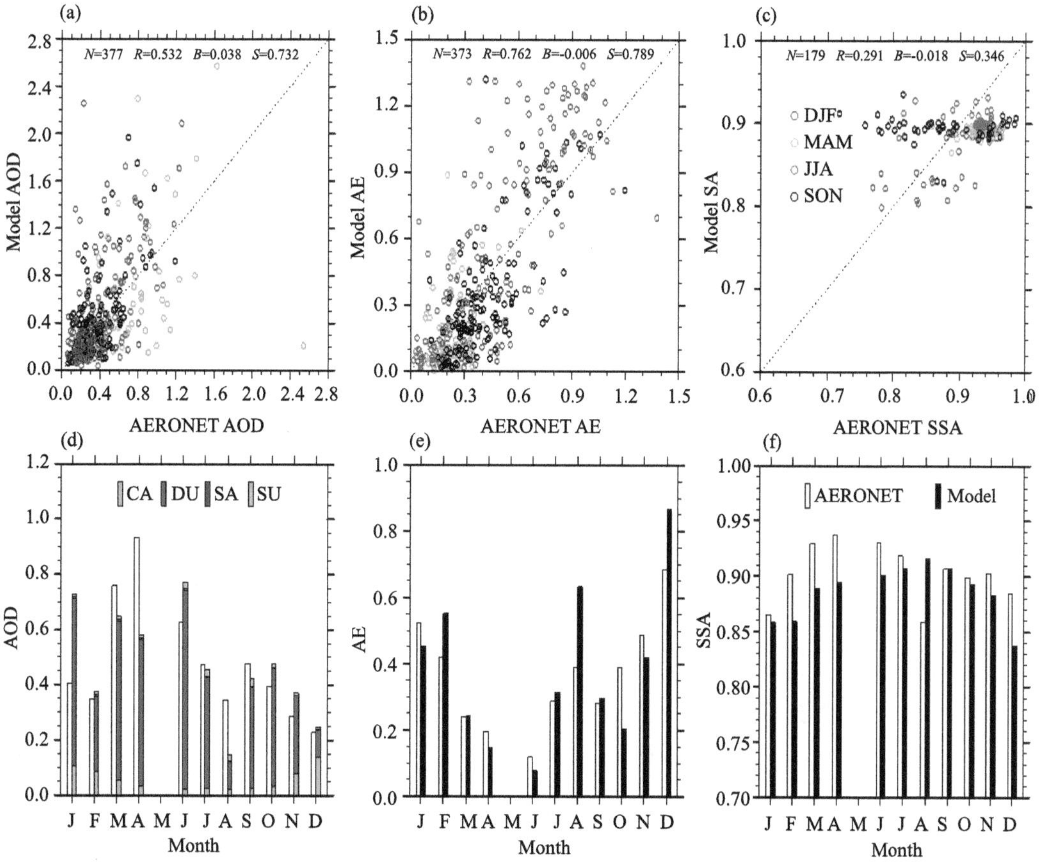

Fig. 3 (Upper panel) Scatter plots of modeled AODs (a), AEs (b), and SSAs (c) against the AERONET retrievals at IER_Cinzana site. Each point represents the matched modeled and retrieved results, and points are colored according to the seasons. N=total number of model−observed pairs, R=correlation coefficient, B=mean bias (model−AERONET), and S=skill score. (Lower panel) Annual cycles (monthly mean) of modeled and AERONET retrieved AODs (d), AEs (e), and SSAs (f). The simulated aerosol components carbonaceous aerosols (CA), dust (DU), sea salt (SA), and sulfate (SU) accumulated to total amount are also shown in panel d.

At the Capo Verde site, located right off the west coast of northern Africa, aerosol is dominated by dust transported from the Sahara desert region. This site represents the aerosol properties on a 300×300 km regional scale (Kinne *et al.*, 2003), thus the local influence is assumed to be small. The model reproduces the seasonal cycles of the retrieved AODs and AEs well with a peak AOD in July and a minimum in December as shown in Figs. 4d and 4e. The correlations between model and AERONET are 0.69 and 0.662 for AOD and AE (Figs. 4a and 4b), respectively. The high skill scores for AOD and AE (0.844 and 0.827, respectively) indicate that the model can generally reproduce the aerosol transport processes from the Sahara desert to the Atlantic. The AODs are low in the winter season resulting from the combination of a shift in the dust transport pathway and weak dust emission from the North African desert. During the summer months, the majority of long-range dust transport is toward the Caribbean, while in winter and spring seasons, the southward shift of the Intertropical Convergence Zone (ITCZ) directs more dust toward South America (Ridley *et al.*, 2012). Compared to other seasons, the higher AEs in winter season are caused by transport of the fine carbonaceous aerosols emitted from biomass burning over the Sahel region. On average, the model underestimates AOD

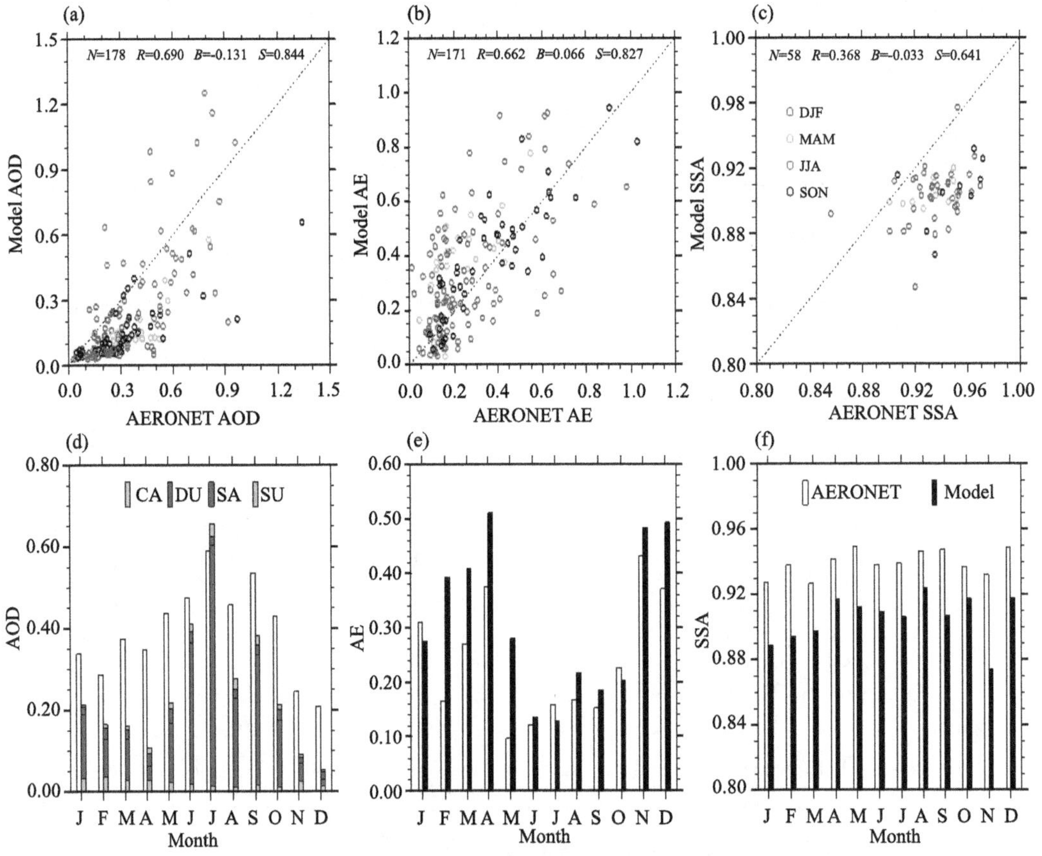

Fig. 4 Same as Fig. 3 but for Capo Verde site

by 0.131 (37.9%) and overestimates AE by 0.066 (23.7%), suggesting an underestimate of dust transported from the northern Africa toward the Atlantic Ocean. With respect to SSA (Fig. 4c), modeled SSAs are mostly lower than those of retrievals and the modeled AEs are generally higher than those of retrievals, suggesting that the model underestimates amount of the non-absorbing sea salt. Although the correlation between modeled and retrieved SSAs is weak, the skill score is high with value of 0.641. This indicates that the amplitude of the retrieved SSA variation is estimated well.

At the Rome Tor Vergata site, located on the eastern Mediterranean basin, aerosols from different sources converge in this area: urban/industrial aerosols and seasonal biomass burning from central and Eastern Europe, dust from North Africa, and sea spray from the Mediterranean Sea (De Tomasi *et al.*, 2006; Santese *et al.*, 2007). As shown in Figs. 5a and 5b, the correlations between the model and AERONET are 0.469 and 0.677, respectively, for AOD and AE. A few outliers in AOD results may result in the lower correlation coefficient, for example, the largest overestimate in summer season. This overestimate could be a result of the large dust transport process caused by the incorrectly nudging the wind field because the modeled AOD is mostly derived from the dust. The

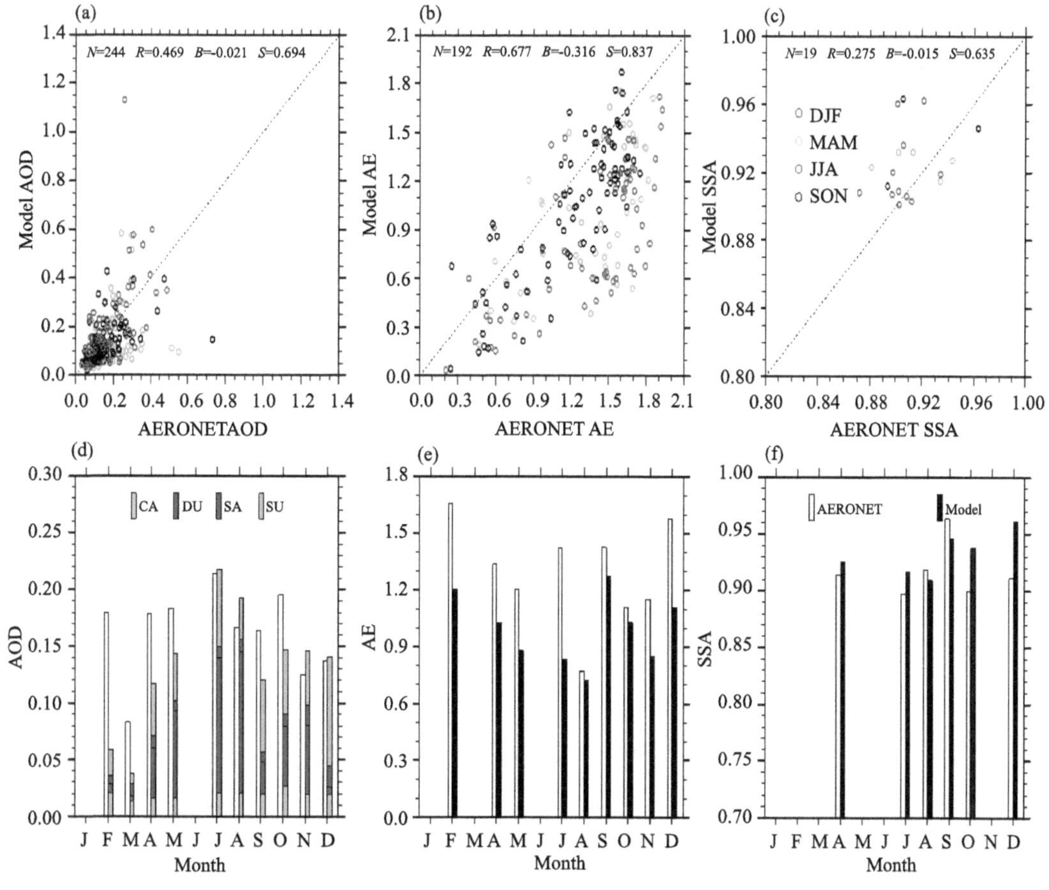

Fig. 5 Same as Fig. 3 but for Rome Tor Vergata site

retrieved AEs vary over a broad range reflecting the frequent variation in aerosol composition. Though closely reproducing the broad and frequent variation of AE with skill score of 0.837, the modeled AEs are generally much lower than the retrieved ones especially during the Sahara desert outbreak seasons. The latter reflects the overestimation of dust transport from the northern Africa toward the Mediterranean basin and Europe. However, the modeled mean AOD is still lower than the retrieved one. One reason could be that the fine aerosols (i.e., carbonaceous and sulfate) are underestimated in the model. Meanwhile, the coarse model resolution could also be a cause of underestimation of AOD especially when the local urban pollution is large. With respect to SSA (Fig. 5c), there are only a few retrievals because the instantaneous AODs are low. AERONET only retrieves the SSA under the conditions that the AOD at 440nm is larger than 0.4 along with a solar zenith angle that exceeds 50o (Dubovik and King, 2000).

At the Taihu site, situated in the Yangtze Delta region of China, aerosol is dominated by the anthropogenic sources from big cities in the Yangtze Delta Region (Pan *et al.*, 2010; Logan *et al.*, 2013). As shown in Fig. 6a, the model severely underestimates the mean AOD by 0.33 (52.9%). One reason could be a result of underestimation of the

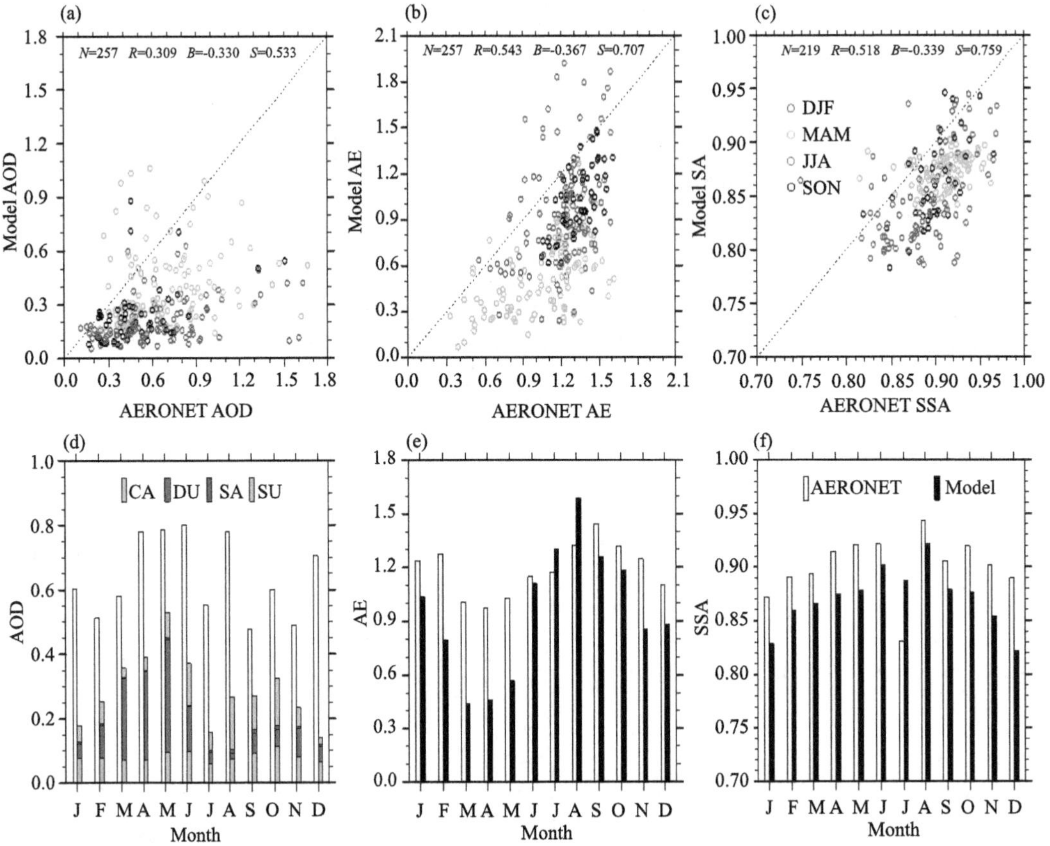

Fig. 6 Same as Fig. 3 but for Taihu site

sulfate possibly due to insufficient emission of anthropogenic SO_2 and gas-to-particle production of sulfate aerosol (Suzuki *et al.*, 2008). The other reasons could be SPRINTARS does neither treat secondary organic aerosols (SOAs) from volatile organic compounds (VOCs) nor nitrate aerosols. These processes can all increase the AOD. The coarse model resolution could also contribute to the underestimation of AOD especially during the winter or summer season when local anthropogenic sources are active. The large variability of AE makes the model simulation a challenge since aerosol components change frequently. The modeled AE variability is comparable to that of the retrievals with a correlation coefficient of 0.543 and a skill score of 0.707, however, the modeled AEs are generally lower than the retrievals, especially during Asian dust outbreak season (Figs. 6b and 6e). These underestimations of AE are likely the result of an underestimation of fine aerosols (sulfate, nitrates, and SOAs) and/or an overestimation of the dust transported from the Gobi Desert between Mongolia and north-central China and the Taklimakan Desert in western China. With respect to SSA (Fig. 6c), correlation and skill score between the modeled and the retrieved results are high, although the modeled SSAs are mostly lower than the retrieved ones. The latter result points to an overestimation of the absorption aerosols or an underestimation of the non-absorption aerosols over eastern Asia.

Over the GSFC site in the eastern coast of the United States, local biases are believed to be small (Kinne *et al.*, 2006). Aerosols are mainly from pollution sources and only sporadically perturbed by dust from the long-range transport (Chin *et al.*, 2009). As shown in Figs. 7a and 7b, the model generally underestimates both the AODs and AEs. The mean underestimates for AOD and AE are 0.062 (43.5%) and 0.476 (29%), respectively. These results illustrate the model also underestimates the fine aerosols over Northern America. The high correlations between the model and the AERONET for both AOD and AE indicate that main aerosol processes are captured by the model, however the severely underestimation of AOD over the summer season causes an underestimation of the amplitude of AOD variation and as a consequence has a low skill score for AOD. The underestimation of AE during the spring season demonstrates that the model overestimates the relative contribution of coarse aerosols, possibly due to an overestimation of the Asian dust transport and an underestimation of the fine aerosols. We further find that the modeled AEs are still lower than the retrievals even when dust contribution is quite small (such as July and August). These underestimates of AE point out the uncertainties of the sulfate optical properties caused by hygroscopic growth, because the prescribed AE values for sulfate decrease sharply over the large humidity conditions as shown in Fig. 1. With respect to SSA (Figs. 7c and 7f), we find the available retrieved SSAs are confined to the period from May to September when higher AODs are observed. The modeled SSAs and AEs are mostly both lower than the retrievals, indicating the model underestimates the fine mode non-absorption aerosols.

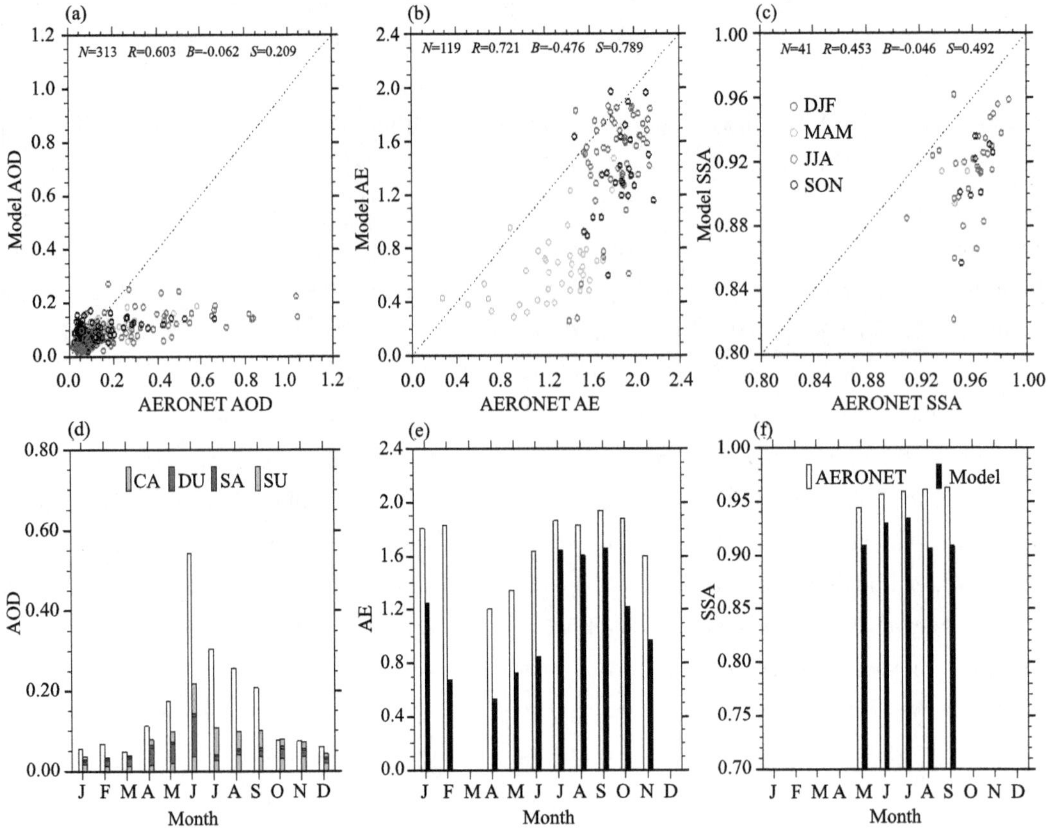

Fig. 7 Same as Fig. 3 but for GSFC site

The Sao Paulo site located in the west portion of Sao Paulo city in Brazil, is mainly dominated by biomass burning aerosols. As shown in Fig. 8d, the model reproduces the observed seasonal variation of AOD reasonably well with a peak during the dry season (August and September), when the Amazonian biomass burning plumes are transported over long distance and can influence Sao Paulo at its higher altitude (Artaxo *et al.*, 2002). The medium correlation, small mean bias, and high skill score for AOD (Fig. 8a) suggest that the used emission inventories for the biomass burning regions are quite acceptable. Both the modeled and observed AEs vary over a narrow range with mean bias value of −0.087 (−6.3%) and skill score value of 0.634, indicating that the model simulates the relatively simple aerosol components well. With respect to SSA (Fig. 8c), the modeled values on average coincide with the retrievals and the skill score is 0.581, suggesting that the assumed mass ratios of OC and BC for the carbonaceous aerosols are suitable. However, more studies are needed to further investigate the reasons of the low correlations between model and AERONET for the AE and SSA.

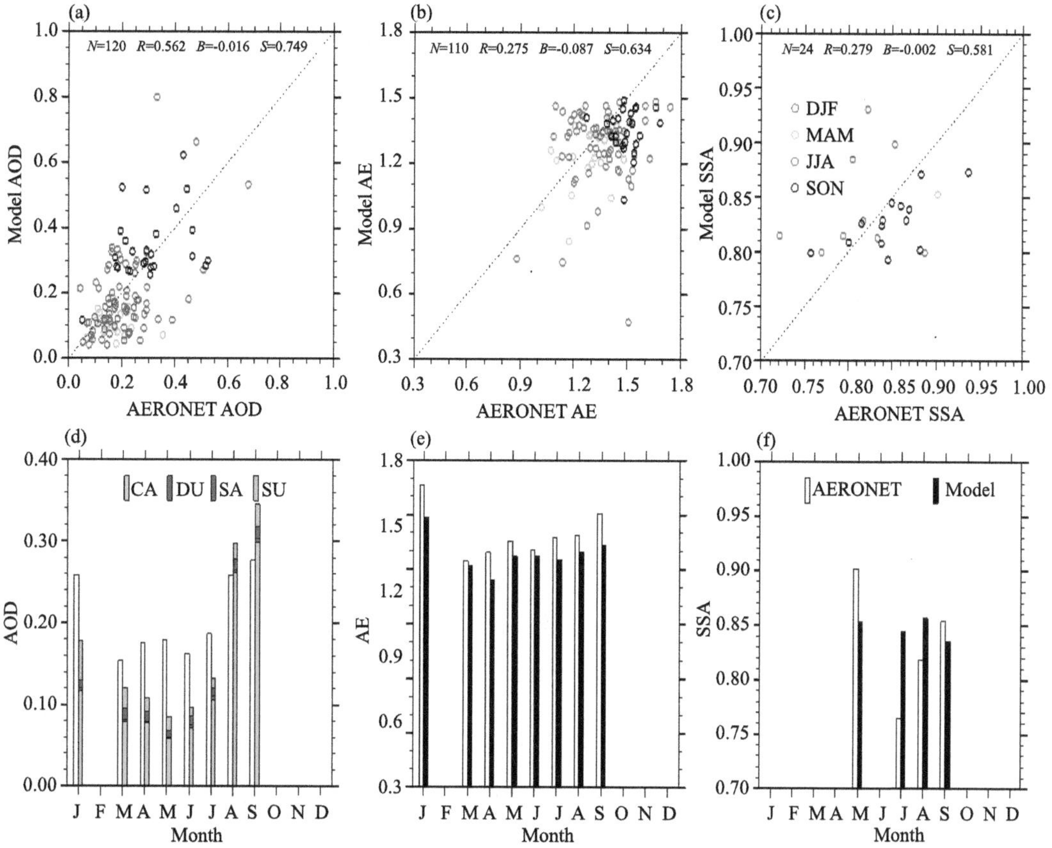

Fig. 8　Same as Fig. 3 but for Sao Paulo site

3.2　Evaluation for All Sites Using 3-year Averaged Observations

Fig. 9 shows the comparisons of the site-specific temporally collocated 3-year mean AODs, AEs and SSAs as simulated by the model and retrieved by the AERONET. With respect to AOD, we find a strong correlation with value of 0.753, and 82.1% of the simulated AODs agree within a factor of two with the AERONET retrieved AODs. The mean bias and the skill score are −0.029 and 0.875, respectively. Furthermore, the largest discrepancy (−0.581) is in Beijing, China, and this could be the result of the unresolved emissions of urban aerosol sources because of the coarse model resolution. Investigating the spatial distribution of the relative mean biases between the modeled and the retrieved AODs, we find that the model tends to underestimate AODs over the regions dominated by anthropogenic aerosols, such as eastern China, North America, and India. These results reflect the underestimation of the fine mode aerosols in the model, as was also noted in Fig. 6 and Fig. 7. The mean AODs generally agree with the retrievals over the dust source region (northern Africa) except the Tamanrasset site, indicating that the parameterization of the dust emission (Takemura *et al.*, 2009) is reasonable. The combination of low mean retrieved AOD and erroneously large mean dust AOD caused by high wind speed leads to the significantly high relative mean bias at the Tamanrasset site. With respect to AE, the correlation and the skill score are high (0.735 and 0.822, respectively), although the modeled mean AE is lower than that of AERONET by 0.388. The model also tends to underestimate the AEs over the pollution regions where mean AEs are high, such as in North America, Europe, and eastern Asia, further pointing to the underestimation of the burden of fine mode aerosols over these regions. With respect to SSA, the model has a lower skill score (0.721) than that of AOD and AE. The simulated SSAs on average coincide with the retrievals, however the correlation is low (0.45). We also find a large variability in the relative mean bias over the biomass burning regions with the largest underestimation in the Ji_Parana_SE site (Brazil) and almost no bias in the Sao Paulo site.

3.3 Evaluation for All Sites Using Instantaneous Observations

The overall statistical comparisons betweenthe model and the AERONET for all the available model-observed pairs over all the AERONET sites during 2006 to 2008 are shown in Table 3. The model can generally reproduce the retrieved AODs, AEs, and SSAs with correlation coefficients of 0.509, 0.661, and 0.446, respectively. On average, the model underestimates the AOD and AE by 0.022 (10.5%) and 0.329 (31.2%), respectively, while the simulated SSA has a good agreement with the retrieval. Among the three optical properties, the model has the highest skill score in reproducing AE and the lowest skill score for SSA. As shown in Fig. 10, the probability distribution functions (PDFs) and cumulative distribution functions (CDFs) of AOD, AE and SSA between the model and AERONET are compared. With respect to AOD, the PDFs are both uni-modal distributions with the same modal value of 0.075, although the model slightly overestimates the occurrence frequencies of small AODs (<0.1). With respect to AE, the AERONET retrievals show a bi-modal distribution with two modal values of 0.3 and 1.5, while the modeled values show only one peak with a modal value of 0.1. The comparisons of the PDF and CDF of AE both illustrate that the model overestimates the occurrence frequencies of small AEs (<1.0) and underestimates those of large AEs (>1.0). With respect to SSA, we find that the model reproduces a similar uni-modal distribution as that

Table 3 Statistical comparisons between modeled and AERONET retrieved AODs, AEs, and SSAs over all selected sites during the period 2006—2008

Optical properties	N	R	B	S
AOD	23496	0.509	−0.022	0.754
AE	14690	0.661	−0.329	0.818
SSA	3971	0.446	−0.030	0.704

Note: N=total number of model-observed pairs, R=correlation coefficient, R=mean bias, and R= skill score.

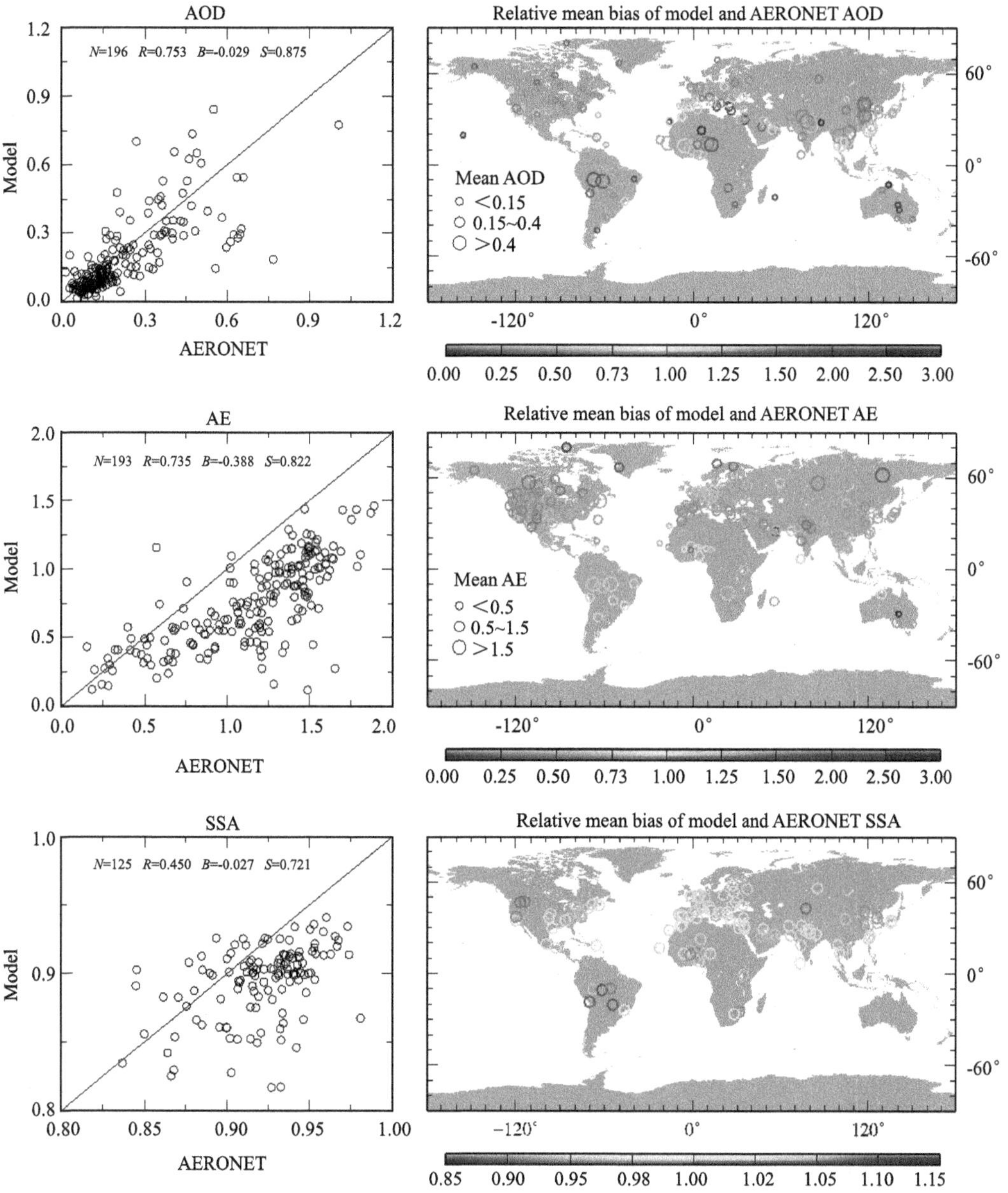

Fig. 9　Comparisons between the modeled and the retrieved AODs (top), AEs (middle), and SSAs (bottom) from a total of 196 AERONET sites (left column). Each point represents the average from modeling and AERONET retrievals at each site during the 3-year period. N=total number of AERONET sites involved in the comparison, R= correlation coefficient, B= mean bias, and S= skill score. The relative mean biases between modeled and AERONET retrieved values are also given using colored open circle (right column). Here, the relative mean bias is calculated as $\sum C_m / \sum C_0$. C_m is the model-estimated value, and C_0 is the observed one.

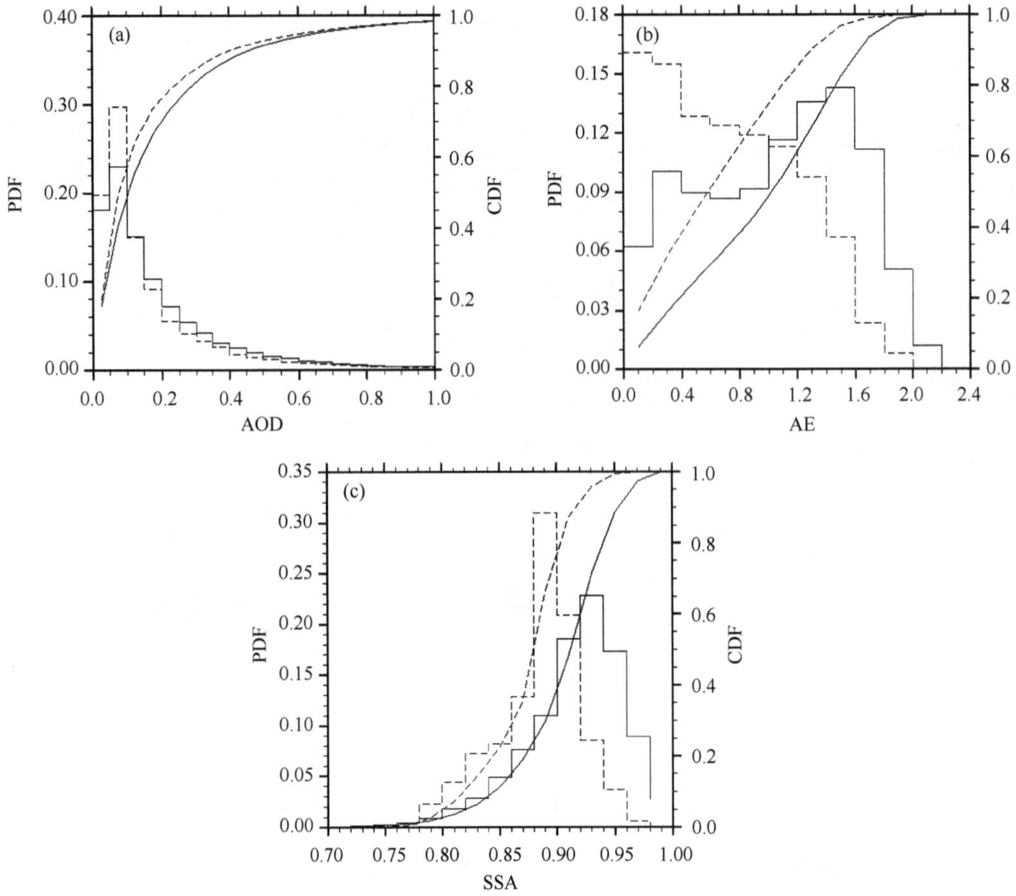

Fig. 10 Probability Distribution Functions (PDFs) and Cumulative Distribution Functions (CDFs) of (a) AOD, (b) AE, and (c) SSA for both the AERONET (solid line) and the model (dashed line) from all matched results over 196 AERONET sites during the period 2006—2008. The bin widths used for histogram calculations are 0.05, 0.2, and 0.02 for AOD, AE, and SSA, respectively.

of the AERONET. However, the peak of the PDF of the modeled SSAs is shifted to smaller value compared to that of the retrieved SSAs, indicating the model overestimates the distribution of small SSAs. Moreover, the PDF of modeled SSAs has a sharper distribution with a peak at 0.89, indicating the modeled SSAs exhibit a smaller variability compared to the retrieved ones.

4 Discussion and Conclusions

The global aerosol key optical properties simulated by a three-dimension aerosol transport module coupled with a new dynamical core for years 2006 to 2008 are evaluated by comparing the simulated AODs, AEs, and SSAs with the corresponding AERONET retrievals. The evaluations reveal that the model can generally capture the seasonal variations of AOD and AE over different aerosol source regions of the world, especially the places where dust or biomass burning aerosol dominates. Meanwhile, the spatial distributions of the AODs and AEs are also reproduced reasonably well. For the all sites 3-year mean AODs and AEs, the correlations between the model and AERONET are 0.753 and 0.735, respectively, and 82.1% of the modeled AODs agree within a factor of two with the retrieved AODs. Compared with the AOD and AE, the model shows a relatively low skill score to reproduce the SSA, however the simulated global 3-year mean SSA coincides with the AERONET retrieval. These results indicate that the model is suitable to be used for the study of the temporal and spatial variations of the aerosol key optical properties and the aerosol associated climate impacts over global scale. The primary model deficiency is an underestimation of fine mode aerosol AOD and a corresponding underestimation of AE over pollution region. The model underestimates the global 3-year mean AOD and AE by 0.022 (10.5%) and 0.329 (31.2%), respectively, compared to the AERONET retrievals. Meanwhile, the model overestimates the frequencies of occurrence for small AEs and SSAs, although the PDF of modeled AODs is comparable to that of the AERONET retrievals. To correct the model underestimations of the AOD and AE over the pollution region, we suggest to improve the prescribed optical parameters for sulfate by performing more sensitivity experiments in order to choose the optimal values, to obtain the better emission inventories by using higher temporal and spatial resolution datasets, to implement the formation of SOAs and nitrates, and to improve the treatment of sulfate chemical processes by introducing more sophisticated chemical dynamics into the model (Goto *et al.*, 2011). We also find the model tends to overestimate the dust transport from the Sahara desert to the Mediterranean and central Europe while underestimating transatlantic transport, although the parameterization for the dust over the source region is acceptable. To eliminate these transport discrepancies, we suggest to assess the transport of dust aerosol by tuning the parameters which influence the dry and wet removal processes and improving the nudging of meteorological fields by using different analysis or reanalysis meteorological data. The comparisons of optical characteristics between our model and AERONET have brought to the forefront the need for model further improvements. We believe such efforts will lead to a considerable improvement of our model to achieve a better agreement with satellite and ground-based measurements, and this is beneficial to accurately assess aerosol climate impacts of the past and project the climate response to the change of aerosols in the future.

Acknowledgements

Some of the authors are supported by projects from JAXA/EarthCARE, MEXT/VL for Climate System Diagnostics, the MOE/Global Environment Research Fund A-1101, NIES/GOSAT, NIES/CGER, MEXT/RECCA/SALSA, National Natural Science Funds of China (41130104), and the National Basic Research Program of China (973 Program, 2013CB955803). We are thankful to the relevant researchers for the AERONET sites, NCEP FNL analysis data, and to the SPRINTARS and NICAM developers. We also acknowledge the useful comments by two anonymous reviewers that helped to improve this paper.

References

Albrecht B A. 1989. Aerosols, Cloud Microphysics, and Fractional Cloudiness. *Science* **245**: 1227-1230.

Arakawa A, Schubert W H. 1974. Interaction of a Cumulus Cloud Ensemble with the Large-Scale Environment, Part I. *Journal of the Atmospheric Sciences*. **31**: 674-701.

Artaxo P, Martins J V, Yamasoe M A, *et al*. 2002. Physical and chemical properties of aerosols in the wet and dry seasons in Rondonia, Amazonia. *Journal of Geophysical Research*. **107**: 8081.

Boylan J W, Russell A G. 2006. PM and light extinction model performance metrics, goals, and criteria for three-dimensional air quality models. *Atmospheric Environment*, **40**: 4946-4959.

Chin M, Diehl T, Dubovik O, *et al*. 2009. Light absorption by pollution, dust, and biomass burning aerosols: a global model study and evaluation with AERONET measurements, Annales geophysicae: atmospheres, hydrospheres and space sciences, pp. 3439-3464.

Coakley J A, Cess R D, Yurevich F B. 1983. The Effect of Tropospheric Aerosols on the Earth's Radiation Budget: A Parameterization for Climate Models. *Journal of the Atmospheric Sciences*. **40**: 116-138.

Crutzen P J, Andreae M O. 1990. Biomass Burning in the Tropics: Impact on Atmospheric Chemistry and Biogeochemical Cycles. *Science*. **250**: 1669-1678.

De Tomasi F, Tafuro A, Perrone M. 2006. Height and seasonal dependence of aerosol optical properties over southeast Italy. *Journal of Geophysical Research*. **111**: D10203.

Diehl T, Heil A, Chin M, *et al*. 2012. Anthropogenic, biomass burning, and volcanic emissions of black carbon, organic carbon, and SO_2 from 1980 to 2010 for hindcast model experiments. *Atmos. Chem. Phys. Discuss.*, **12**:24895-24954.

Dubovik O, King M D. 2000. A flexible inversion algorithm for retrieval of aerosol optical properties from Sun and sky radiance measurements. *Journal of Geophysical Research*. **105**: 20673-20696.

Dubovik O, Smirnov A, Holben B N, *et al*. 2000. Accuracy assessments of aerosol optical properties retrieved from Aerosol Robotic Network (AERONET) Sun and sky radiance measurements. *Journal of Geophysical Research*. **105**: 9791-9806.

Generoso S, Bey I, Labonne M, *et al*. 2008. Aerosol vertical distribution in dust outflow over the Atlantic: Comparisons between GEOS-Chem and Cloud-Aerosol Lidar and Infrared Pathfinder Satellite Observation (CALIPSO). *J. Geophys. Res.*, **113**: D24209.

Goto D, Kanazawa S, Nakajima T, *et al*. 2011. Evaluation of a relationship between aerosols and surface downward shortwave flux through an integrative analysis of modeling and observation. *Atmospheric Environment*, 294-301.

Holben B, Eck T, Slutsker I, *et al*. 1998. AERONET—A federated instrument network and data archive for aerosol characterization. *Remote Sensing of Environment*. **66**: 1-16.

Kinne S, Lohmann U, Feichter J,*et al*. 2003. Monthly averages of aerosol properties: A global comparison among models, satellite data, and AERONET ground data. *Journal of Geophysical Research: Atmospheres*, **108**: 4634.

Kinne S, Schulz M, Textor C, *et al*. 2006. An AeroCom initial assessment - optical properties in aerosol component modules of global models. *Atmos. Chem. Phys.*, **6**:1815-1834.

Logan T, Xi B, Dong X, *et al*. 2013. Classification and investigation of Asian aerosol absorptive properties. *Atmos. Chem. Phys.*, **13**: 2253-2265.

Matthias V. 2008. The aerosol distribution in Europe derived with the Community Multiscale Air Quality (CMAQ) model: Comparison to near surface in situ and sunphotometer measurements. *Atmos. Chem. Phys.*, **8**: 5077-5097.

Miura H, Satoh M, Nasuno T, *et al*. 2007. A Madden-Julian Oscillation Event Realistically Simulated by a Global Cloud-Resolving Model. *Science*. **318**: 1763-1765.

Niwa Y, Patra P K, Sawa Y, *et al*. 2011. Three-dimensional variations of atmospheric CO_2: aircraft measurements and multi-transport model simulations. *Atmos. Chem. Phys.*, **11**: 13359-13375.

Nakajima T, Tsukamoto M, Tsushima Y, *et al*. 2000. Modeling of the Radiative Process in an Atmospheric General Circulation Model. *Applied Optics*. **39**: 4869-4878.

Pan D-M, Randall D D A. 1998. A cumulus parameterization with a prognostic closure. *Quarterly Journal of the Royal Meteorological Society*. **124**: 949-981.

Pan L, Che H, Geng F, *et al*. 2010. Aerosol optical properties based on ground measurements over the Chinese Yangtze Delta Region. *Atmospheric Environment*. **44**: 2587-2596.

Ridley D A, Heald C L, Ford B. 2012. North African dust export and deposition: A satellite and model perspective. *Journal of Geophysical Research*. **117**: D02202.

Santese M, De Tomasi F, Perrone M R. 2007. AERONET versus MODIS aerosol parameters at different spatial resolutions over southeast Italy. *Journal of Geophysical Research*. **112**: D10214.

Satoh M, Matsuno T, Tomita H, *et al*. 2008. Nonhydrostatic icosahedral atmospheric model (NICAM) for global cloud resolving simulations. *Journal of Computational Physics*. **227**: 3486-3514.

Schutgens N A J, Miyoshi T, Takemura T, *et al*. 2010. Applying an ensemble Kalman filter to the assimilation of AERONET observations in a global aerosol transport model. *Atmospheric Chemistry and Physics*. **10**: 2561-2576.

Sekiguchi M, Nakajima T. 2008. A k-distribution-based radiation code and its computational optimization for an atmospheric general circulation model. *Journal of Quantitative Spectroscopy and Radiative Transfer*. **109**: 2779-2793.

Sudo K, Takahashi M, Kurokawa J I, *et al*. 2002b. CHASER: A global chemical model of the troposphere 1. Model description. *Journal of Geophysical Research*. **107**: 4339.

Suzuki K, Nakajima T, Satoh M, *et al*. 2008. Global cloud-system-resolving simulation of aerosol effect on warm clouds. *Geophysical Research Letters*. **35**: L19817.

Suzuki K, Stephens G L, van den Heever S C, *et al*. 2011. Diagnosis of the Warm Rain Process in Cloud-Resolving Models Using Joint CloudSat and MODIS Observations. *Journal of the Atmospheric Sciences*, **68**: 2655-2670.

Takata K, Emori S, Watanabe T. 2003. Development of the minimal advanced treatments of surface interaction and runoff. *Global and Planetary Change*. **38**: 209-222.

Takemura, T. 2012. Distributions and climate effects of atmospheric aerosols from the preindustrial era to 2100 along Representative Concentration Pathways (RCPs) simulated using the global aerosol model SPRINTARS. Atmos. Chem. Phys. 12, 11555-11572.

Takemura T, Egashira M, Matsuzawa K, *et al*. 2009. A simulation of the global distribution and radiative forcing of soil dust aerosols at the Last Glacial Maximum. *Atmospheric Chemistry and Physics*. **9**: 3061-3073.

Takemura T, Nakajima T, Dubovik O, *et al*. 2002. Single-Scattering Albedo and Radiative Forcing of Various Aerosol Species with a Global Three-Dimensional Model. *Journal of Climate*. **15**: 333-352.

Takemura T, Okamoto H, Maruyama Y, *et al*. 2000. Global three-dimensional simulation of aerosol optical thickness distribution of various origins. *Journal of Geophysical Research*. **105**: 17853-17873.

Taylor K E. 2001. Summarizing multiple aspects of model performance in a single diagram. *Journal of Geophysical Research*. **106**: 7183-7192.

Textor C, Schulz M, Guibert S, *et al*. 2007. The effect of harmonized emissions on aerosol properties in global models-an Aero Com experiment. *Atmospheric Chemistry and Physics*. **7**: 4489-4501.

Tomita H, Satoh M. 2004. A new dynamical framework of nonhydrostatic global model using the icosahedral grid. *Fluid Dynamics Research*. **34**: 357-400.

ESTIMATION OF THE ANTHROPOGENIC HEAT RELEASE DISTRIBUTION IN CHINA FROM 1992 TO 2009①

CHEN Bing[1,2], SHI Guangyu[1②], WANG Biao[1], ZHAO Jianqi[1] and TAN Saichun[1]

[1] The State Key Laboratory of Numerical Modeling for Atmospheric Sciences and Geophysical Fluid Dynamics (LASG), Institute of Atmospheric Physics, Chinese Academy of Sciences, Beijing 100029, China

[2] Graduate University of Chinese Academy of Sciences, Beijing 100049, China

Abstract

Stable light data from Defense Meteorological Satellite Program (DMSP) Operational Linescan System (OLS) satellites and authoritative energy consumption data distributed by China's National Bureau of Statistics were applied to estimate the distribution of anthropogenic heat release in China from 1992 to 2009. A strong linear relationship was found between DMSP/OLS Digital Number data and anthropogenic heat flux density. The results indicate that anthropogenic heat release in China was geographically concentrated and distributed and was fundamentally correlated with economic activities. The anthropogenic release in economically developed areas in Northern, Eastern and Southern China was much larger than other regions, whereas that in Northwestern and Southwestern China was very small. The mean anthropogenic heat flux density in China increased from 0.07 W • m^{-2} in 1978 to 0.28 W • m^{-2} in 2008. The results indicate that in the anthropogenic heat-concentrated regions of Beijing, the Yangtze River Delta and the Pearl River Delta area, the levels were much higher than the average. The effect of aggravating anthropogenic heat release on climate change warrants further investigation.

Key words: DMSP/OLS data, estimation, distribution, anthropogenic heat flux, China

1 Introduction

Energy consumption and greenhouse gases from fossil fuel combustion have increased sharply since the Industrial Revolution. Aerosols, anthropogenic heat and greenhouse

---

① The paper pubished in *Acta Meteor. Sinica*, **26**(4), 507-515, doi: 10.1007/s13351-012-0409-y. 2012.

Supported by the National Natural Science Foundation of China (No. 40775008 and No. 41075015) and the Ministry of Science and Technology of China (2010DFA22770).

② Corresponding author: shigy@mail.iap.ac.cn

gases such as CO_2, H_2O, and CH_4, have been released into the atmosphere and exert profound influence on the climate. Changes in the atmosphere's composition, land use and land cover change (LUCC) are considered the main aspects of the climate effect caused by human factors. The impacts of human activities contribute to changes in the composition of the atmosphere and disrupt the energy balance of the earth-atmosphere system, which lead to a warmer and changeable climate. As we know, "urban heat islands" are a common phenomenon partly caused by the release of heat into the environment from energy use for human activities (IPCC, 2007). The economic prosperity of the modern world is based on the vast consumption of different energy resources, which eventually turn into anthropogenic heat released into the atmosphere. Anthropogenic heat is essential for the climate in urban areas (Block *et al*, 2004; Fan and Sailor, 2005). Anthropogenic heat may increase turbulent fluxes in sensible and latent heat, which can result in the atmosphere containing more energy (Oke, 1988). The flux of anthropogenic heat caused by human activities can have a significant influence on the dynamics and thermodynamics of the urban boundary layer (Ichinose *et al.*, 1999; Block *et al.*, 2004; Fan and Sailor, 2005). It can also increase the temperature and height of the boundary layer, especially at night, which affects reaction rates and chemical processing of emitted species (Markar *et al.*, 2006). The climate effect due to anthropogenic heat release is important to climate change. Clearly, the energy balance of the surface can be broken due to anthropogenic heat. Anthropogenic heat energy forms have been considered as both latent heat and sensible heat emissions in different climate models (Oleson *et al.*, 2010; Block *et al.*, 2004). Estimating the anthropogenic heat flux is essential for determining the effect of anthropogenic heat in climate models. Previous research has generally focused on a single city or a small area (Lee *et al.*, 2009; Hamilton *et al.*, 2009) rather than a larger scale and has not focused on the continuously changing trends. Contrary to the methods used in urban areas, where anthropogenic heat is considered from all major sources, including buildings, industry and vehicles (Lee *et al.*, 2009; Hamilton *et al.*, 2009), energy from additional heating resources are ultimately converted into anthropogenic heat. Considering the development of urban agglomeration on a global scale in the near future, the flux of anthropogenic heat release will increase sharply with the development of the world economy. The role anthropogenic heat plays in the climate effect of urban agglomerations is essential in global climate change. Contrary to previous research focused on anthropogenic heat flux in a single area or city (Lee *et al.*, 2009; Hamilton *et al.*, 2009) and the climate effect limited to a small area (Tong *et al.*, 2004; Block *et al.*, 2004), in the present study, the anthropogenic heat flux distribution in China from 1992 to 2009 was estimated using DMSP/OLS satellite data, which provide a useful parameterization for climate models and make further research on the climate effect of anthropogenic heat and urban agglomerations in climate models feasible.

The U.S. Air Force Defense Meteorological Satellite Program (DMSP) has been in operation since the mid-1960s. The Defense Meteorological Satellite Program (DMSP) Operational Linescan System (OLS) is an oscillating scan radiometer with a broad field of view (about 3000 km swath) and captures images at a nominal resolution of 0.56 km. The images are smoothed onboard into 5×5 pixel blocks to 2.8 km to reduce the amount of memory required onboard the satellite. The OLS sensor has two broadband sensors, one is visible/infrared (VNIR, 0.4－1.1 μm) and the other is thermal infrared (TIR, 10.5－12.6 μm), which is able to detect lights from cities and towns and gas flares as well as ephemeral events such as fires or clouds illuminated by moonlight and lightening. The OLS sensor is not only able to detect visible light band sources down to 10^{-9} Watts/cm^2/sr, but also produces visually consistent imagery of clouds at all scan angles. The sensitivity of the OLS sensor is four orders of magnitude greater than other sensors, such as NOAA-Advanced Very High Resolution Radiometer (AVHRR) or Landsat Thematic Mapper (Elvidge *et al.*, 1997a), which provide the OLS with a unique ability to detect low levels of visible and near-infrared radiance at night. The DMSP/OLS data have been widely applied in many research fields, such as in estimating population (Amaral *et al.*, 2006), energy consumption (Husi *et al.*, 2010), economic activity (Elvidge *et al.*, 1997b; Tilottama *et al.*, 2009), human settlement and urban extension (Small *et al.*, 2005). Additionally, the DMSP/OLS data are applied to analyze greenhouse gas emissions (Doll *et al.* 2000) and light pollution (Cinzano *et al.*, 2001). Recent research (Tilottama *et al.*, 2009; Husi *et al.*, 2010) indicates that the DMSP/OLS satellite data are closely correlated with economic development levels and energy consumption, which provide a good way to estimate these levels. Generally speaking, the areas where the DMSP/OLS data are high are often developed economies with high energy consumption (Elvidge *et al.*, 1997; Husi *et al.*, 2010; Tilottama *et al.*, 2009) and, therefore, more anthropogenic heat emissions. The world economy is based on the vast consumption of various energy resources, which turn into anthropogenic heat that is eventually released into the atmosphere. In this paper, a simple model based on DMSP/OLS satellite data provided by the National Geophysical Data Center (NGDC) of the National Oceanic and Atmospheric Administration (NOAA) and energy consumption statistics published by China's National Bureau of Statistics is developed to analyze the distribution of anthropogenic heat flux in China, which is closely correlated with economic activity and energy consumption.

2 Data and Methods

Assuming that all energy consumption by humans is eventually converted into heat, the anthropogenic heat flux density (Q_a) of a specific area and over a specified time can be approximated by the following equation:

$$Q_a = \frac{energy\ consumption[\mathrm{J}]}{time[\mathrm{s}] \cdot area[\mathrm{m}^2]} \tag{1}$$

The effect of anthropogenic heat in different areas and at different scales can be obtained by equation (1). Using equation (1) along with global energy consumption data for the year 2009 provided by BP (www. bp. com/statisticalreview), we found a mean anthropogenic energy flux density (Q_a) of about 0. 10 $W \cdot m^{-2}$ for all global land area and 0. 28 $W \cdot m^{-2}$ for China.

The authoritative statistics come from the *China Statistical Yearbook* 2009 and *China Energy Statistical Yearbook* 2009, the latest publications compiled by China's National Bureau of Statistics. The *China Statistical Yearbook* is published annually by China's National Bureau of Statistics and contains detailed statistics on all aspects of society, including the economy, population, energy data and resources in China, with statistics provided separately by provinces or districts. The *China Energy Statistical Yearbook* contains detailed data on energy production and consumption among the provinces. Stable light data from the DMSP/OLS Nighttime Lights Version 4 Time Series were applied to the research. These data are cloud-free composites made using all the available archived DMSP/OLS smooth resolution data by calendar year. The products are 30 arc second grids that span −180 to 180 degrees longitude and −65 to 75 degrees latitude. The stable light data contain the light from cities, towns, and other sites with persistent lighting. Glare, sunlight and moonlight data, as well as lighting from auroras, have been excluded. Ephemeral events, such as fires, have been discarded. The background noise was identified and replaced with zero values. Each grid of the stable light data from the DMSP/OLS Nighttime Lights Version 4 Time Series has a Digital Number from 0 to 63. The Digital Number, which is a positive integer assigned to the response of a sensor relative to the intensity of the signal received by the sensor, depends on the number of bits assigned to the quantifying sensor response. For the analysis presented in this work, continuous stable light data from 1992 to 2009 from DMSP satellites were provided by NGDC, including the following four satellites: F10 (from 1992 to 1994), F12 (from 1994 to 1999), F14 (from 1997 to 2003) and F16 (from 2004 to 2009). The mean anthropogenic heat release flux for all of China, as well as the districts in China, from 1992 to 2008 was obtained from China's National Bureau of Statistics. The mean Digital Number for all of China and its districts was obtained by Arc GIS. The relationship between the mean anthropogenic heat flux density and the mean Digital Number was analyzed. The coefficients of determination (R^2) produced by the linear regression between the stable light data from four satellites and the anthropogenic heat flux data were all more than 0. 9, which indicates a strong linear correlation as shown in Table 1 and Figure 1. Considering the differences among the various satellites due to varying instruments and observation conditions, different satellite data obtained in the same year were analyzed using the T-test. Analysis of the satellite data for China obtained by F10 and F12 in 1994 indicates that the correlation coefficient between the two datasets is 0. 996 and the T-test result shows no significant difference between the two satellites. The same result was obtained with the data for China from F12 and F14

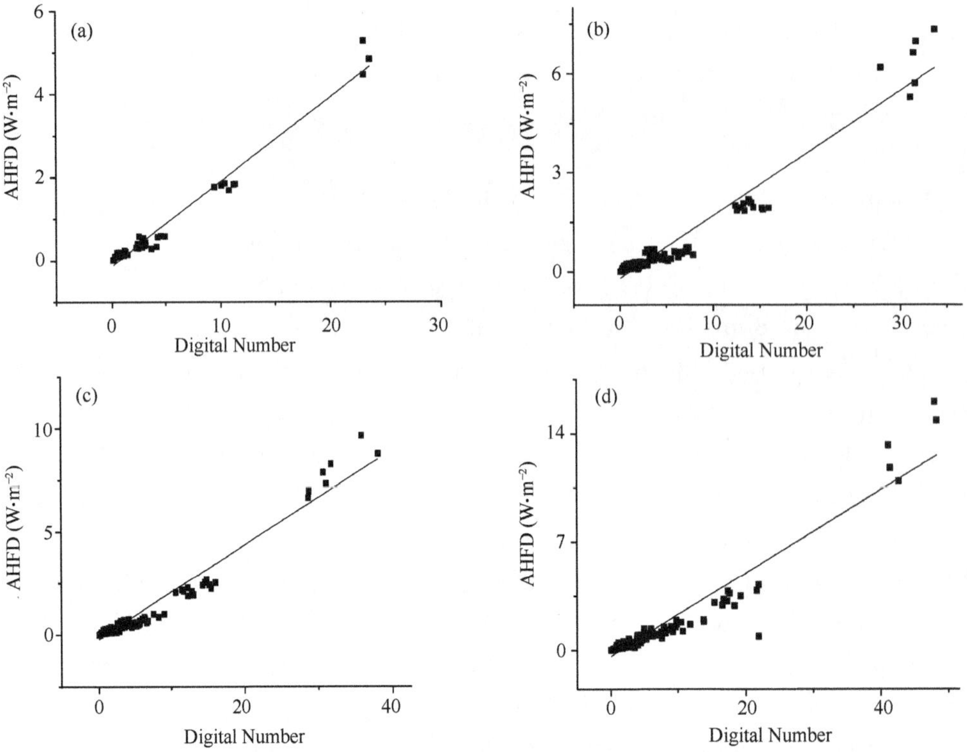

Fig. 1 Relationshipe between Digital Number of DMSP/OLS stable light and Anthropogenic Heat Flux Density(AHFD) (W·m^{-2}) in China

(a) the F10 satellite from 1992 to 1994; (b) the F12 satellite from 1994 to 1999; (c) the F14 satellite from 1997 to 2003; (d) the F16 satellite from 2004 to 2009

between 1997 and 1999. The correlation coefficient between them is 0.997, whereas the *T*-test result shows that the satellite data are quite similar. The results allow us to conclude that there is no significant difference between the various satellite data. It is therefore feasible for the data from different satellites to be applied in this paper. One curve-fitting model requires the following criteria: (1) when the Digital Number (DN) of the satellite data is 0, the anthropogenic heat flux density should be 0 and (2) when the DN becomes large, the anthropogenic heat flux density should be at a reasonable level. A curve fit model was used to estimate the anthropogenic heat flux density where the anthropogenic heat flux density was expected to increase with increasing Digital Number (DN) and the anthropogenic heat flux would stay at a reasonable level when the Digital Number is at its minimum or maximum. The areas where the DN is 0 are probably desolate areas, such as forests, crop land, ice land, desert regions or oceans. In these areas, the anthropogenic heat flux is very low (the probable value is 0). For this reason, the DMSP/OLS data are suitable for estimating anthropogenic heat flux in the model. Based on this assumption, different curve-fitting models, including many nonlinear models, were used to estimate the model. Moreover, the statistical hypothesis test was applied to the simulation model. Based on analysis of the different model outcomes, the simulation results indicate that the linear model, $y = k \cdot x$ (where k is the slope of the linear function), is the best among the various models.

Table 1 The simulated result between the linear function, $y=k \cdot x$, and the stable light data from four Satellites

Satellite series number	Coefficient of determination (R^2)
F10	0.977
F12	0.944
F14	0.949
F16	0.914

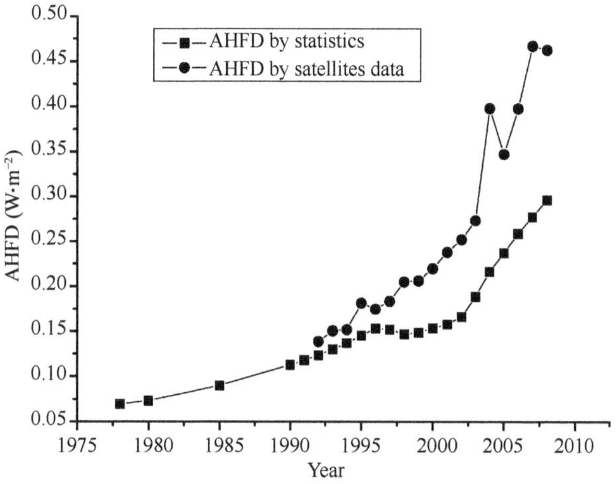

Fig. 2 The mean Anthropogenic Heat Flux Density (AHFD) in China from 1978 to 2008 by statistics and the estimated result from DMSP/OLS satellite data from 1992 to 2008 ($W \cdot m^{-2}$)

3 Statistical Analysis of Anthropogenic Heat Release

Energy consumption in China has been recorded from 1978 to 2009 by China's National Bureau of Statistics; energy consumption data by districts and provinces are available from 1990 to 2008. In the past 30 years, the mean anthropogenic heat release flux in China has increased sharply. The flux was 0.07 $W \cdot m^{-2}$ in 1978, whereas it was 0.28 $W \cdot m^{-2}$ in 2008, indicating a threefold increase over 30 years.

As shown in Figure 3, the distribution of anthropogenic heat flux density in China is generally non-uniform, which is consistent with the economic development in different districts of China. The mean anthropogenic heat flux density in Beijing is 3.5 $W \cdot m^{-2}$, where as it is 4.4 $W \cdot m^{-2}$ in Tianjin and about 15.0 $W \cdot m^{-2}$ in Shanghai. The anthropogenic heat flux density is greater than 1.0 $W \cdot m^{-2}$ in most areas of China, including most provinces of Northern and Eastern China, as well as coastal regions. Overall, the anthropogenic heat flux is very low in Northwestern China, whereas the levels

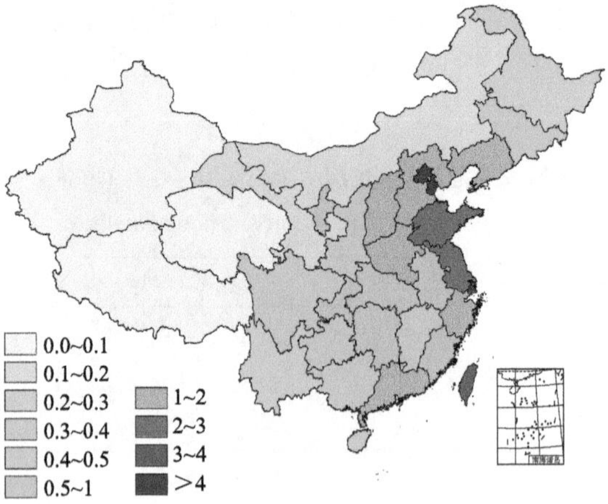

Fig. 3 The distribution of mean anthropogenic heat flux density in provinces and districts of China in 2008 (W · m^{-2})

in Northern, Central, Eastern and Southern China are much higher than other regions. With the development of urban agglomerations and the booming economy in China, the anthropogenic heat release in Northern, Eastern and Southern China is becoming increasingly serious and becoming a large atmospheric heat source. Attention should be paid to the possible regional climate effects caused by anthropogenic heat, which deserve further research.

4 Estimating the Anthropogenic Heat Flux Distribution in China by DMSP/OLS Data

The estimation of the anthropogenic heat release distribution for China from 1992 to 2009 obtained by DMSP/OLS data is shown in Figure 4. A chart of the anthropogenic heat flux density distribution in China was obtained for each year during the period from 1992 to 2009. The estimation of anthropogenic heat release by satellite data is consistent with the statistical results. Obviously, the distinct result that the developing process of anthropogenic heat release in China can be obtained from the satellite data, which is compatible with the development of the economy andurban agglomerations in China. Generally, anthropogenic heat is concentrated in Northeastern, Northern and Eastern China, as well as the coastal cities of Southeast China, especially in Beijing, the Yangtze River Delta and the Pearl River Delta area. The results indicate that the anthropogenic heat flux density distribution in China's inland cities appear as spots on a map. The anthropogenic heat flux density distribution map for 1992 in China suggests that thermal pollution is not serious, except in the economically developed areas. Some years later, the results show that the anthropogenic heat distribution increases on a larger scale in the developed areas. The anthropogenic heat release has deteriorated for all of China; significant development of anthropogenic heat release in Central and Southwestern China is

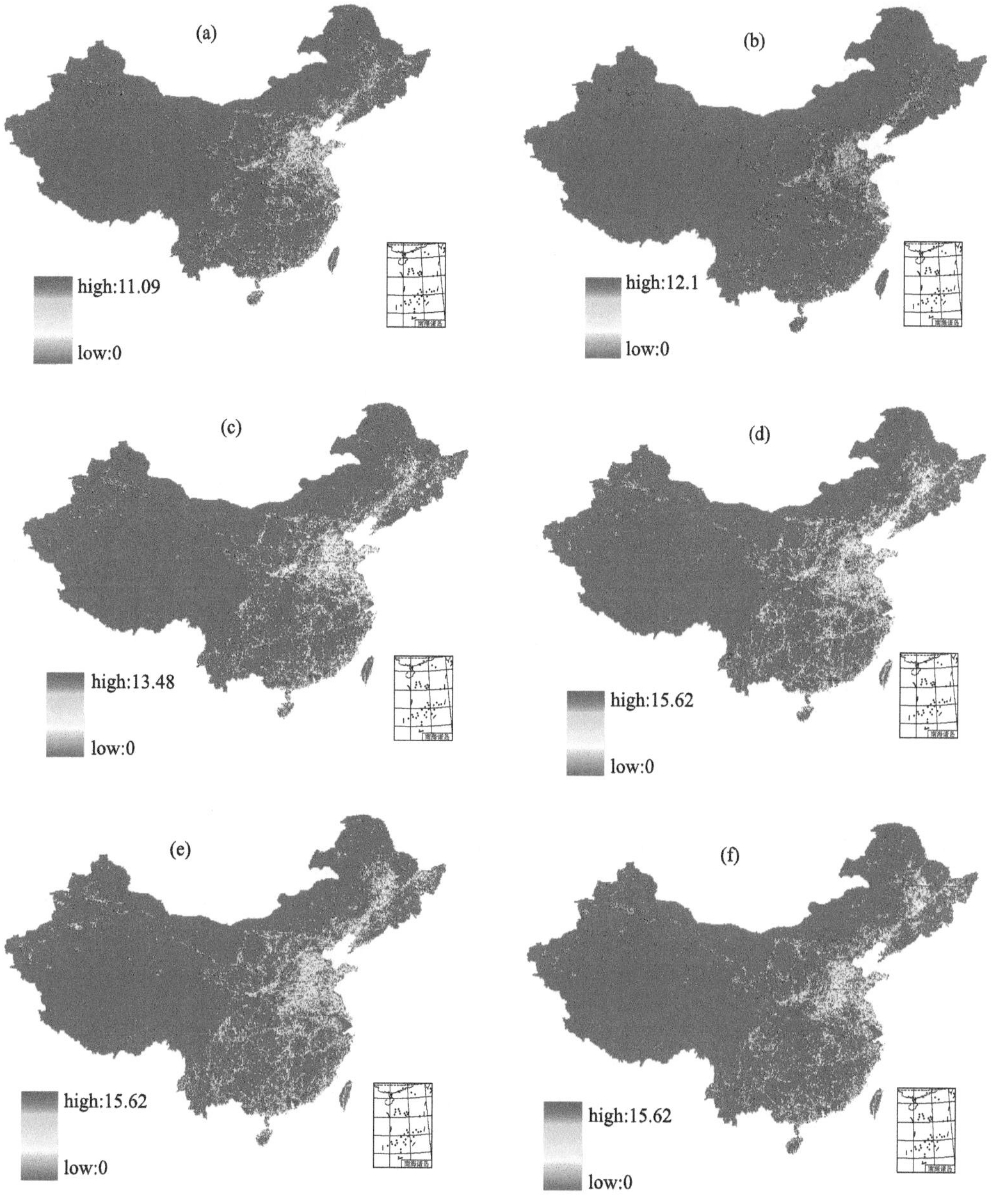

Fig. 4 The estimated anthropogenic heat flux density distribution in China from 1992 to 2009 by DMSP/OLS data ($W \cdot m^{-2}$).

(a) in the year 1992 by F10; (b) in the year 1995 by F12; (c) in the year 2000 by F14; (d) in the year 2005 by F16; (e) in the year 2008 by F16; (f) in the year 2009 by F16

shown by satellite data from 2000. A vast area, including Northeastern, Northern, Central and Southern China, appears seriously affected by thermal pollution in 2005. The latest anthropogenic heat flux density distribution map in China suggests that the anthropogenic heat is still growing. The trend toward increasing anthropogenic heat is clearly presented by satellite data. The DMSP/OLS data provide an effective way to analyze anthropogenic heat release as a developing process in China.

5 Analysis of the Model Results

The mean anthropogenic heat flux density for each grid of China was obtained by the simple model. The total anthropogenic heat flux density in a district was calculated as the total of the grids within its boundary using Arc GIS. By using Equation (1), the total anthropogenic heat emission of an area over one year was obtained. Then, the anthropogenic heat emission was converted into the energy consumption unit of 10000 tons of coal equivalents (tce) to assess the model results (1 tce=29.2 GJ) and compare them with the official energy consumption statistics. The results are presented in Table 2.

Table 2 Comparison between official energy consumption statistics and model results

Districts of China	Official energy consumption statistics in 2008 (10000 tce)	Model results (10000 tce)	Residual percentage
Beijing	6327.0	7791.2	23
Tianjin	5364.0	6619.7	23
Hebei	24322.0	29997.1	23
Shanxi	15675.0	18877.4	20
Inner Mongolia	14100.0	15381.8	9
Liaoning	17801.0	19218.8	8
Jilin	7221.0	11834.4	64
Heilongjiang	9979.0	23450.6	135
Shanghai	10207.0	8136.9	−20
Jiangsu	22232.0	37827.6	70
Zhejiang	15107.0	24590.2	63
Anhui	8325.0	16498.6	98
Fujian	8254.0	12839.6	56
Jiangxi	5383.0	8074.5	50
Shandong	30570.0	40206.2	32
Henan	18976.0	29819.4	57
Hubei	12845.0	13447.1	5
Hunan	12355.0	10753.4	−13
Guangdong	23476.0	39385.4	68
Guangxi	6497.0	11994.5	85
Hainan	1135.0	3624.7	219
Chongqing	6472.0	4876.2	−25
Sichuan	15145.0	13056.0	−14
Guizhou	7084.0	5360.9	−24
Yunnan	7511.0	13352.9	78
Shaanxi	7417.0	15508.3	109
Gansu	5346.0	7776.0	45
Qinghai	2279.0	2278.0	0
Ningxia	3229.0	4520.6	40
Xinjiang	7069.0	14138.0	100
Hong Kong	641.1	1391.6	117
Taiwan	503.1	509.1	1

As shown in Table 2, the modeling results seem generally reasonable in most regions, while significant errors occur in some regions. Errors made by the model ranging from 0 to 30 percent occurred in 14 of the 32 districts, whereas the model resulted in significant errors for 7 districts in China. Possible explanations for these errors will be addressed in the discussion section. Although there are errors in the modeling results, the DMSP/OLS data still provide us with a useful way to estimate the anthropogenic heat flux density distribution in China.

The mean anthropogenic heat flux density (AHFD) estimated by DMSP/OLS data is shown in Fig. 2. The variation in the anthropogenic heat flux density with time is quite similar to the statistical results. We can also clearly observe AHFD's increasing trend from the DMSP/OLS satellite data. The mean anthropogenic heat flux has increased sharply in the past 20 years. However, the chart shows a low oscillating trend during the period from the late 1990s to the early 2000s. A much more rapid increase in AHFD is shown since the beginning of 21st century, which is compatible with the data published by China's National Bureau of Statistics. The probable reason for the oscillation at the end of 20th century may be the Asian financial crisis, which broke in 1997. Obvious effects were imposed by the financial crisis on the global economy that were clearly observed from the satellite data, which shows that satellite data outcomes are quite compatible with authoritative statistics. At the beginning of the 1990s, the AHDF obtained by statistical and satellite data were very similar; however, after this period, the results from the satellite data are generally higher than the statistical data. The possible reason for this discrepancy could be due to innovations in energy technology in industrial production, which caused a continual decline in the energy consumption per unit of GDP. The increase in energy consumption is not proportional to economic development, which results in significant model errors.

6　Discussions and Conclusions

The DMSP/OLS stable light data from 1992 to 2009 and the authoritative energy consumption statistics were applied to estimate the anthropogenic heat flux distribution in China. Compared with authoritative statistics, the modeling results of the anthropogenic heat distribution generally prove credible. Although different models were applied to estimate the anthropogenic heat flux distribution in China, there are also errors in the model adopted in this paper. A linear model was adopted because the results of this model resulted in minimal error. Generally speaking, the DMSP/OLS data are closely correlated with economic activities, such as the GDP and energy consumption, for a given area (Elvidge *et al.*, 1997b). However, significant differences exist with regards to economic development in different districts. For the regions where economic development is dependent on energy consumption, the model results seem generally credible. Energy consumption increases as the economy grows. However, for the districts where economic development does not depend as much on energy consumption, significant errors occurred in the model. For example, the districts for which the model results showed an overestimate were usually the districts known for tourism, where fewer energy resources are consumed in the economic development process, such as Hong Kong, Hainan, and Shaanxi. The model underestimated the results for the districts within developing areas where vast energy resources are needed for the industrial production process and daily life, such as Northwestern China. Furthermore, regional differences in energy efficiency for economic development may be a major reason for the errors. For some regions, energy consumption growth has not been proportional to economic growth due to energy technology innovation. Generally, the anthropogenic heat flux is closely correlated with economic development and energy consumption, but the relationship between energy consumption and economic development is complex.

There are various explanations that may account for the errors produced in the linear model. Errors produced by the satellite data should not be ignored. The model assumes that the Digital Number value is proportional to the radiance in each grid of DMSP/OLS data, but this only occurs when DN is not very large. When the DN is close to the maximum value, the radiance is no longer proportional. As a result, the assumptions of the model are not correct when DN is large. Additionally, the DMSP/OLS satellite data should account for errors in the model outcomes. Many factors may contribute to errors in the satellite data, such as the sensor saturation effect and errors produced by the instruments and in processing the satellite data. Different line functions were applied in the model by different satellites. Furthermore, different satellite data were applied in the estimation of the anthropogenic heat distribution. Differences in the various satellite data may be the main reason for the discrepancy between Fig. 2 and Fig. 4 during the period from 1992 to 2008. Slight differences exist among the data from the four satellites used in this paper. There were no significant differences between the model results using data from two different satellites for the same year. Analysis of the modeling results from the satellite data obtained by F10 and F12 in 1994 indicates that the correlation coefficient between them is more than 0. 99, and the *T*-test result shows that there is no significant difference between them. The same results were obtained from the data for F12 and F14 from 1997 to 1999. The results show that the modeling results by different satellites in the same year are quite similar. The data differences will not influence the final estimation results on the whole. The data from the four DMSP/OLS satellites were applied separately in 4 different linear models. There is a slight difference between the data from the various satellites, but using the data from one satellite, which were produced by the same instruments and data processing, the model results appear credible. Despite errors in the model results, there is an obvious advantage in estimating the anthropogenic heat flux distribution in China by satellite data. The anthropogenic heat release distribution map formed by the satellite data seems much more credible when compared to the statistical maps. Moreover, the process of anthropogenic heat development can be clearly obtained from the DMSP/OLS data.

The DMSP/OLS provide an effective way to estimate the anthropogenic heat flux distribution on a large scale. From the estimation results using DMSP/OLS data, the aggravating trend was easily obtained. Additionally, a strong linear relationship was found between anthropogenic heat flux density and the satellite data. The regional characteristics of the anthropogenic heat distribution could also be clearly seen, which is much better than the statistical results shown in Fig. 3. The modeling results indicate that the anthropogenic heat release in China was geographically concentrated and distributed and fundamentally correlated with economic activities. The release in economically developed areas in Northern, Eastern and Southern China was much larger than other regions, whereas that in Northwestern and Southwestern China was very small. A vast area, including Beijing, the Yangtze River Delta and the Pearl River Delta, in which anthropogenic heat is concentrated, has an anthropogenic heat flux density larger than 10 $W \cdot m^{-2}$, which is high enough to affect the local climate. With economic development and sharp increases in the global population, anthropogenic heat pollution will become increasingly serious. With the boom in urban agglomerations on a global scale, large areas where anthropogenic heat flux is serious are bound to turn up in the near future. Anthropogenic heat release is an important aspect of human influence on climate change and should not be ignored in climate change research. Further research should be focused on the climate effect of anthropogenic heat release in regional and global climate models. The distribution of anthropogenic heat flux in China from 1992 to 2009 was estimated using DMSP/OLS satellite data in this paper, which provides useful parameterization for further research on climate models. The results indicate that the anthropogenic heat flux is very large in some regions, reaching a level that is high enough to influence the regional climate. Large scale areas which have the ability to heat the lower atmosphere on a global scale are significant to research on climate change. This paper focuses on estimating the anthropogenic heat flux distribution in China from 1992 to 2009 using a model based on DMSP/OLS stable light data. However, this method is still in the exploratory stage. The initial results suggest that more accurate satellite data and more practical models should be applied in future research for better modeling results. The climate effect caused by anthropogenic heat should be considered in climate models in further research. The climate effect by anthropogenic heat is still considered at the regional scale, but it deserves further research attention at the global climate change scale.

Acknowledgement

Special thanks to the Earth Observation Group, National Geophysical Data Center, and National Oceanic and Atmospheric Administration of the United States.

References

Amaral S, Monteiro A M V, Camara G, *et al.*, 2006. DMSP/OLS night-time imagery for urban population estimates in Brazilian Amazon, *Int. J. Remote Sensing*. **27**:855-870.

Block A, Keuler K, Schaller E. 2004. Impacts of anthropogenic heat on regional climate patterns, *Geophys. Res. Lett.*, **31**: L 12211, doi: 10.1029/2004GL019852.

Cinzano P, Falchi F, Elvidge C D. 2001. The first world atlas of the artificial night sky brightness. *Monthly Notices of the Royal Astronomical Society*. **328** (3): 689-707.

Compiled by China's National Bureau of Statistics. 2010. *China Energy Statistical Yearbook* 2009. China Statistics Press, Beijing, 1-536pp. (in Chinese).

Compiled by China's National Bureau of Statistics. 2010. *China Statistical Yearbook* 2010, China Statistics Press, Beijing, 1-1072pp. (in Chinese).

Doll C N H, Muller J P, Elvidge C D. 2000. Night-time imagery as a tool for global mapping of socio-economic parameters and greenhouse gas emissions. *Ambio*, **29** (3): 157-162.

Elvidge C D, Baugh K E, Kihn E A, *et al*. 1997a. Mapping city lights with nighttime data from the DMSP operational linescan system, *Photogrammetric Engineering and Remote Sensing*. **63**(6):727-734.

Elvidge C D, Baugh K E, Kihn E A, *et al*. 1997b. Relation between satellite observed visible-near infrared emissions, population, economic activity and electric power consumption, *Int. J. Remote Sensing.*, **18**: 1373-1379.

Fan H, Sailor D J. 2005. Modeling the impacts of anthropogenic heating on the urban climate of Philadelphia: a comparison of implementations in two PBL schemes, *Atmos. Environ.* **39**:73-84.

Hamilton I G, Davies M, Steadman P, *et al*. 2009. The significance of the anthropogenic heat emissions of London's buildings: A comparison against capuured shortwave solar radiation, *Building and Environment*. **44**: 807-817.

Husi L, Masanao H, Hiroshi Y, *et al*. 2010. Estimating energy consumption from night-time DMSP/OLS imagery after correcting for saturation effects, *Int. J. Remote Sensing*. **31**: 4443-4458.

Ichinose T, Shimodozono K, Hanaki K, 1999. Impact of anthropogenic heat on urban climate in Tokyo, *Atmos. Environ.* **33**: 3897-3909.

IPCC, 2007. *Climate change* 2007. The Physical Science Basis, Contribution of Working Group I to the Fourth Assessment Report of the Intergovernmental Panel on Climate Change [Solomon S, Qin D, *et al*. (eds.)], Cambridge University Press, Cambridge, United Kingdom and New York, NY, USA, 996pp.

Lee S H, Song C K, Baik J J, *et al*. 2009. Estimation of anthropogenic heat emission in the Gyeong-In region of Korea, *Theor. Appl. Climatol.* **96**: 291-303.

Makar P A, Gravel S, Chirkov V, *et al*., 2006. Heat flux, urban properties, and regional weather, *Atmos. Environ.* **40**: 2750-2766.

Oke T R. 1988. The urban energy balance, *Progerss in Physical Geography*. **12**: 471-580.

Oleson K W, Bonan G B, Fedema J, *et al*. 2010. An examination of urban heat island characteristics in a global climate model, *Int. J. Climatol.*, DOI: 10.1002/joc.2201 (in press).

Small C, Francesca P, Elvidge C D. 2005. Spatial analysis of global urban extent from DMSP-OLS night lights, *Remote Sensing of Environment*, **96**:277-291.

Tilottama G, Sharolyn A, Rebecca L P, *et al*. 2009. Estimation of Mexico's Informal Economy and Remittances Using Nighttime Imagery, *Remote Sens.* **1**: 418-444.

Tong H, Liu Zhihua, Sang Jianguo, *et al*. 2004. The impact of Urban Anthropogenic Heat on the Beijing Heat Environment, *Climatic and Environmental Research*., **9**(3): 410-421 (in Chinese).

阜康大气气溶胶中水溶性无机离子粒径分布特征研究①

苗红妍[1,2]②　温天雪[1]③　王跃思[1]　刘子锐[1]　王　丽[1]　兰中东[3]

1)中国科学院大气物理研究所大气边界层物理和大气化学国家重点实验室，北京 100029；
2)中国科学院大学，北京 100049；
3)中国科学院新疆生态与地理研究所新疆阜康荒漠生态国家野外科学观测研究站，阜康 830011

摘要

为了解阜康大气气溶胶中水溶性无机离子的浓度水平、来源以及粒径分布，本研究于2011年2月—2012年2月利用8级惯性撞击式分级采样器采集了阜康大气气溶胶样品，使用离子色谱测定了其中水溶性无机离子含量。分析比较了非采暖期和采暖期主要离子的变化趋势、浓度水平、构成、来源以及粒径分布，在此基础上选取特殊采样日分析了重污染、秸秆燃烧以及春耕期的离子组成以及粒径分布的差异。结果显示，阜康细粒子、粗粒子中总水溶性无机离子(TWSI)在非采暖期和采暖期的浓度分别为11.17 $\mu g \cdot m^{-3}$、12.68 $\mu g \cdot m^{-3}$和35.98、22.22 $\mu g \cdot m^{-3}$；非采暖期的SO_4^{2-}主要来自盐碱土扬尘，NO_3^-和NH_4^+主要来自农田土壤扬尘，而采暖期的SO_4^{2-}、NO_3^-和NH_4^+主要来自煤炭等化石燃料燃烧。八种离子在非采暖期和采暖期均呈现双峰分布，相对于非采暖期，采暖期的SO_4^{2-}、NO_3^-和NH_4^+在细粒径段的峰值发生了粒径增长，SO_4^{2-}和NH_4^+在粗粒径段的峰值出现在3.3～4.7 μm处。重污染期间二次污染严重，离子主要分布在1.1～2.1 μm处；秸秆燃烧期受生物质燃烧影响大，离子主要分布在<0.65 μm粒径段；春耕期土壤扬尘较多，离子主要分布在>3.3 μm粒径段。

关键词：粒径分布；采暖期；非采暖期；水溶性无机离子；重污染；农业活动；阜康

我国西北地区的地理和气候环境与中东部地区存在较大的差异，气溶胶的质量浓度和化学组成也与中东部地区存在一定的差别(Chen *et al.*，2012；Zhang *et al.*，2012；Hu *et al.*，2010；郑小波等，2012)。但是以往关于西北气溶胶的研究主要集中于兰州(张宁等，2008)、乌鲁木齐(杨浩，2001)等城市，并且绝大多数研究关注可吸入颗粒物和细粒子，对气溶胶粒径分布研究很少。已有研究显示，认知气溶胶中化学组分粒径分布对气溶胶的气候和环境效应研究具有重要作用(Spinder *et al.*，2012)。

阜康荒漠生态国家野外科学观测研究站(简称阜康站)(图1)位于新疆维吾尔自治区阜康市北亭镇，准噶尔盆地南缘，阜康县的三工河流域，往南距阜康市中心18 km，往西南距乌鲁木齐市76 km。属于典型的大陆性干旱气候，夏季炎热干燥，冬季酷寒潮湿，每年的采暖时间长达6个月(当年10月15日—次年4月15日)。阜康站是绿洲与荒漠之间的过渡区，

---

① 本文将发表于环境科学，2014

基金资助项目：中国科学院战略性先导科技专项(XDA05100100)；中国科学院战略性先导科技专项(B类)(XDB05020000)，国家自然科学基金项目(41230642，41375128)

② 作者简介：苗红妍(1985—)，女，博士研究生，主要研究方向为大气化学和大气环境，E-mail：mhy@dq.cern.ac.cn

③ 温天雪，通讯联系人，E-mail：wtx@dq.cern.ac.cn

对该站点的观测不仅可以补充城市的观测，也有助于全面了解过渡区的大气状况。站点周边以农田为主，除施肥、犁地、秋割等农业活动外无其他明显的人为污染源。本研究于2011年2月—2012年2月对阜康大气气溶胶进行了分级采样，分析了大气气溶胶中水溶性无机离子的年均浓度水平、构成、来源以及粒径分布，探究了重污染、秸秆燃烧以及春耕期间大气气溶胶中水溶性无机离子的特征。

1 材料与方法

采样点位于阜康站(87.92°E,44.28°N)实验楼二楼楼顶，使用安德森8级撞击式分级采样器(Andersen impacter serial L-8996)采集大气气溶胶样品，流速为28.3 L·min^{-1}，粒子50%切割等效空气动力学粒径为9.0、5.8、4.7、3.3、2.1、1.1、0.65和0.43 μm。2011年2月—2012年2月，每两周采样一次，连续采集48 h，采样时间从当日10:00至第三日上午10:00，使用混合纤维素酯膜采样。采集后的膜样品置于冰箱内冷冻避光保存至分析，共采集有效样品30组。

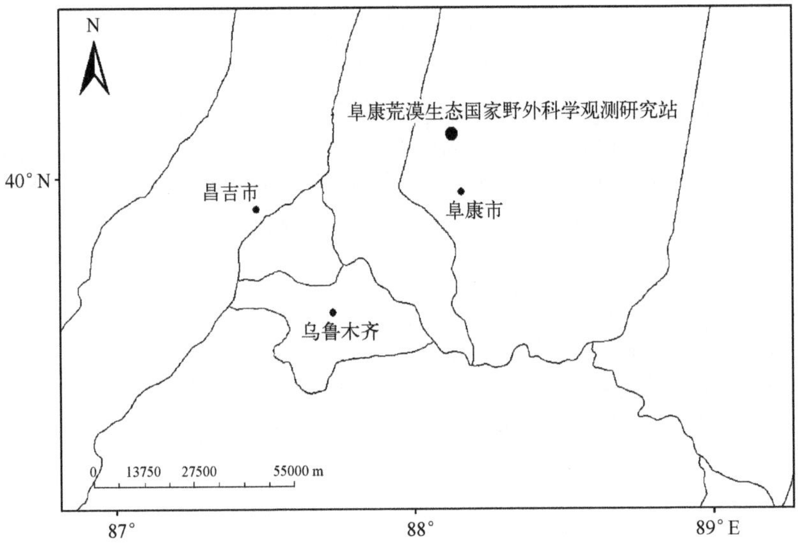

图1 采样点示意

Fig. 1 Location of the sampling station

取1/2张膜置于PET瓶中，加入25 mL去离子水，恒温超声提取30 min，经0.22 μm滤膜过滤后用ICS-90离子色谱测定其中阳离子Na^+、NH_4^+、K^+、Mg^{2+}、Ca^{2+}和阴离子Cl^-、NO_3^-、SO_4^{2-}的含量。阳离子检测采用Ionpac CS12A 4×250 mm分离柱、CSRS 300-4 mm抑制器，淋洗液为22 mmol·L^{-1}的MSA；阴离子检测采用Ionpac AS14A 4×250 mm分离柱，ASRS 300-4 mm抑制器，淋洗液为3.5 mmol·L^{-1}的Na_2CO_3和1.0 mmol·L^{-1}的$NaHCO_3$混合溶液。进样量为10 μL时，各离子的最低检测限均小于0.12 μg·m^{-3}。

2　结果与讨论

2.1　水溶性离子浓度变化趋势

由于 Andersen 采样器没有 2.5 μm 的切割粒径，因此本研究把空气动力学直径 2.1 μm 作为粗、细粒子的分界，$D_P \leqslant 2.1$ μm 的粒子称为细粒子，$D_P > 2.1$ μm 的粒子称为粗粒子。

如图 2 所示，阜康大气气溶胶中水溶性离子受采暖影响较大，离子浓度水平和变化幅度在非采暖期和采暖期明显不同，细粒子表现更为突出。非采暖期水溶性无机离子浓度较低，变化小，细粒子和粗粒子中的总水溶性离子(TWSI)分别为 11.17 μg·m^{-3} 和 12.68 μg·m^{-3}。采暖期水溶性无机离子浓度高，多次出现污染过程，细粒子和粗粒子中的 TWSI 分别为 35.98 μg·m^{-3} 和 22.22 μg·m^{-3}，是非采暖期的 3.2 和 1.8 倍。并且采暖期水溶性离子浓度水平变化幅度大，细粒子中 TWSI 的全年最高值和最低值均出现在采暖期，最低值(2012 年 1 月 18 日)仅有 4.83 μg·m^{-3}，最高值(2012 年 1 月 10 日)达到了 115.70 μg·m^{-3}，是最低值的 24 倍。与西北的乌鲁木齐(刘新春等，2012)、西安(张碧云等，2012)、兰州(张宁等，2008)等城市相比，阜康大气气溶胶中水溶性无机离子含量较低，但是高于北部的阿克达拉

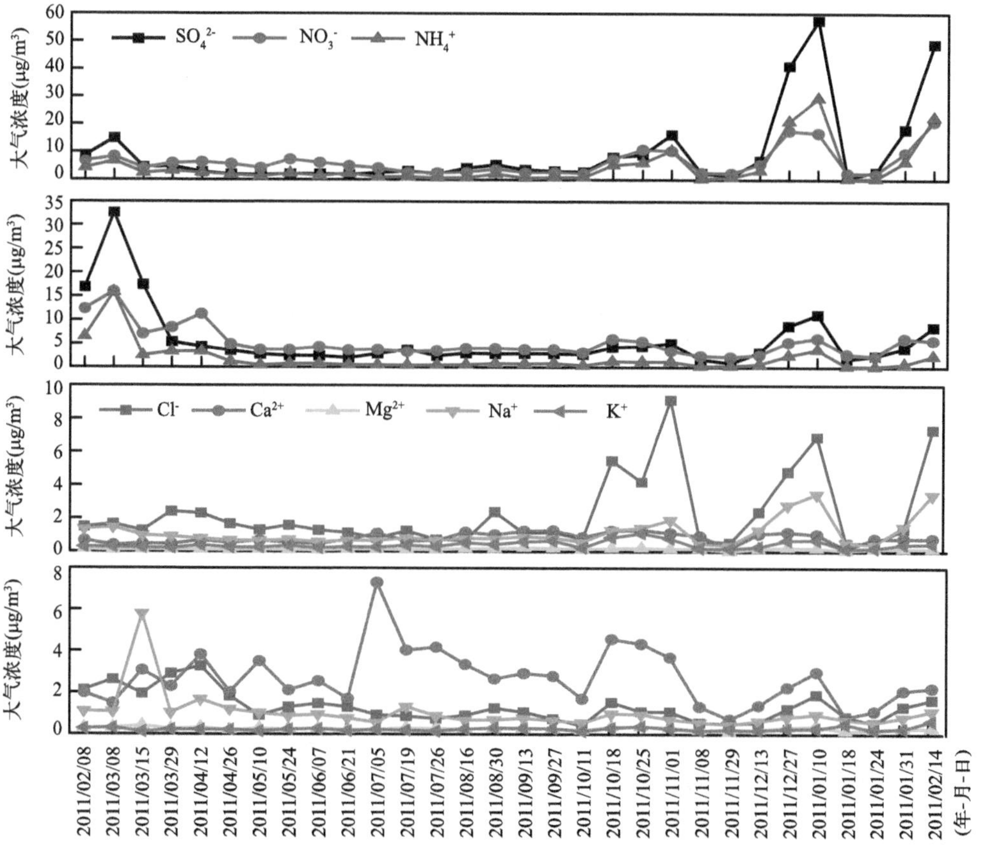

图 2　阜康气溶胶中水溶性无机离子的时间变化

Fig. 2　Temporal variation of water soluble inorganic ions in atmospheric aerosols in Fukang

(Zhang *et al.*, 2012)背景区，表明阜康仍然受到部分人为污染，可能与周边的农业活动及采暖期化石燃料燃烧有关。

除 Ca^{2+} 以外，各水溶性离子的浓度变化与 TWSI 总体一致，采暖期明显高于非采暖期。而 Ca^{2+} 无论是粗粒子还是细粒子，均是非采暖期高于采暖期，特别是在6—11月期间粗粒子浓度较高，并且变化幅度更大，土壤尘来源的 Mg^{2+} 在此期间变化较小，周边也没有明显的建筑活动，表明非采暖期 Ca^{2+} 可能有其他未知的来源。

虽然除 Ca^{2+} 外其他离子浓度变化趋势基本一致，但部分离子在某些特定时期呈现一定的独特变化。每年4—5月是阜康的春耕时期，在这期间水溶性离子浓度在粗粒子中显著升高，而细粒子中无明显变化，如2011年4月12日的样品，粗粒子中 NO_3^-、NH_4^+、Cl^-、Ca^{2+}、Mg^{2+} 和 Na^+ 的浓度分别是前后两次采样的1.0～2.8倍，而细粒子中这几种离子浓度仅是前后两次的0.8～1.5倍。此外，在10中旬到11月初农民集中整地焚烧秸秆期间，生物质燃烧的示踪离子 K^+ 和 Cl(祝斌等，2005)浓度较平时出现了显著增加，如2011年11月1日样品，细粒子中的 Cl^- 浓度为全年最高值，K^+ 也显著高于非秸秆燃烧期，S-N-A（S:SO_4^{2-}，N:NO_3^-，A:NH_4^+）浓度水平也相对较高。

2.2 无机离子构成及来源

非采暖期和采暖期气溶胶中水溶性无机离子的构成以及在粗、细粒子中的分布存在一定的差别。由表1可知，非采暖期细粒子中含量最高的3种离子依次是 $NO_3^- > SO_4^{2-} > NH_4^+$，粗粒子中为 $NO_3^- > Ca^{2+} > SO_4^{2-}$，说明 NO_3^- 对当地大气质量的影响很大，下文将进一步探究 NO_3^- 的来源；非采暖期气溶胶中水溶性离子在粗粒子中分布较多，气溶胶中 SO_4^{2-}、NO_3^-、NH_4^+ 和 Cl^- 在粗粒子中的分布比例分别为50%、49%、33%和47%，高于我国其他地区(赵亚楠，2009)。采暖期细粒子和粗粒子中含量最高的三种离子都依次为：$SO_4^{2-} > NO_3^- > NH_4^+$，这说明采暖期煤炭燃烧对粗、细粒子都造成了一定的影响，致使与煤炭燃烧密切相关的 SO_4^{2-} 成为气溶胶中含量最高的离子。采暖期除了 Ca^{2+} 和 Mg^{2+} 之外的6种离子都主要分布在细粒子中，细粒子中S-N-A以及 Cl^- 的浓度分别是非采暖期的5.2、2.1、5.9和2.6倍，与其他地区类似。

鉴于气溶胶中的三种主要成分S-N-A在非采暖期和采暖期的构成比例存在一定的差别，对三者进行了相关分析。非采暖期气溶胶中 SO_4^{2-} 与 NO_3^- 和 NH_4^+ 相关性差，与 Na^+、Ca^{2+} 和 Mg^{2+} 显著相关($R^2 \geqslant 0.64$，$n=13$)，考虑到西北盐渍区离子主要以 SO_4^{2-}-Na^+ 为主(顾峰雪等，2003)，Ca^{2+} 和 Mg^{2+} 又是沙尘的主要离子，加上 SO_4^{2-} 在粗粒子中分布较多，因此推测非采暖期的 SO_4^{2-} 主要来自盐碱土扬尘，而不是二次转化；气溶胶中的 NO_3^-、NH_4^+、Cl^- 三者相关性很好($R^2 \geqslant 0.74$，$n=13$)，这3种离子正是肥料的主要成分，推测非采暖期这3种离子主要来自肥料残留的农田土壤扬尘；可见非采暖期阜康气溶胶中S-N-A与其他地区二次转化同源性的观测结果不同(胡敏等，2005；Chen *et al.*, 2011；苗红等，2013)。采暖期气溶胶中S-N-A三者之间呈现显著相关($R^2 \geqslant 0.80$，$n=17$)，同源性强，而且三者主要分布在细粒子中，因此推测这3种离子与乌鲁木齐(亚力昆江—吐尔逊等，2010；Li *et al.*, 2009)、北京(张凯等，2006)、西安(韩月梅等，2009)等地一样，主要来自煤炭等化石燃料燃烧。

表 1 阜康气溶胶中水溶性无机离子浓度($\mu g \cdot m^{-3}$)

Table 1 Concentration of water-soluble inorganic ions in atmospheric aerosols in Fukang ($\mu g \cdot m^{-3}$)

项目	非采暖期(年-月-日)(2012-04-26—2012-10-11)		采暖期(年-月-日)(2011-02-08—2011-04-12，2011-10-18—2012-02-14)		重污染期(年-月-日)(2012-01-10—2012-01-13)		秸秆燃烧期(年-月-日)(2011-11-01—2011-11-03)		春耕期(年-月-日)(2011-04-12—2011-04-15)	
离子	细粒子	粗粒子	细粒子	粗粒子	细粒子	粗粒子	细粒子	粗粒子	细粒子	粗粒子
TWSI	11.17	12.68	35.98	22.22	115.69	27.28	49.95	15.69	15.72	28.12
SO_4^{2-}	2.79	2.82	14.65	7.82	57.54	11.15	16.20	5.05	2.82	4.39
NO_3^-	3.91	3.77	8.16	6.16	16.61	6.05	10.32	3.47	6.18	11.20
NH_4^+	1.24	0.61	7.30	2.67	29.29	3.66	10.37	1.14	2.51	3.30
Cl^-	1.21	1.05	3.10	1.50	6.89	1.88	9.11	1.07	2.28	3.25
Ca^{2+}	0.83	3.14	0.78	2.36	1.04	2.98	1.11	3.68	0.68	3.81
Mg^{2+}	0.14	0.25	0.15	0.28	0.20	0.37	0.17	0.32	0.12	0.32
Na^+	0.70	0.83	1.42	1.16	3.44	0.94	1.89	0.71	0.76	1.65
K^+	0.35	0.21	0.42	0.27	0.68	0.25	0.78	0.25	0.37	0.20

NO_3^-/SO_4^{2-}比值通常用来判断固定源和移动源对大气质量的相对贡献。阜康非采暖期粗、细粒子中NO_3^-/SO_4^{2-}分别为1.3和1.4，采暖期比值为0.8和0.6。从比值看，采暖期比值小于1，这符合采暖期煤烟型污染的特点。非采暖期的比值大于1，与我国北京(越普生等，2011)、上海(徐旭，2010)等大城市的观测结果接近，但是与交通发达的城市不同，本研究采样点周边受机动车影响很小。因此推测NO_3^-显著高于SO_4^{2-}主要有以下两点原因：首先，阜康站位于乡镇，采样点周边无明显SO_2排放源，非采暖期SO_4^{2-}量较低；其次，非采暖期的肥料挥发以及肥料残留的农田土壤扬尘致使大气中NO_3^-的含量相对较高，两点共同导致NO_3^-/SO_4^{2-}高值的出现。由此可知，阜康的非采暖期用NO_3^-/SO_4^{2-}来判断固定源和移动源的贡献并不恰当，NO_3^-/SO_4^{2-}比值并不总能判断固定源和移动源的相对贡献，需要根据采样点的情况具体分析。

2.3 气溶胶中水溶性无机离子粒径分布

图3给出了八种离子在非采暖期以及采暖期的平均粒径分布。非采暖期气溶胶中的水溶性无机离子都呈双峰分布，其中NH_4^+以细粒径段的峰值为主，SO_4^{2-}、NO_3^-、Cl^-、Na^+以及K^+在粗细粒径段分布相当，而Ca^{2+}和Mg^{2+}以粗粒径段的峰值为主。细粒径段中除了NO_3^-之外的七种离子都在0.43～0.65 μm处出现峰值，而NO_3^-在0.43～2.1 μm处出现峰值；八种离子在粗粒径段的峰值都出现在4.7～5.8 μm处。相比较而言，非采暖期阜康气溶胶中水溶性离子在粗粒径段的分布比例高于我国中东部地区(王丽等，2013；于阳春等，2011；耿彦红等，2010)，推测这主要与当地距离荒漠较近有关。

采暖期部分离子粒径分布的峰值和出峰位置较非采暖期存在一定的差别。采暖期的SO_4^{2-}、NO_3^-、NH_4^+、Cl^-、Na^+虽然也呈现双峰分布，但五种离子在细粒径段的峰值均显著

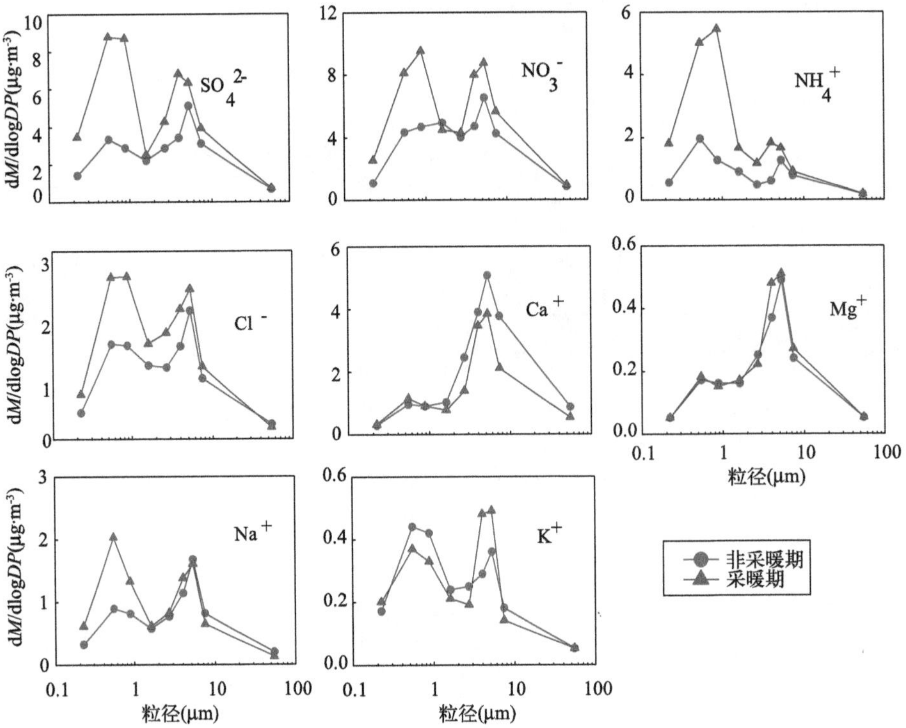

图 3 气溶胶中水溶性无机离子粒径分布

Fig. 3 Mass size distributions of water-soluble inorganic ions

高于粗粒径段，而且这五种离子在细粒径段峰值较非采暖期的增加量显著高于粗粒径段峰值的增加量；采暖期的 K^+ 也呈现双峰分布，其在采暖期粗粒径的峰值略高于非采暖期；Ca^{2+} 和 Mg^{2+} 的粒径分布受采暖影响较小，Ca^{2+} 在粗粒径段的峰值略低于非采暖期。采暖期的 SO_4^{2-}、NO_3^- 和 NH_4^+ 在细粒径段的峰值均出现在 0.43～1.1 μm 处，较非采暖期发生了向较大粒径段移动的趋势，以往在京津冀(胡敏等，2006；Cheng *et al.*，2011；苗红妍等，2013)、珠三角(Liu *et al.*，2008)湿度增加的情况下也都观测到了离子向较大粒径转移的现象，只是这些地区均是夏季湿度大，冬季湿度小，所以吸湿增长往往发生在夏季，而阜康则是采暖期湿度大、非采暖期湿度小，而且采暖期气体污染物增加迅速，气粒转化强烈，导致阜康水溶性离子在采暖期出现明显粒径增长。粗粒径段中除了 SO_4^{2-} 和 NH_4^+ 之外的六种离子均在 4.7～5.8 μm 处出现峰值，而 SO_4^{2-} 和 NH_4^+ 的峰值出现在 3.3～4.7 μm 处，这两种离子受采暖影响较大，推测该峰值可能是受到煤炭粉尘以及飞灰颗粒的影响。乔佳佳在青岛的采暖期也观测到了颗粒物在 3.3～4.7 μm 出现峰值的现象(乔佳佳，2010)。

2.4 重污染、秸秆燃烧和春耕期的水溶性离子特征

根据 2.1 节的描述，选择 2012 年 1 月 10 日、2011 年 11 月 1 日和 2011 年 4 月 12 日这三个样品作为重污染、秸秆燃烧期以及春耕期的代表，在此基础上对 3 个特定采样日水溶性离子特征进行比较分析。

2.4.1 离子构成

表 1 给出了冬季重污染、秸秆燃烧以及春耕三段时期水溶性无机离子的浓度水平。从

离子构成来看，在细粒子中，重污染期和秸秆燃烧期三种主要离子都是 S-A-N，但是两段时期主要离子的构成比例存在较大的差别，重污染期间 S-A-N 总和高达 TWSI 的 89%，其他离子含量均较低，而秸秆燃烧期的 S-A-N 总和占 TWSI 的 74%，除了这 3 种主要离子外，秸秆燃烧期的 Cl^- 含量较高，占 TWSI 的 18%，而且 K^+ 的绝对浓度和所占比例也均高于重污染期，可见重污染期二次离子污染十分严重，与其他地区污染期的观测结果一致（Li *et al.*，2009；Tan *et al.*，2009；Shen *et al.*，2009），除二次离子外，Cl^- 与 K^+ 也是影响秸秆燃烧期质量的主要因素。春耕期离子构成与重污染和秸秆燃烧期不同，三种主要离子是 N-S-A，而且 Cl^- 含量较高，与 NH_4^+ 接近。

在粗粒子中，重污染期的三种主要离子是 S-N-A，说明化石燃料燃烧对粗粒子也造成了一定的影响，而秸秆燃烧期和春耕期分别是 S-Ca^{2+}-N 和 N-S-Ca^{2+}，并且两段时期 Na^+ 比例较高，因此推测这两段时期粗粒子中的离子组成主要与农业活动有关，田间耕作和整理秸秆过程中会将土壤扬尘带至大气中，进而影响大气质量，赵鹏等（赵鹏等，2006）关于北京郊区菜地的研究中也发现了类似的现象。

2.4.2　粒径分布

重污染、秸秆燃烧以及春耕期水溶性无机离子在不同粒径段的分布比例存在较明显的差别。如图 4 所示，重污染和秸秆燃烧期的离子除了 Ca^{2+} 和 Mg^{2+} 之外都主要分布在细粒子中，而春耕期除 K^+ 外都主要分布在粗粒子中，并且 K^+ 在粗粒径段的分布与其他两段时期相似，由此可见，重污染期和秸秆燃烧期主要是细粒子污染，而春耕期主要是粗粒子污染。

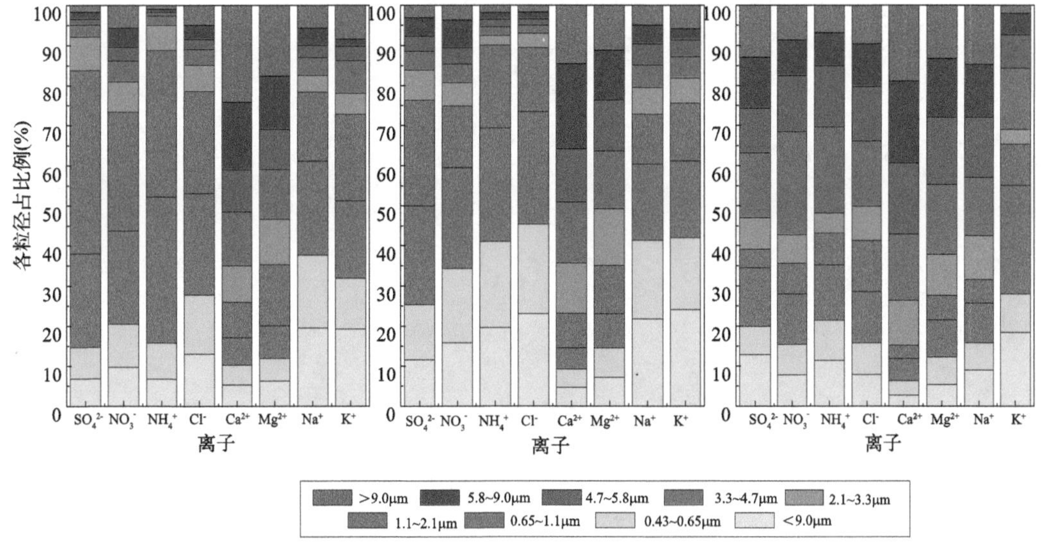

分别是 2012 年 1 月 10 日、2011 年 11 月 1 日以及 2011 年 4 月 12 日的样品

图 4　重污染、秸秆燃烧以及春耕期水溶性无机离子在不同粒径段的分布比例

Fig. 4　Size distributions of ions during heavy pollution, straw burning and spring planting periods

从各个粒径段的分布来看，重污染期的离子主要集中在 1.1～2.1 μm 处，三种主要离子 S-N-A 在该段的比例分别达到了 46%、30%和 37%。这是因为一方面重污染期间大气相对稳定，已经生成的粒子不断的积累老化，进而增大了离子粒径，另一方面，重污染期大气湿度高于 75%，离子极易吸湿增长，或者与非降水的云滴或雾滴结合进而增加了其在大粒径段的

含量。秸秆燃烧期的离子主要集中在<0.65 μm粒径段，除Ca^{2+}和Mg^{2+}外其他6种离子在该粒径段的比例均超过了25%，Cl^-和K^+分别达到了45%和42%，而且二者在<0.43 μm的比例均超过了20%。已有研究报道，凝结模态的粒子主要来自燃烧源的直接排放或二次物种的气-粒转化反应，包括大气中燃烧产生的热蒸汽的凝结以及成核的气态反应产物，加上本次秸秆燃烧期间大气的相对湿度高于70%，离子会进一步吸湿增长，因此，秸秆燃烧过程中的离子多集中在<0.65 μm粒径段也就不足为奇。春耕期的离子主要分布在>3.3 μm粒径段，除K^+之外的七种离子在该粒径段的分布比例均超过了50%，这是由于相对于其他两个季节，春季风速较大，湿度较小($RH \approx 40\%$)，加上农田耕作对土壤的扰动，沙尘、农田土壤更容易进入大气中。

3 结论

(1)阜康气溶胶中水溶性无机离子受采暖影响较大，非采暖期离子含量较低，浓度水平变化小，采暖期离子含量较高，浓度水平变化大。

(2)非采暖期和采暖期气溶胶中主要离子构成比例存在一定的差别，非采暖期的SO_4^{2-}主要来自盐碱土，NO_3^-和NH_4^+来自农田土壤扬尘，采暖期的SO_4^{2-}、NO_3^-和NH_4^+都主要来自煤炭等化石燃料燃烧。

(3)水溶性无机离子在非采暖期和采暖期均呈现双峰分布，采暖期SO_4^{2-}、NO_3^-、NH_4^+、Cl^-和Na^+较非采暖期在细粒径段峰值的增加量大于粗粒径段，并且SO_4^{2-}、NO_3^-、NH_4^+在细粒径段的峰值较非采暖期发生了位移，而Ca^{2+}和Mg^{2+}受采暖影响很小。

(4)重污染期以二次污染为主，离子主要集中在1.1～2.1 μm处，秸秆燃烧期除S-A-N外，还受到Cl^-和K^+的影响，离子主要集中在<0.65 μm粒径段，春耕时期Ca^{2+}和Na^+的含量较高，离子主要分布在>3.3 μm粒径段。

参考文献

耿彦红，刘卫，单键，等. 2010. 上海市大气颗粒物中水溶性离子的粒径分布特征. 中国环境科学，**30**(12)：1585-1589.

顾峰雪，张远东，刘永强，等. 2003. 阜康绿洲土壤盐渍化特征及其与肥力的相关性分析. 干旱区资源与环境，**17**(2)：78-82.

韩月梅，沈振兴，曹军骥，等. 2009. 西安市大气颗粒物中水溶性无机离子的季节变化特征. 环境化学，**28**(2)：261-266.

胡敏，赵云良，何凌燕，等. 2005. 北京冬、夏季颗粒物及其离子成分质量浓度谱分布. 环境科学，**26**(4)：1-6.

刘新春，钟玉婷，何清，等. 2012. 乌鲁木齐市不同天气条件下TSP水溶性离子特征分析. 沙漠与绿洲气象，**6**(5)：60-66.

苗红妍，温天雪，王丽，等. 2013. 唐山大气颗粒物中水溶性无机盐的观测研究. 环境科学，**34**(4)：1225-1231.

乔佳佳，祁建华，刘苗苗，等. 2010. 青岛采暖期不同天气状况下大气颗粒态无机氮分布研究. 环境科学，**31**(1)：29-35.

王丽，温天雪，苗红妍，等. 2013. 保定大气颗粒物中水溶性无机离子质量浓度及粒径分布. 环境科学研究，**26**(5)：516-521.

徐昶. 2010. 中国特大城市气溶胶的理化性质、来源及其形成机制. 上海：复旦大学.

亚力昆江·吐尔逊，迪丽努尔·塔力甫，阿不力克木·阿布力孜，等. 2010. 乌鲁木齐市冬季大气$PM_{10\text{-}2.5}$、

$PM_{2.5}$中水溶性无机离子的化学特征. 环境工程, **28**(S1): 196-199.

杨浩. 2011. 乌鲁木齐市 TSP 浓度、形貌及水溶性离子特性分析[D]. 乌鲁木齐:新疆大学.

于阳春. 2011. 济南市大气颗粒物中水溶性无机离子的粒径分布研究. 济南:山东大学.

张碧云,张承中,周变红,等. 2012. 西安采暖期 $PM_{2.5}$及其水溶性无机离子的时段分布特征. 环境工程学报, **6**(5):1643-1646.

张凯,王跃思,温天雪,等. 2006. 北京大气重污染过程 PM_{10}中水溶性盐的研究. 中国环境科学, **26**(4): 385-389.

张宁,李利平,王武功,等. 2008. 兰州市城区与背景点冬季大气气溶胶中主要无机离子的组成特征. 环境化学, **27**(4): 494-498.

赵鹏,朱彤,梁宝生,等. 2006. 北京郊区农田夏季大气颗粒物质量和离子成分谱分布特征. 环境科学, **27**(2): 193-199.

赵普生,张小玲,孟伟,等. 2011. 京津冀区域气溶胶中无机水溶性离子污染特征分析. 环境科学, **32**(6):1546-1549.

赵亚楠,王跃思,温天雪,等. 2009. 贡嘎山大气气溶胶中水溶性无机离子的观测与分析研究. 环境科学, **30**(1):9-13.

郑小波,罗宇翔,赵天良,等. 2012. 中国气溶胶分布的地理学和气候学特征. 地理科学, **32**(3): 265-272.

祝斌,朱先磊,张元勋,等. 2005. 农作物秸秆燃烧 $PM_{2.5}$排放因子的研究. 环境科学研究, **18**(2): 29-33.

Cheng M, You C, Cao J, *et al*. 2012. Spatial and seasonal variability of water-soluble ions in $PM_{2.5}$ aerosols in 14 major cities in China. *Atmospheric Environment*, **60**:182-192.

Cheng S, Yang L, Zhou X, *et al*. 2011. Size-fractionated water-soluble ions, situ pH and water content in aerosol on hazy days and the influences on visibility impairment in Jinan, China. *Atmospheric Environment*, **45**(27): 4631-4640.

Hu H, Yang Q, Lu X, *et al*. 2010. Air pollution and control in different areas of China. *Critical Reviews in Environmental Science and Technology*, **40**(6): 452-518.

Li J, Zhuang G, Huang K, *et al*. 2009. The chemistry of heavy haze over Urumqi, Central Asia. *Journal of Atmospheric Chemistry*, **61**(1): 57-72.

Liu S, Hu M, Slanina S, *et al*. 2008. Size distribution and source analysis of ionic compositions of aerosols in polluted periods at Xinken in Pearl River Delta (PRD) of China. *Atmospheric Environment*, **42**(25): 6284-6295.

Quinn P, Overt D, Bates T, *et al*. 1993. Dimethylsulfied cloud condensation nuclei climate system relevant size-resolved measurements of the chemical physical-properties of atmospheric aerosol-particles. *Journal of Geophysical Research-Atmospheres*, **98**(D6): 10411-10427.

Shen Z, Cao J, Arimoto R, *et al*. 2009. Ionic composition of TSP and $PM_{2.5}$ during dust storms and air pollution episodes at Xi′an, China. *Atmospheric Environment*, **43**(18): 2911-2918.

Spindler G, Gnauk T, Gruner A, *et al*. 2012. Size-segregated characterization of PM_{10} at the EMEP site Melpitz (Germany) using a five-stage impactor: a six year study. *Journal of Atmospheric Chemistry*, **69**(2): 127-157.

Tan J, DuanJ, Chen D, *et al*. 2009. Chemical characteristics of haze during summer and winter in Guangzhou. *Atmospheric Research*, **94**(2): 238-245.

Wang M, Penner J. 2009. Aerosol indirect forcing in a global model with particle nucleation. *Atmospheric Chemistry and Physics*, **9**(1): 239-260.

Zhang X, Wang Y, Niu T, *et al*. 2012. Atmospheric aerosol compositions in China: spatial/temporal variability, chemical signature, regional haze distribution and comparisons with global aerosols. *Atmospheric Chemistry and Physics*, **12**(2): 779-799.

SIZE DISTRIBUTIONS OF WATER-SOLUBLE INORGANIC IONS IN ATMOSPHERIC AEROSOLS IN FUKANG

MIAO Hongyan[1,2], WEN Tianxue[1], WANG Yuesi[1], LIU Zirui[1], WANG Li[1], LAN Zhongdong[3]

1) State Key Laboratory of Atmospheric Boundary Layer Physics and Atmospheric Chemistry, Institute of Atmospheric Physics, Chinese Academy of Sciences, Beijing 100029, China;

2) University of Chinese Academy of Sciences, Beijing 100049, China;

3) Xinjiang Institute of Ecology and Geography, Chinese Academy of Sciences, Xinjiang Fukang Desert Ecological State Field Scientific Observation Station, Fukang 830011, China

Abstract

To investigate the levels, and size distributions of water soluble inorganic components, samples were collected with Andersen cascade sampler from Feb. 2011 to Feb. 2012, in Fukang and were analyzed by IC. The variation trend, concentration level, composition, sources and size distribution of major ions during non-heating were compared with heating period. Based on the specific samples, we analyzed the ionic compositions and size distributions during heavy pollution, straw burning and spring planting periods. The results showed that the inorganic components in Fukang were severely affected by heating. The total water soluble ions in fine and coarse particles during non-heating and heating period were 11.17, 12.68 $\mu g \cdot m^{-3}$ and 35.98, 22.22 $\mu g \cdot m^{-3}$, respectively. SO_4^{2-} was mainly from saline-alkali soil, and NO_3^- and NH_{4+} were from resuspension of farmland soil during non-heating period, while SO_4^{2-}, NO_3^- and NH_{4+} were all from the fossil fuel consumption during heating period. All ions were bimodal distribution during non-heating and heating periods. During heating period, the particle size growth of SO_4^{2-}, NO_3^- and NH_4^+ in fine mode was found and SO_4^{2-} and NH_4^+ peaked at 3.3－4.7 μm in coarse particles. Secondary pollutions were serious during heavy pollution days with high levels of secondary ions between 1.1 and 2.1 μm. Biomass burning obviously affected the size distribution of ions during straw burning period and ions focused on smaller than 0.65 μm, while there were more soil dusts during spring planting periods and ions concentrated in larger than 3.3 μm.

Key words: size distribution, heating and non-heating periods, water-soluble inorganic ions, heavy pollution, agricultural activity, Fukang

QUASI-DISTRIBUTED REGION SELECTABLE GAS SENSING FOR LONG DISTANCE PIPELINE MAINTENANCE①

LU Mifang[1], Koji Nonaka[1], Hirokazu Kobayashi[1], YANG Jun[2]②, YUAN Libo[2]

1) Kochi University of Technology, Kochi, Japan

2) Harbin Engineering University, Harbin, China

Abstract

In this work, a novel optical gas tele-monitoring concept is proposed. By following this concept, we construct long distance region selectable gas sensing system, which can address gases of single and/or different types at multi-locations. This approach is based on optical spectroscopy of selected absorption lines of gasleakage. A gas lines spectrum can address from long distance monitoring center using optical fiber, gas sensing region, and region selector. Region selecting technique monitors the selected gas absorption and identifies the location on it simultaneously. The technique has potential to be applied to long distance light weight fiber optic wide-region gas sensing.

Multi-regions tele-monitoring experiment using FBG monitor as spectroscopy unit, long propagation fibers, gas cells as leakage sensing regions and FBGs as region selectors is demonstrated. Available numbers and coverage of multi-sensing regions are estimated using loss of sensing unit and propagation.

Key words: pipeline maintenance; region selectable sensing; region selector; gas absorption spectroscopy; FBG

1 Introduction

Pipelines are playing an increasingly important role in energy transportation, especially in the process of shipping hydrocarbons over long distances. Pipelines are popular infrastructure due to their safety and economy. However, being laid over long distances in remote areas, pipelines are typically affected by geohazards and harsh environmental conditions which may cause failures. Pipeline failure can lead to large business losses and

---

① The paper published in *Meas. Sci. Technol.*, **24** 095104 doi:10.1088/0957-0233/24/9/095104. 2013.

② Corresponding author: E-mail: 158010m@gs.kochi-tech.ac.jp

environmental damage. Consequently, there is a growing demand for the maintenance of pipeline over long distances. A system is required that has the ability to detect and locate problems such as leakage and emission along the pipeline. A technique that can perform multipoint and real-time monitoring is expected(Nikes,2009).

A lot of electrical gas sensing systems have been explored in order to satisfy the requirements of the gas industry. Gas sensing electrodes are free of the liquid junction potential problems associated with pH and ion sensitive electrodes, and are free of redox interferences, but the technique suffers from the limitation of dissolved gas(Ross *et al.* ,). Homogeneous semiconductor gas sensor array simply detect the resistivity from variety of gases. The sensors can convert the concentrations of gases into an electrical signal. However, these electrical gas sensors are basically onsite point sensors with power supply (Clifford *et al.* ,1985).

Gas absorption spectroscopy(Rothman *et al.* ,2009) offers direct, accurate and highly selective means of gas measurement. Optical gas sensors can realize high sensitive detection of gas concentration(Patterson *et al.* ,2007;Patterson *et al.* ,2007;Lancaster *et al.* ,1999). However, they are not applicable in a long distance monitoring system, since they are usually composed of separate optical components. Optical fibre sensors are proven to be an effective method for environmental monitoring in various environmental processes(Stewart *et al.* , 1998). When we apply optical fibre sensing technique to gas spectroscopy, it is probable to realize long distance optical gas tele-monitoring technique.

This paper explains a new concept, which is capable of addressing multiple gas sensors in one single fibre, by inserting optical gas sensors into our region selectable system(Xu *et al.* ,2010;Xu *et al.* ,2010). Wavelength scanning LD and wavelength analyser connect with optical fibres and region selectable reflectors to insert the light source and receive the results of spectroscopy of multi regions.

To distinguish multi regions, gas spectroscopy combines with region selectable technique by synchronizing Fibre Bragg Grating (FBG)reflector with the target absorption line in each sensing region. Spectroscopy performance under different temperatures, pressures and different leakage occurrences are measured. We discuss possible optical loss and estimated available sensing length of the system.

The remainder of this paper is organized as follows. In Chapter 2, the concept of quasi-distributed region selectable gas sensing is explained on one of its applications, long distance pipeline maintenance. Chapter 3 talks about basic tele-spectroscopy technique and its performance. Impact factors such as absorption line broadening and FBG instability have been discussed. Chapter 4 talks about multi-region gas sensing. Multi sensing situation is discussed. In Chapter 5, the performance of multi-region gas sensing over one km long distance is demonstrated and discussed. Finally, the summary of this study is given in Chapter 6.

2 Concept of Region Selectable Gas Sensing by Tele-Spectroscopy

In our region selectable concept, we distinguish various sensing regions by different reflection wavelengths, which are selected by bandpass optical filters named region selectors in this work. When a gas absorption line center wavelength is selected in one sensing region, gas absorption spectrum can be detected when gas exists in the region. In this paper, we would like to explain and demonstrate this concept of spectroscopy-based region selectable gas sensing on one of its potential applications, gas leakage monitoring of gas pipeline. As shown in Fig. 1, a pipeline with gas flowing inside is put in the suburb. Gas monitoring system is supposed to be set aside the pipeline. The entire monitoring area is separated into several sensing regions by region selectors, i. e., FBGs in this study. In each sensing region, the gas absorption spectrum can be reflected back by the region selector. Wavelength scanning light source is needed to provide wavelength scanning light for the monitoring system. Analyser is required to detect and record the reflected leakage gas spectrum and their distance. Leakage gas can be caught by the corresponding sensing region as shown by the inset figure of Fig. 1. Components which are able to capture the leaks, such as open gas cells or special sensing fiber(Guan *et al.*, 2011) are used to work as sensing regions.

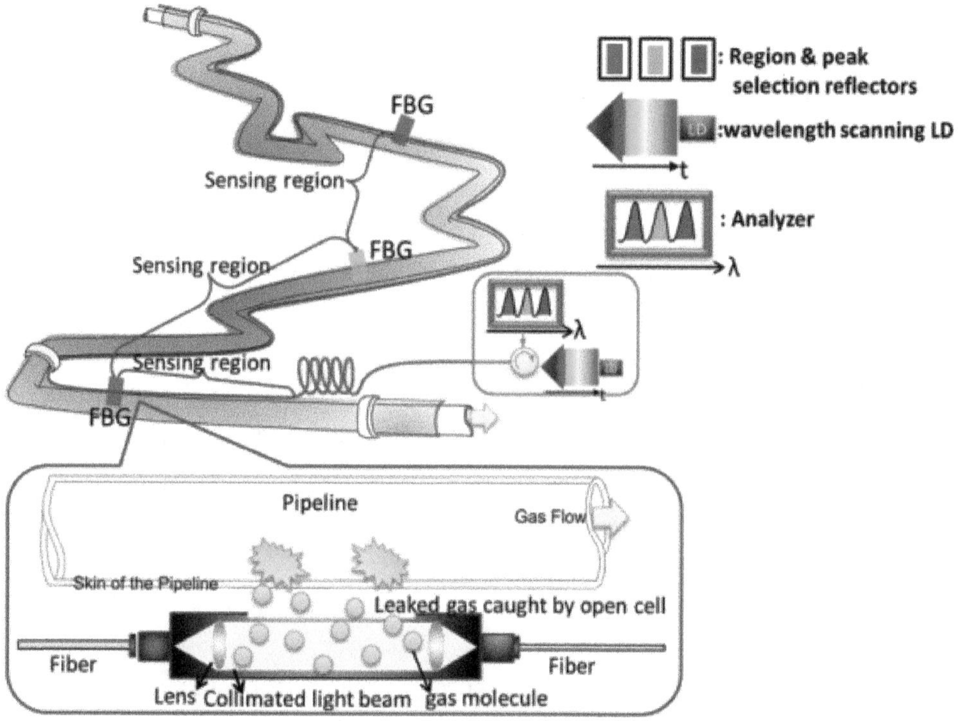

Fig. 1 Conceptual schematic of the gas sensing for long distance pipeline.

To directly detect the leakage, this gas sensing system should be set aside gas pipeline. When leaked gas flows out of the pipeline, gas cell inside sleeve picks up nearby gas leakage as shown in the inset of Fig. 1. This process could be imitated using gas cell with fibres, as shown in Fig. 2. We use agas tube cell filled with hydrogen cyanide gas (HCN13) to work as leakage imitation. Cell is launched to scanning LD light through fibres. Scanning light wave propagates through the gas cell and is partially reflected back by FBG. Round trip absorption profile is monitored.

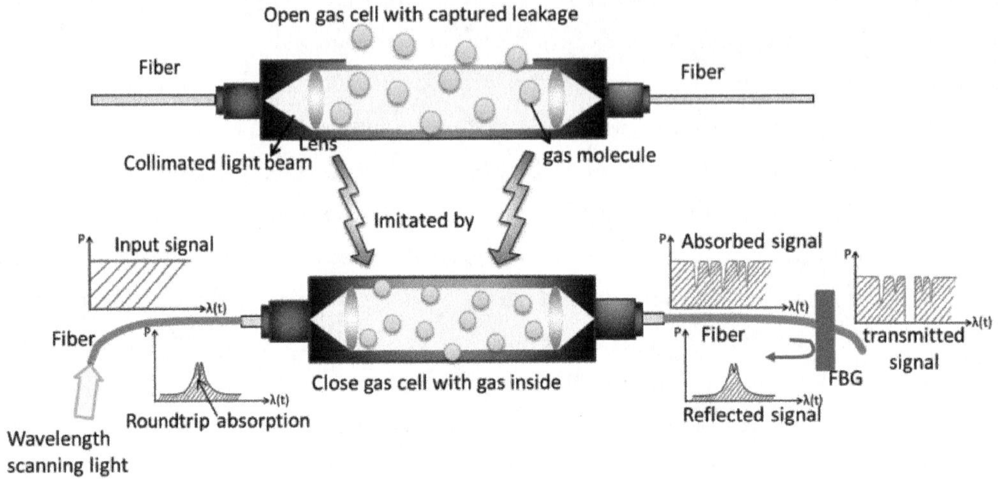

Fig. 2 Imitation of gas leakage in pipeline using gas cell with fibres.

3 Tele-Spectroscopy by Scanning-LD, FBGs and Fibers

The concept of basic tele-spectroscopy technique and an experimental demonstration are explained in this section, including the schematic and signal analysis. Potential impacts on the technique such as the absorption line broadening, FBG vibration are also discussed here.

3.1 Schematic of Tele-Spectroscopy

Schematic of tele-spectroscopy is shown in Fig. 3a. The technique uses long propagation fibers, FBG monitor as spectroscopy unit, gas cells as leakage sensing regions and FBGs as region selectors. The FBG, which has almost complete reflection at selected wavelength range, adds to the sensitivity of the system. It is laid behind the sensing region to reflect the absorption spectrum from which the location and species of the leakage can be analyzed. An FBG sensor monitor (Anritsu SF3011A) is composed of a wavelength analyser with 0.01 nm resolution and a wavelength scanning LD with the scanning range from 1520 nm to 1570 nm. Wavelength scanned light (Reid *et al.*, 1981) propagates through the sensing region in fibre. Selected wavelength range is reflected back by FBG into the analyser. If gas is located in the sensing region, round trip absorption will be measured and recorded as spectrum in this demonstration.

Hydrogen Cyanide gas cell with fibres, as shown in Fig. 3b, is used to simulate the gas leakage on the sensing cell along in pipeline. 16.5 cm-long HCN13 gas packaged cell under weak pressure of 10 Torr was used for demonstration.

There are many absorption lines for each type of gas. The absorption spectrum of HCN13 gas can be seen from Fig. 3c. The wavelength range of the absorption lines of HCN13 gas, from 1525 nm to 1565 nm, is within the wavelength range of the scanning LD. It has more than 46 absorption lines (Rothman *et al.*, 2009; Krips *et al.*, 2010). According to gas spectroscopy, when a light wave propagates through gas, the intensity of the light will reduce due to gas absorption. If we ensure the sensing region focus on one special absorption line, and select the target absorption line by region selector (FBG), the gas leakage could be detected by analysing the reflected absorption spectrum. As shown in Fig. 3c, we choose one of the absorption lines, λ_{gas}, for detection, and synchronize the peak reflection wavelength of FBG, λ_{FBG}, with the target absorption line. The intensity of the spectrum at λ_{gas} can be monitored in this way. It needs to be noted that the bandwidth of the reflection spectrum of FBG, $\Delta\lambda_{FBG}$, is required to be broader than that of gas absorption lines, $\Delta\lambda_{gas}$.

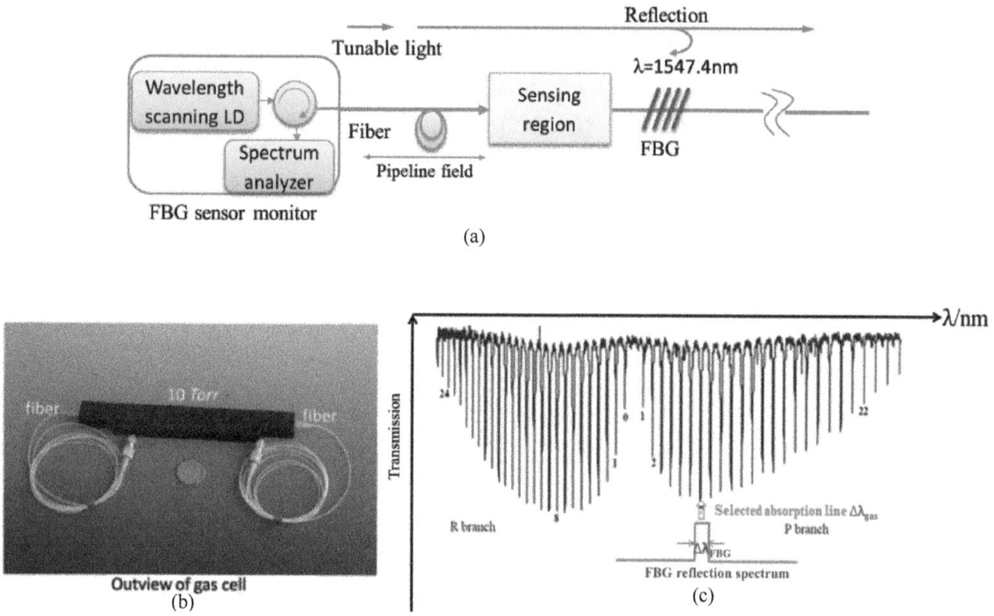

Fig. 3　Schematic of tele-spectroscopy gas sensing.

(a) Demonstration set-up; (b) Outview of gas cell(Manua);

(c) Synchronization of FBG and HCN13 gas absorption line.

In this demonstration, the target absorption line is P_7 of Ref. (Rothman *et al.*, 2009; William *et al.*, 2005), 1547.4 nm. The 3 dB linewidth of the absorption line, which is less than 0.01 nm, is within the 3 dB bandwidth of the FBG reflection spectrum ($\Delta\lambda_{FBG}$ = 0.2 nm). Therefore, absorption line of HCN13 gas, 1547.4 nm, can be monitored and analyzed.

3.2 Tele-spectroscopy Experiment and Analysis

The detected reflection spectrum and their signal analysis are shown in Fig. 4a and 4b, respectively. Fig. 4a expresses the intensity distribution of the reflected spectrum. Curve A illustrates the reflected power distribution when there is no gas in the objective region, i. e. gas cell doesn't exist in the system. Curve B illustrates the reflected power distribution when gas exists in the same sensing region. By comparing the intensity distribution of the two figures, we can see that the peak intensity and waveform of the reflected signal is obviously reduced when gas cell exists in the region because of link loss with the gas cell. In order to analyze the absorption intensity more clearly, we divided the intensity of curve B spectrum by curve A at around wavelength range of $\Delta\lambda$ FBG and normalized the ratio as shown in Fig. 4b. The normalized transmissivity demonstrates the ratio of the cell transmitted power P_{trans} and FBG reflected power P_{in}, referred to as P_{trans}/P_{in}. Additionally, link loss has been eliminated in the normalization process. The normalized figure shows that the minimum value appears at the wavelength of 1547.4 nm, which is well matched to the central wavelength of absorption line. The absorption depth is 11% when the gas pressure is 10 Torr, length of gas cell is 16.5 cm.

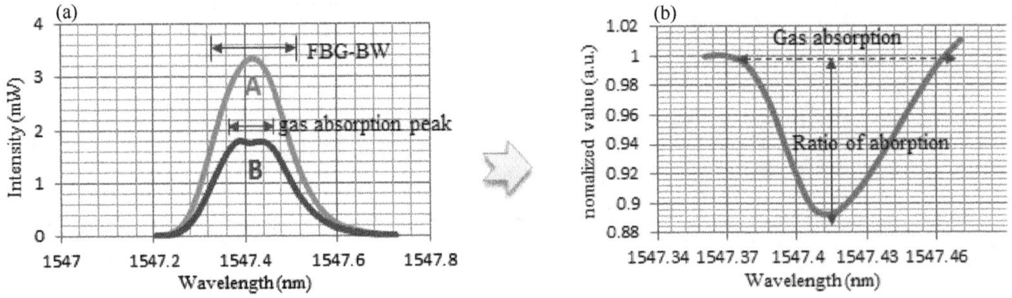

Fig. 4 Absorption intensity of reflected signal and normalized figure.
(a) Intensity distribution of detected signal; (b) Normalization intensity curve.

3.3 Discussion of Tele-spectroscopy Technique

The region selection technique by absorption peak wavelength selection suffers from the impact of the potential absorption line broadening in various temperatures and pressures. Additionally, the FBG profile in various temperatures affects the performance of the system as well. We measure the effects and discuss as follows.

3.3.1 Impact of absorption line broadening under varied temperatures and pressures

The transmitted power P_{trans} through gas leakage, is related to the absorption coefficient α and the absorption path length L by Equation (3.1):

$$P_{trans} = P_{in} \cdot \exp(-\alpha \cdot L) \tag{3.1}$$

where P_{in} is the power of the incident light.

We get the absorption ratio when we divide the absorption power $(P_{in} - P_{trans})$ by the incident light power P_{in}:

$$R = \frac{P_{in} - P_{trans}}{P_{in}} = 1 - \exp(-\alpha \times L) \approx \alpha \cdot L \tag{3.2}$$

where we approximate $\alpha \cdot L \ll 1$.

The magnitude of absorption coefficient α is determined by absorption line of the gas, it also related to ambient temperature T and pressure P.

Linewidth broadening $\Delta\lambda_{gas}$ mainly caused by the interaction of the molecules during elastic collisions, which result in Gaussian broadening (caused by Doppler broadening, for example) and Lorentzian line shape, known as temperature broadening or pressure broadening.

The broadening of hydrogen cyanide gas absorption line P_7 (Rothman *et al.*, 2009; William *et al.*, 2005) under different pressures and temperatures are measured in following experiment. Two different HCN13 gas cells, 10-Torr, 5.5 cm-long and 100-Torr, 5.5 cm-long, are used for the demonstration under 290 K and 320 K, respectively. Absorption depths and full widths at half maximum (FWHM) of the absorption line are measured. Absorption areas are calculated. During this demonstration, we only vary the ambient temperature of the gas cells, while the temperature of FBG are kept in the same range because the line broadening is negligibly small comparing with the resolution of spectroscopy.

The experimental results show that temperature change did not impact the absorption linewidth under 100 Torr. HCN gas under 10 Torr, by contrary, suffered a FWHM broadening of 0.02 m from 290 K to 320 K. Fortunately, such weak broadening does not give impact to the performance of the sensing system since the FWHMs of absorption line are within the 0.2 nm-FWHM of the reflection spectrum of FBG, which is much wider than the 10 Torr absorption line. And the temperature dependence of other lines of HCN spectrum is not expected to be significantly different (William *et al.*, 2005).

At each temperature, the detected FWHM of line P_7 under 100 Torr is broader than that under 10 Torr. The absorption under 10 Torr is less than that under 100 Torr. However, gas leakages in both cases are within the detectable range.

Table 1 lists the measured and estimated pressure broadening of line P_7 of HCN13 under various pressures at 290 K. We estimated the linewidth of line P_7 under 100 Torr and 760 Torr using the data of reference 15.

Compared with the estimated linewidth under 100 Torr, the measured data are lager. It is reasonable as the measured value includes the error of measurement device. Therefore, the measured result has good agreement with estimated value.

In case of leakage with 760 Torr pressure, that is 1 atm-pressure, 100% leakage, differential technique cannot catch an entire absorption line because the transition linewidth becomes wider than a typical FBG linewidth (0.2 nm). We can only tell the rough amount of gas leakage by directly reviewing the spectrum with selected FBG line disappearance.

Table 1 Measured and estimated pressure dependence of line P_7 of HCN13 under varied pressures

Pressure	10 Torr	100 Torr		760Torr(1atm)
FWHM (nm)	0.05	Measured	Estimated	Estimated
		0.09	≈0.065	≈0.5

In summary, we have demonstrated the temperature dependence and pressure dependence of line P_7 of HCN13 gas as well as the absorption under various temperatures and pressures. The slightly temperature broadening of the linewidth of HCN13 gas is tolerable for this technique because the bandwidth of FBG is much larger than that of the absorption line. Compared to temperature broadening, pressure broadening is more obvious and leads to apparently increase of absorption depth, consequently, the absorption ratio has been improved greatly.

Additionally, this system focuses on fuel gases, which possess pure absorption spectroscopy with narrow absorption lines. In case of fuel gas mixed with water vapour, we would like to treat water vapour as brief background of fuel gas measurement as H_2O has broad absorption band with line center wavelength of 1.38 μm, from 1.32 μm to 1.56 μm, which is not sensitive chemical phenomena.

3.3.2 Impact of FBG instability

The peak reflection wavelength of typical FBG at 1550 nm band has an approximately 1-nm shift over 100℃ temperature range which may cause some serious problems in a sensing system.

Fortunately, an athermal packaged FBG with passive or active temperature compensation(Nikles, 2009) is more reliable. Therefore, it can be applied in an ambient temperature range. An athermal packaged FBG is applicable for −5～+70℃ in case of $\Delta\lambda_{FBG}\sim$ 0.2 nm due to the small shift (～0.07 nm) of FBG peak. Consequently, the target gas can be detected.

As to realistic pipeline monitoring, in order to avoid the fluctuation of FBG peak caused by the ambient temperature of the pipeline, commercial available athermal packaged FBG can be employed in the system.

4 Multi-Region Gas Sensing Demonstration

In this section, we would like to explain the concept of wide-region optical gas tele-monitoring by inserting tele-spectroscopy into region selectable system. The performance of tele-spectroscopy applied to multi region gas sensing has been demonstrated by constructing a three-region gas sensing using FBG selectors. Absorption spectrums of different occurrences of gas leakage have been detected. Leaks in different regions have been distinguished. Multi sensing situation has been discussed in this section.

4.1 Setup of Three-Region Gas Sensing Demonstration

The schematic of three-region gas sensing is shown in Fig. 5. The implementation of three-region selectionis explained in the bottom half of Fig. 5. Different absorption lines λ_1, λ_2, and λ_3 are selected from the P-branch of gas spectrum in Fig. 3c for sensing region 1, region 2, and region 3, respectively. The peak reflection wavelength of $\lambda_{FBG\,i}$ (i= 1, 2, 3) for region i should be synchronized with $\lambda_{gas\,i}$, respectively.

The three-region gas sensing setup is shown in the top half of Fig. 5. Three FBGs, whose peak reflection wavelength range $\Delta\lambda_{FBG\,i}$ includes the target absorption lines $\lambda_{gas\,i}$, are used to separate the sensing regions. HCN13 gas cells under pressure of 10Torr are used to simulate the leakage at the sensing regions.

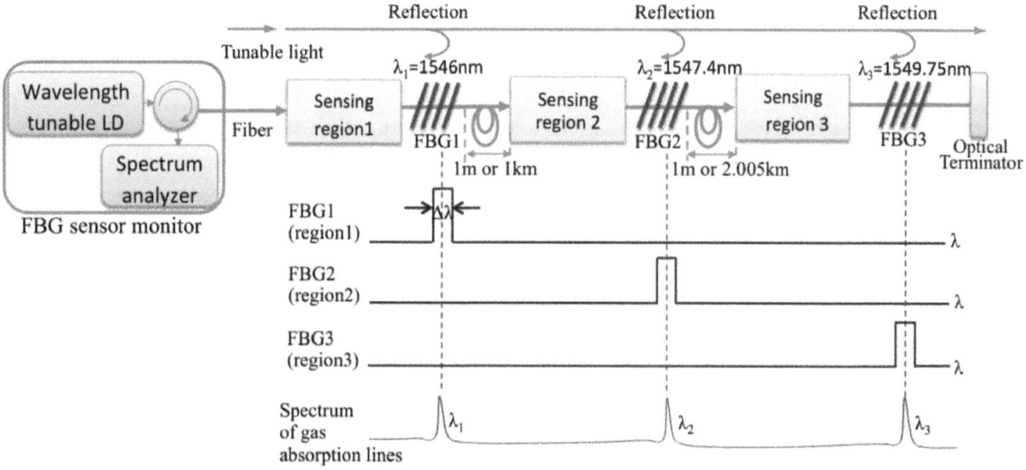

Fig. 5　Schematic of three-region gas sensing demonstration. Setup of three-region gas sensing demonstration and synchronization of the FBG peak and absorption line.

In this demonstration experiment, major absorption lines of HCN13 gas in Fig. 3c at the wavelength of P_5=1546 nm, P_7=1547. 4 nm, and P_{10}=1549. 75 nm are selected. The reflection peak wavelength of FBG1 is at 1546 nm, FBG2 is at 1547. 4 nm, and FBG3 is at 1549. 75 nm. The absorption sensitivity can be enhanced due to reflected round trip propagating and detection set-up.

The reflected spectrums of gas under four different situations are demonstrated. Firstly, we detect the spectrum when there is no gas absorption of leakage at any sensing region, define it as situation (a). Secondly, we measure the signal when only sensing region 2 has gas by connecting 16.5cm-long, 10-Torr gas cell in sensing region 2. This is named as situation (b). Thirdly, the reflected spectrum when gas only exists in region 3 is measured. Gas cell is connected in sensing region 3 to simulate gas leakage. It is referred to as situation (c). Finally, we demonstrate the performance when both sensing region 2 and sensing region 3 have leakage. In this situation, 10-Torr, 5.5 cm-long gas cell is connected in sensing region 2, and 10-Torr, 16.5 cm-long hydrogen cyanide gas cell is connected in sensing region 3 simultaneously. This case is known as situation (d). The spectrum of the four situations are detected and analyzed as follows.

4.2 Results of Three-Region Gas Sensing Demonstration

The results of the demonstration experiment of the muti-regiongas sensing system are shown in Fig. 6. Reflected absorption spectrums of situation (a), (b), (c) and (d) are shown in Fig. 6a, b, c, and d, respectively. The blue curve of situation (a) is due to the fact that there is no gas in the sensing regions, which becomes reference of measurement. The peak reflected spectrum of FBG2 has an absorption peak at the wavelength of 1547.4 nm, as shown in Fig. 6b because gas exists in sensing region 2. The amplitude reduction of peak reflected spectrum of FBG3 is caused by the introduction of link loss. The normalized curve of case 2 in the small inset figure shows that the detected absorption depth is 11%. The absorption peak occurs in the reflected spectrum of FBG3 in case 3, as shown in Fig. 6c, due to the fact that gas exists in sensing region 3. The normalized figure shows that the absorption depthin this situationis 12%. In case 4, the absorption depths are 2.6% in sensing region 2 and 10.6% in sensing region 3, respectively, as shown in Fig. 6d.

4.3 Analysis of Multi sensing Situation

One of the important issues for multi-region gas sensing system is the discrimination of the locations of each gas leakage when problems occur. Due to the fact that gas absorbs light at all absorption lines, in multi regions, power reduction will present at the backward sensing regions even leakage only exists in one sensing region. Using this three-region sensing model, we discuss about the difference between situations (b), (c) and (d), gas leakages occur in both region 2 and region 3. The absorption R_{3d} in the peak of FBG3 contains the absorption in both sensing region 2 R_{32} and sensing region 3 R_{33}. Therefore, it is important to estimate the origin of power reduction in each region.

For multi-region gas sensing, the equation of absorption ratio Eq. (3.2) for ith absorption line($i=1,2,3$: which means ith sensing region) takes the form of

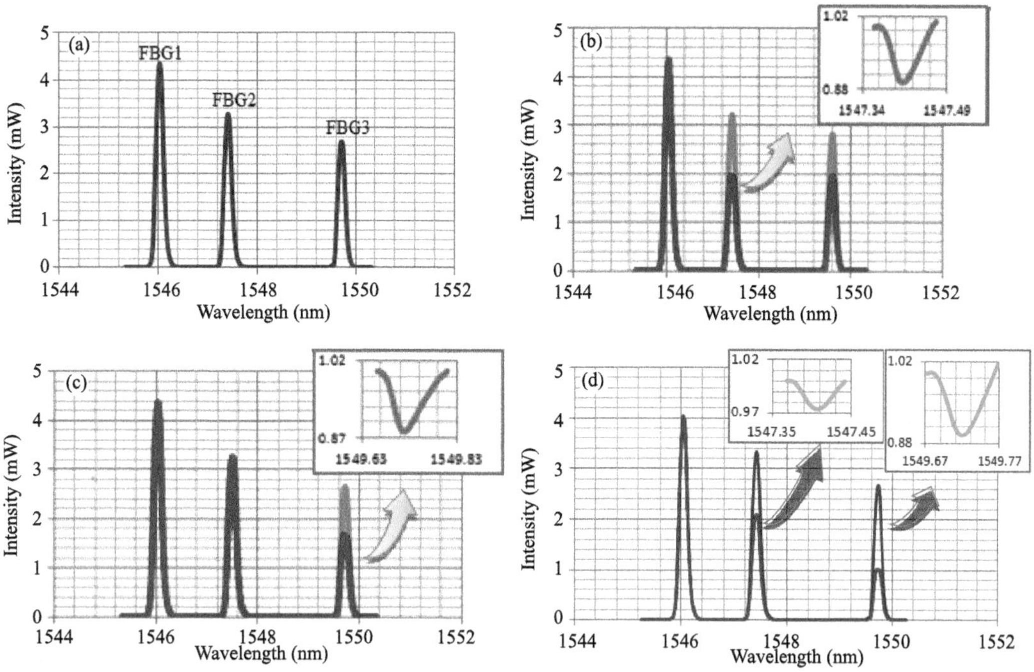

Fig. 6　Experimental results of reflected spectrum of three-region gas sensing system (a), (b), (c) and (d) show the absorption spectrum of situation(a), (b), (c), and (d), respectively.

$$R_{ij} = \frac{P_{\text{in}} - P_{\text{trans}}}{P_{\text{in}}} = 1 - \exp(-\alpha_i \cdot L_j) \approx \alpha_i \cdot L_j, \tag{4.1}$$

where α_i shows the absorption coefficient of ith absorption line, j shows absorption path length of jth sensing region, $j=1,2,3$. In situation (d) of this demonstration, $L_2 = 5.5$ cm, $L_3 = 16.5$ cm.

The detected absorption ratio R_{3d} of 3rd absorption line is given by:

$$R_{3d} = 1 - \exp(-\alpha_3 \cdot L_2) \times \exp(-\alpha_3 \cdot L_3) \approx \alpha_3 \cdot L_2 + \alpha_3 \cdot L_3 = R_{32} + R_{33}. \tag{4.2}$$

In this demonstration experiment, λ_1, λ_2, and λ_3 are near to each other, α_1, α_2, and α_3 have almost the same value.

Consequently, the detected ratio of 3rd absorption line is:

$$R_{3d} = R_{33} + R_{32} \approx R_{33} + R_{22}. \tag{4.3}$$

The absorption depth in sensing region 2 is 2.6%, sensing region 3 is 10.6%, so the depth of absorption caused by the leakage in sensing region 3 would be about 10.6% − 2.6% = 8%.

To sum up, when gas absorption presents in sensing region 3, we should compare the detected absorption R_{3d} with forward sensing region R_{2d}.

If $R_{3d} - R_{2d} \approx 0$, it indicates the absorption in 3rd absorption line comes from the leaks of forward sensing region.

If $R_{3d} - R_{2d} \gg 0$, it suggests that the leaks also exist in region 3.

5 Long Distance Multi-Region Gas Sensing

In this section we would like to explain the concept of long distance multi-region gas sensing by reporting both its schematic and performance. Available numbers and coverage of sensing regions are limited by the loss of sensing unit and fiber propagation. The possible sensing length is evaluated.

5.1 Demonstration of Multi-Region Tele-Monitoring

We demonstrated the performance of three-region gas sensing over 3km. The schematic of the demonstration experiment is shown in Fig. 5. To construct sensing regions far from the monitoring center, long fibers with length of 1 km and 2.005 km are connected in front of sensing region 2 and sensing region 3, respectively. The same as in multi-region gas sensing demonstration, this experiment uses FBG sensor monitor as wavelength scanning LD and spectrum analyzer, FBGs as region selectors and gas cell as sensing region.

Three different occurrences of leakage are detected and analyzed. Firstly, we measure the reflected spectrum when there is no leakage at any region, named as situation (a). Then, we measure the response when leakage only exists in sensing region 2, called situation (b). Finally, we measure the absorption when leakage only exists in sensing region 3, referred to as situation (c). The recorded reflection spectrum and normalized absorption depth of the three situations are displayed in Fig. 7.

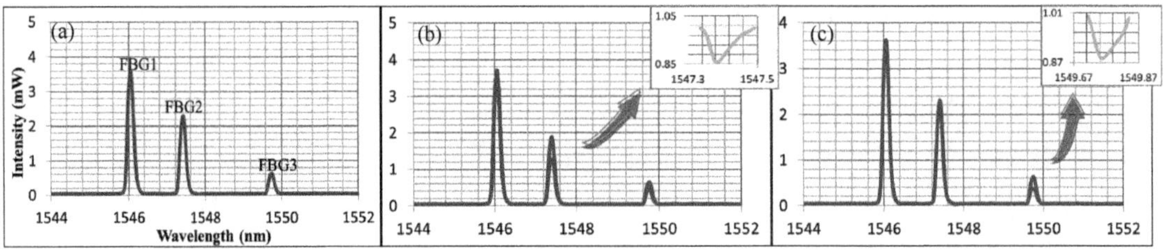

Fig. 7 Experimental results of reflected spectrum of long distance gas sensing system
(a), (b), (c) show the absorption spectrum of situation (a)~(c), respectively.

When leakage only exists in sensing region 2, the absorption peak appears at the peak reflection wavelength of FBG2, the analyzed absorption depth is 14.2%. It is 12% when the leakage only occurs at sensing region 3, the absorption peak is at the peak reflection wavelength of FBG3 in this situation.

5.2 Impact of Optical Loss in the Design of Gas Sensing

The sensing length of this technique is limited by optical loss such as numbers of link loss for each sensing region and length of propagation in fiber.

When we take link loss into account, the transmissivity T_i of transmitted power and input power for sensing region i is given by

$$T_i = C^i \cdot \exp[-\alpha_F \cdot L_i], \quad (5.1)$$

where C is the coefficient of transmitted power related to coupling loss, L_i represents the distance between the monitor center and the ith sensing region. α_F is the loss parameter of optical fiber. It includes not only material absorption but also other sources of power attenuation. A typical optical fiber has a 0.2dB/km transmission loss, consequently, $\alpha_F \approx 0.046$.

We substitute $C = 0.71$ and $\alpha_F = 0.046$ in equation (5.1) and obtain

$$T_i = 0.71^i \cdot \exp[-0.046 \cdot L_i]. \quad (5.2)$$

The output power and minimum detectable power of FBG sensor monitor accept system loss within 30dB. The maximum number of sensing region N is required to satisfy

$$10^{-3} \leqslant 0.71^N \cdot \exp[-0.046 \cdot L_N]. \quad (5.3)$$

It means available number of sensing regions and tele-monitoring length of this demonstration system can be explained as follows

$$-3 \leqslant N \cdot (-0.1487) + (-0.02) \cdot L_N. \quad (5.4)$$

Assuming the coverage of each sensing region is 1km, the maximum number of sensing regions would be 19, the available sensing length is less than 20 km;

If the designed coverage of each sensing region is 10 km, the maximum number of sensing region is 8, the corresponding sensing length approaches 80 km.

The total length and sensitivity of one system can easily enhance by inserting optical fiber amplifier before detection.

6 Summary

In this paper we proposed an innovative concept of gas sensing which can be used to detect the types and locations of gas at multiple regions over long distances. The technique is based on optical tele-spectroscopy and the fact that one kind of gas has numerous absorption lines. Basic concept of tele-spectroscopy was explained. Potential impact factors such as absorption line broadening under varied temperatures and pressures were demonstrated and discussed. Low pressure leakage around 10 Torr is detectable.

Multi-regions tele-monitoring experiment using FBG monitor as spectroscopy unit, long propagation fibers, gas cells as leakage sensing regions and FBGs as region selectors is demonstrated. Four different situations of the existence of gas were measured. Leakage distinguish in multi sensing situation was also discussed based on the three-region gas sensing demonstration. The location of the gas leakage was estimated by the absorption profile in the reflected spectrum.

The performance of multi-region tele-monitoring was demonstrated in this study as well. Long fibers were used to connect long distance absorption sensing regions. The results indicated that the technique worked well in long distance gas monitoring. Optical loss was analyzed and evaluated based on link loss and fiber propagation loss. Relationship between available region numbers and tele-monitoring length were estimated based on designed coverage of sensing region.

The advantage of the approach is that it is simple, safe, low cost and can be easily extended to long distance, multipoint sensing system. The technique is capable of being used for long distance distributed pipeline maintenance.

Acknowledgement

This work was partially supported by the China-Japan international corporation project No. 2010DFA32920, JST Application Research No. 1513, JSPS No. 18360180 and No. 22657062.

References

Clifford P, Alto P. 1985. Selective gas detection and measurement system. *United States Patent*, **4** 542-640, Sep. 24.

Guan C Y, Tian F J, Dai Q, *et al*. 2011. Characteristics of embedded-core hollow optical fiber. *Optics Express*, 20069-20078.

Krips M, Crocker A F, Bureau M, *et al*. 2010. Molecular gas in SAURON early-type galaxies: detection of 13CO and HCN emission. *Mon. Not. R. Astron. Soc.*, **407**:2261-2268.

Lancaster D G, Richter D, Tittel F K. 1999. Portable fiber-coupled diode-laser-based sensor for multiple trace gas detection. *Appl. Phys.* B **69**:459-465.

Manual. C-Band calibrator hydrogen cyanide gas cell H13C14N.

Nikles M. 2009. Long-distance fiber optic sensing solutions for pipeline leakage, intrusion and ground movement detection. *Proc.* SPIE7316, 02-1-02-13.

Patterson C S, McMillan L C, Longbottom C, *et al*. 2007. Portable optical spectroscopy for accurate analysis of ethane in exhaled breath. *Meas. Sci. Technol.*, **18**:1459-1464.

Reid J, Labrie D. 1981. Second-harmonic detection with tunable diode lasers-comparison of experiment and theory. *App. Phys.* B, **26**:203-210.

Ross J W, Riseman J H, Krueger J A. Potentiometric gas sensing electrodes. *Gas Electrodes*, 473-487.

Rothman L S, Gordon I E, Barbe A, *et al*. 2009. The HITRAN 2008 molecular spectroscopic database. *JQSRT*, **110**:533-572.

Stewart, Stewart G, Moodie D, *et al*. 1998. Design of a fibre optic multi-point sensor for gas detection. *Sensors and Actuators B*, **51**:227-232.

William Swann C, Sarah L. Gilbert. 2005. Line centers, pressure shift, and pressure broadening of 1530—1560 nm hydrogen cyanide wavelength calibration lines. *Opt. Am.* B, **22**(8).

Xu X J, Nonaka K. 2010a. Fiber strain measurement for wide region quasi-distributed sensing by optical correlation sensor with region separation techniques. *Journal of Sensors*, 839803.

Xu X J, Nonaka K. 2010b. The regional selectable distributed fibre-optic sensing system based on pulse correlation and partial reflectors. *Meas. Sci. Technol.*, **21**.

Yu Lung Lo, Chih Ping Kuo. 2003. Packaging a fiber Bragg grating with metal coating for an athermal design. *Journal of Lightwave Technology*, **21**(5).

水稻根系内生细菌对未来大气 CO_2 浓度升高的响应①

任改弟[1,2]　张华勇[1]　林先贵[1]　朱建国[1]　贾仲君[1]②

1）土壤与农业可持续发展国家重点实验室，中国科学院南京土壤研究所，南京 210008；
2）中国科学院大学，北京 100049

摘要

针对中国 FACE(Free Air CO_2 Enrichment)平台的镇籼 96、扬稻 8 号、II 优 084 和扬两优 6 号四种水稻品种，采用新一代高通量测序技术，研究了水稻根系内生菌的整体微生物群落对未来大气 CO_2 浓度升高的响应。结果表明，水稻内生菌群落中 γ-变形菌纲的肠杆菌科相对丰度最高，占整体微生物群落的 30.8%～59.8%。对于镇籼 96、扬稻 8 号和 II 优 084 三种水稻品种，大气 CO_2 浓度升高可能抑制了数量上占优势的微生物菌群（优势菌群）生长，而促进了数量上不占优势的微生物菌群（稀少菌群）繁殖。例如，对于 II 优 084 品种，相对丰度高于 14.6% 的 4 种水稻内生菌为肠杆菌科、假单胞菌科、黄单胞菌科和气单胞菌科，大气 CO_2 浓度升高，这些优势菌群的相对丰度由 74.8% 降为 67.2%；相反，稀少菌群主要由鞘脂杆菌科、丛毛单胞菌科、黄杆菌科及草酸杆菌科组成，其相对丰度则由 4.13% 增至 16.9%，其中，与对照处理相比，鞘脂杆菌科相对丰度增加比例高达 344 倍，是大气 CO_2 浓度升高的最敏感微生物类群。但对于水稻品种扬两优 6 号，根系内生菌对大气 CO_2 浓度升高的响应模式与其他三种品种不完全一致。这些研究结果表明，微生物的相对丰度可能是影响水稻根系内生菌对大气 CO_2 浓度升高响应的重要因素，为研究全球变化下整体微生物结构与功能的演变规律提供了一定的依据。

关键词：高通量测序；微生物群落；植物内生菌；FACE

植物内生菌是一类生活在植物组织内部、对植物组织不引起明显病害症状的微生物（Dennis，1995），这些微生物在植物生长和健康等方面发挥了重要作用。近年来，植物内生菌因其在固氮（Landha *et al.*，1983；Muthukumarasarry *et al.*，2007）、防病（Berg *et al.*，2005）、促生（Singh *et al.*，2006）、强化植物修复污染环境（马莹等，2013）等方面的优势而受到广泛关注。事实上，植物组织内部具有丰富的微生物资源。目前已经从甘蔗（*Saccharum officinarum* L.）（Dong *et al.*，1994）、玉米（*Zay mays*）（Palus *et al.*，1996）、水稻（*Oryza sativa* L.）（Barraquio *et al.*，1997）等禾本科农作物中发现了多种具有固氮功能的内生细菌。然而，由于研究手段的限制，目前国内外主要集中于特定内生菌对植物的固氮、抗病、促生等作用的研究，在整体微生物群落水平的研究报道较少。

微生物是地球物质和能量流动的重要引擎，广泛参与了多种营养元素的生物地球化学循环。全球气候变化（比如大气 CO_2 浓度升高下）可能对生态系统中微生物驱动的生态过程

① 本文发表于土壤学报．**50**(6)：1162-1171，2013．科技部国际合作专项项目“大气组成变化及其影响与对策研究”(2010DFA22770)和中国科学院应用微生物研究网络项目(KSCX2-EW-G-16)资助

② 作者简介：任改弟(1984-)，女，博士研究生，主要从事环境微生物生态学研究。E-mail：gdren@issas.ac.cn
通讯作者，E-mail：jia@issas.ac.cn

产生重要影响。然而目前研究大多集中于土壤微生物群落对全球气候变化的响应与适应(Lipson,2006;He *et al.*,2012),全球变化下植物内生菌的响应与适应规律研究较少。水稻是世界上最重要的粮食作物之一,是世界上50%以上人口的粮食之源。研究水稻根系部内生菌的微生物群落结构和组成及其对未来大气CO_2浓度升高的响应是农业可持续发展的重要内容之一。目前已有利用传统的可培养的方法(Ladha *et al.*,1983;Muthukumarasamy *et al.*,2007;Singh *et al.*,2006;Barraquio *et al.*,1997)、克隆文库方法(Sun *et al.*,2008)、变性梯度凝胶电泳(Denaturing gradient gel electrophoresis,DGGE)方法(Hardoim *et al.*,2011)对水稻根系内生菌的功能或多样性进行研究。但这些方法存在信息量小、费时费力、分辨率低等不足,限制了对水稻内生菌多样性的认识。新一代高通量测序技术为解决上述方法的局限性提供了新的机遇,该技术因信息量大、分辨率高、快速、准确、高效等优点而被越来越多地应用于微生物生态学研究中。

未来大气CO_2浓度升高将对农业生态系统产生重要影响。已有研究表明,大气CO_2浓度升高改变了土壤微生物群落组成和结构(He *et al.*,2012),而水稻根系内生菌的响应以及不同品种的水稻根系内生菌的整体微生物群落响应模式鲜见报道。本研究依托中国FACE(Free Air CO_2 Enrichment)平台,采集对照圈(即对照处理,浓度为当前大气CO_2浓度)和FACE圈(即大气CO_2浓度升高处理,其浓度比对照圈高(200±40)μmol/mol)的四种水稻(包括两个籼稻品种和两个杂交稻品种)根系样品,利用新一代454高通量测序技术对水稻根系内生菌的16S rRNA基因进行深度测序,旨在揭示水稻根系内生菌的整体微生物群落组成,以及不同水稻品种根系内生菌对未来大气CO_2升高的响应规律。

1 材料与方法

1.1 研究区概况

中国FACE平台位于江苏省扬州市小纪镇(119°42′0″E, 32°35′5″N),建于2004年。处于亚热带地区,年降雨量900～1000 mm,年均温度16℃,日均光照12.3 MJ/m^2,年均日照时间超过2000 h,无霜期大于230 d,该实验站为当地典型的单季稻-麦轮作方式,水稻耕种历史大于50年,是我国代表性的水稻产区。土壤类型为下位砂姜土。供试土壤基本理化性质如下:pH 6.8,有机碳18.4 g/kg,全氮1.5 g/kg,全磷0.63 g/kg,全钾14.0 mg/kg,容重1.16 g/cm^3,砂粒(2～0.02 mm)578 g/kg,粉粒(0.02～0.002 mm)285 g/kg,黏粒(<0.002 mm)137 g/kg。土壤详细理化特性已有报道(苑学霞等,2006)。

1.2 试验设计

FACE系统的田间实验区由6个对边距为12.5 m的八角形实验圈构成,具体设计和布局已有详细报道(刘钢等,2002)。其中对照圈(Ambient圈)包括三个重复,其大气CO_2浓度为当前大气CO_2浓度,即(355±15)μmol/mol,本处理为对照处理,用aCO_2表示;另三个圈为FACE圈,其大气CO_2浓度较对照圈高(200±40)μmol/mol,即大气CO_2浓度升高处理,用eCO_2表示。FACE圈和对照圈随机分布且在相邻两圈之间设有缓冲区以防止各个圈中大气CO_2浓度相互影响。

水稻根系样品采集于2010年10月10日(水稻成熟期)进行,共采集了四种水稻品种的根系样品。对照处理的根系样品共计10个,其中镇籼96、扬稻8号、II优084、扬两优6号的根系样品重复分别为3、2、2、3;FACE处理的根系样品共计9个,其中镇籼96、扬稻8号、II优084、扬两优6号的根系样品重复分别为2、2、2、3。用冰盒将水稻植物植株根部带回实验室后,反复小心抖落土壤并用去离子水洗涤根系备用。镇籼96和扬稻8号属于籼稻品种,II优084、扬两优6号属于杂交稻品种。

1.3 水稻根系内生微生物DNA提取

水稻根系内生菌的微生物基因组DNA提取根据已报道的方法(Garbeva *et al.*,2001)并稍作改进。首先将清洗后的新鲜植物根系浸入pH 8.0的TE缓冲液(1 mol/L Tris-HCL,500 mmol/L EDTA)中(根重量:TE缓冲液体积=1:10),以200 r/min的转速摇床振荡1 h使根表(Rhizoplane)微生物与根分离,收集根表微生物菌悬液,如此共收集4次,前3次收集的根表微生物菌悬液合并。将前3次和第4次收集到的菌悬液离心沉淀后分别进行DNA提取,然后通过1.2%琼脂糖凝胶电泳检测DNA提取量。本电泳结果表明,前3次提取的DNA电泳条带亮度远大于第4次(图未列出),说明经过3次洗脱后绝大部分根表微生物已经从根表释放,将不会对根内微生物DNA造成实质性严重污染。

根系经过上述4次TE缓冲液洗脱后,将根系剪碎研磨至糊状,浸泡在灭菌的pH 8.0的磷酸盐缓冲液中,以200 r/min的转速摇床振荡1 h,用玻璃棉对植物悬液进行过滤,收集过滤液(即菌体细胞液)。再将过滤后的植物渣浸泡于磷酸盐缓冲液,经过摇床振荡一玻璃棉过滤后,对菌体细胞进行再次收集。反复共收集3次以便最大程度地回收水稻根系内生菌的微生物细胞。3次收集的细胞悬液合并,离心沉淀并获得微生物菌体后,加入溶解酶裂解微生物细胞,然后加入十二烷基磺酸钠(Sodium dodecyl sulfonate,SDS)和蛋白酶K使得DNA从核蛋白中游离出来,再加入5 mol/L的NaCl使蛋白沉淀。再经氯仿-异戊醇反复抽提蛋白,异丙醇离心沉淀DNA和70%冰乙醇洗涤,最后用TE缓冲液对DNA进行溶解。将DNA保存于−20℃备用。

1.4 新一代高通量测序

利用通用引物(515F-907R)扩增微生物16S rRNA基因的V4区域,修饰后的通用引物含有不同的Tag标签用以区分不同样品。每对引物中同时包含adaptor和Key序列以提高下游测序效率。具体引物组成如下:5'-adapter A or B + Key Sequence + (Tag) + (template- specific sequence) -3'。例如,515F端的引物组成为5'-CGTATCGCCTCCCTCGCGCCA+TCAG+(6 bp tag)+(GTGCCAGCMGCCGCGG)-3';907R端的引物组成为5'-CTATGCGCCTTGCCAGCCCGC + TCAG + (6 bp tag) + (CCGTCAATTCMTTTRAGTTT)-3'。聚合酶链式反应(PCR)体系如下:0.25 μl的TaKaRa Taq HS(5 Uμl^{-1}),5.0 μl的10×PCR Buffer(Mg^{2+} Plus),4.0 μl的dNTP Mixture(各2.5 mmol/L),1.0 μl的引物(20 μmol/L),1.0 μl的DNA模板,37.75 μl的无菌水,总反应体系50 μl。PCR扩增的反应条件如下:94℃,5.0 min;32×(94℃,30 s;55℃,30 s;72℃,45 s);72℃,5 min。PCR产物经过切胶纯化、等摩尔数混合后进行高通量测序。

1.5 高通量数据分析

高通量数据分析采用 Quantitative Insights Into Microbial Ecology(QIIME)(http://qiime.sourceforge.net)。主要步骤如下:(1)对原始数据进行质量控制,过滤掉低质量的16S rRNA 基因序列。低质量序列的标准为:序列长度< 200 bp,平均质量得分< 25,序列模糊碱基 $N>1$,同聚物中的寡核苷酸个数>6;(2)根据 Tag 标签,将所有的序列分配至对应的水稻根系样品;(3)在 97%的序列相似度将 16S rRNA 基因序列归为不同的 OTUs(Operational taxonomic units,操作分类单元)(Edgar,2010)。OTU 产出后,统计各个样品含有 OTUs 总量及每个 OTU 的序列条数,并从每个 OTU 中选取一个代表性序列,在 80%的置信度水平采用 RDP Classifier(Wang *et al.*,2007)对序列进行分类鉴定,得到每个 OTU 的微生物物种分类学信息。利用 R 软件 Vegan 程序包(R v. 2.15.0),采用主成分分析(Principal component analysis,PCA)方法研究微生物群落结构变化。

2 结果

2.1 新一代高通量测序

对四个水稻品种根系内生菌的 16S rRNA 基因序列进行了高通量测序,结果表明,删除掉低质量序列后,所有 19 个样品共得到了 120193 条高质量的 16S rRNA 基因序列(表 1),每个样品的平均序列数为 6326,平均序列长度 402 bp,几乎所有序列均被鉴定为细菌,仅 1 条序列为古菌。在 97%的序列相似度水平上,所有 120193 条序列可聚类为 9503 个不同的 OTUs(表 2),平均每个样品的 OTUs 个数是 1091。进一步对每个 OTUs 中的代表性序列进行分类鉴定,结果表明 9503 个不同的 OTUs 可归为 10 个门(表 2),18 个纲,37 个目,62 个科,133 个属。在所有 9503 个 OTUs 中,高达 8 424 个 OTUs 被鉴定为变形菌门,占所有 OTUs 的比例高达 88.6%(表 2)。其次为拟杆菌门和厚壁菌门,OTUs 数目分别为 617 和 227(表 2)。在变形菌门的 8424 个 OTUs 中,6039 个 OTUs 属于 γ-变形菌纲,占变形菌门 OTUs 总数的 71.7%,而 α-、β-和 δ-变形菌纲仅分别占整个变形菌门 OTUs 总数的 5.05%、14.6%和 0.43%(表 2)。因此,在门的分类水平上,变形菌门是最为优势的微生物菌群;在纲的水平上 γ-变形菌纲是最为优势的微生物菌群。

表 1 高通量序列数及其微生物分类水平概述

Table 1 High throughput sequence number and taxonomic classification of endophytic microbes

水稻品种 Rice variety	CO_2处理 CO_2 treatment	高质量序列数 High quality reads number	序列被鉴定到各个分类水平的百分比 Percentage of the sequence identified at different taxonomic levels (%)				
			门 Phylum	纲 Class	目 Order	科 Family	属 Genus
ZX-96	aCO_2	7 343±1 979	99.8±0.1	98.5±1.1	88.6±3.6	87.1±2.5	34.6±9.3
	eCO_2	5 000±1 767	99.9±0.1	98.1±0.0	86.7±0.1	85.3±0.1	33.2±10.4
YD-8	aCO_2	7 101±1 024	99.7±0.1	98.2±0.9	92.2±5.3	90.5±5.9	24.7±23.2
	eCO_2	7 005±267	99.7±0.1	98.2±0.1	89.0±3.1	87.8±3.2	33.0±2.2

（续表）

水稻品种 Rice variety	CO_2处理 CO_2 treatment	高质量序列数 High quality reads number	序列被鉴定到各个分类水平的百分比 Percentage of the sequence identified at different taxonomic levels (%)				
			门 Phylum	纲 Class	目 Order	科 Family	属 Genus
TY-084	aCO_2	7 654±848	99.8±0.1	98.3±1.0	89.5±4.8	88.4±4.3	29.1±3.6
	eCO_2	7 633±750	99.8±0.1	98.5±1.2	88.6±1.8	86.5±0.9	35.8±8.2
YLY-6	aCO_2	4 020±1 782	99.3±0.7	97.8±1.1	90.0±2.3	87.6±4.0	36.3±8.0
	eCO_2	5 775±680	100±0	98.1±1.2	91.2±2.1	89.5±2.4	44.8±11.9
总计 Total	120 193	99.7	98.1	89.5	87.7	34.5	

注：aCO_2表示 ambient CO_2对照处理；eCO_2表示 CO_2浓度升高处理。ZX-96 表示水稻品种镇籼 96；YD-8 表示水稻品种扬稻 8 号；TY-084 表示水稻品种 II 优 084；YLY-6 表示水稻品种扬两优 6 号。下同

Note: "aCO_2" and "eCO_2" denote ambient CO_2 and elevated CO_2, respectively. ZX-96, YD-8, TY-084, and YLY-6 denote rice variety of ZhenXian96, YangDao8, IIYou084, and YangLiangYou6, respectively. The same below。

表 2 样品中检测到的所有门的 OTUs 数量及隶属于变形菌门的 OTUs 数量

Table 2 Number of OTUs in all the phyla and number of OTUs in Proteobacteria detected in samples

门 Phylum	纲 Class	OTUs 数目 OTU No.	占总 OTU 的百分比 Percentage of the total OTUs (%)	占变形菌门总 OTUs 的百分比 Percentage of the total OTU in Proteobacteria (%)
变形菌门 Proteobacteria		8 424	88.6	
	γ-变形菌纲 Gammaproteobacteria	6 039	63.5	71.7
	β-变形菌纲 Betaproteobacteria	1 231	13.0	14.6
	α-变形菌纲 Alphaproteobacteria	425	4.47	5.05
	δ-变形菌纲 Deltaproteobacteria	36	0.38	0.43
	未确定变形菌门 Unclassified-Proteobacteria	693	7.29	8.23
拟杆菌门 *Bacteroidetes*		617	6.49	
厚壁菌门 *Firmicutes*		227	2.39	
放线菌门 *Actinobacteria*		24	0.25	
浮霉菌门 *Planctomycetes*		11	0.12	
酸杆菌门 *Acidobacteria*		5	0.05	
疣微菌门 *Verrucomicrobia*		4	0.04	
绿弯菌门 *Chloroflexi*		3	0.03	
广古菌门 *Euryarchaeota*		1	0.01	
螺旋体门 *Spirochaetes*		1	0.01	
未确定细菌 *Unclassified-bacteria*		186	1.96	
总计 Total		9503		

注：OTUs 表示分类操作单元；OTUs 是在 97%的序列相似水平计算得到的；门或纲根据 OTUs 的数目降序排列

Note: "OTUs" denotes operational taxonomic units. OTUs are calculated out on the basis of 97% sequence similarity. The Phylum or Class is presented in descending order on the basis of their OTUs numbers

2.2 不同水稻品种根部内生微生物群落组成

高通量数据结果表明，≥87.7% 的 16S rRNA 基因序列能够鉴定到科以上的分类水平，但是只有 34.5%的序列能鉴定到属的水平(表 1)，因此，后续详细分析将在科水平进行。图 1 结果显示，肠杆菌科是水稻根系内生菌群落的优势微生物，占细菌总量的 30.8%～59.8%。其次是假单胞菌科、黄单胞菌科和气单胞菌科，分别占整个微生物群落的 3.40%～27.8%、6.63%～14.5%和 1.11%～14.6%。鞘脂杆菌科、丛毛单胞菌科、黄杆菌科和草酸杆菌科的相对丰度较低，而且在各个处理中相对丰度的变异性较大。

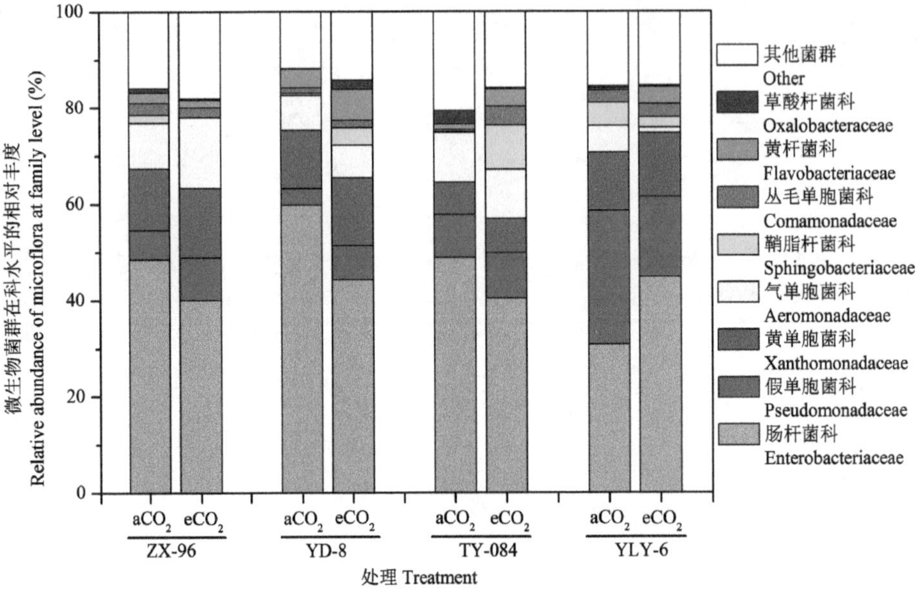

图 1 不同水稻品种根系内生菌群的相对丰度

Fig. 1 Relative abundance of endophytic microflora in roots of different rice varieties

2.3 水稻根系内生菌对大气 CO_2 浓度升高的响应

根据水稻根系内生菌群的相对丰度高低，综合考虑微生物菌群对大气 CO_2 浓度升高的响应模式(正响应或负相应)，将微生物划分为优势菌群和稀少菌群(图 2)。此外，因为微生物在各个品种中的相对丰度差异较大，对于每一个水稻根系内生菌群，优势菌群和稀少菌群判定阈值因品种而异。例如，对于水稻品种镇籼 96(ZX-96)和扬稻 8 号(YD-8)，相对丰度分别>12.8%和>12.0%的微生物菌群被认为是优势菌群，同时这些根系内生菌相对丰度随大气 CO_2 浓度升高而降低；相对丰度小于等于此百分比的微生物菌群被称为稀少菌群，其相对丰度较低且随 CO_2 浓度升高而增加(图 2a，图 2b)。同理，对于水稻品种 II 优 084(TY-084)，相对丰度>2.69%的微生物菌群被认为是优势菌群，而相对丰度≤2.69%的微生物被认为是稀少菌群(图 2c)。

对于水稻品种镇籼 96(ZX-96)、扬稻 8 号(YD-8)和 II 优 084(TY-084)，与对照相比，FACE 圈水稻根系优势菌群的相对丰度较低，稀少菌群的相对丰度较高(图 2，图 3)。例如，对于品种II优084(TY-084)，数量上占优势的微生物菌群(相对丰度>14.6%)主要包括：肠杆菌科、假单胞菌科、黄单胞菌科和气单胞菌科(图2c)。与对照处理的水稻根系优势菌群相对丰度67.2%相比，FACE圈处理下高达74.8%(图3c)。而稀少菌群(相对丰度≤14.6%)主要包括鞘脂杆菌科、丛毛单胞菌科、黄杆菌科和草酸杆菌科(图2c)，其相对丰度由对照处理的4.13%显著增加至FACE圈的16.9%($p<0.05$)(图3c)。对于水稻品种镇籼96和扬稻8号，尽管优势菌群仅肠杆菌科一种(图2a，图2b)，但整体微生物群落对大气CO_2浓度升高的响应模式与水稻品种II优084相似(图2c，图3c)。

对于水稻品种扬两优6号，其根系内生菌对大气CO_2浓度的响应模式与其他品种不完全一致。例如，大气CO_2浓度升高下根系内生菌优势菌群(肠杆菌科和假单胞菌科)的相对丰度高于对照处理，而稀少菌群的相对丰度则低于对照处理。其根系内生菌优势菌群分别为肠杆菌科和假单胞菌科，在对照圈中相对丰度分别为30.8%和27.8%，远低于其他三种品种(镇籼96、扬稻8号和II优084)的肠杆菌科优势菌相对丰度(48.4%~59.8%)，表明扬两优6号根系内生菌群的结构组成较其他品种更为复杂，并表现出明显不同的响应模式(图2d，图3d)。

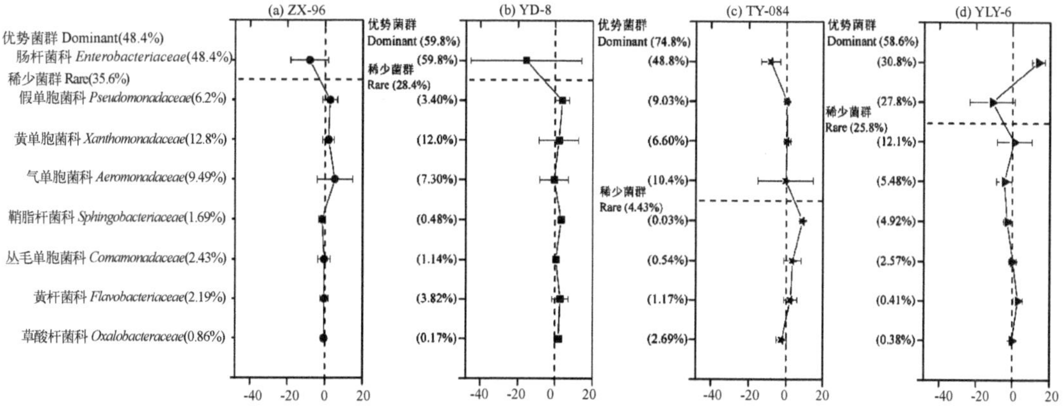

图2 不同水稻品种微生物菌群的相对丰度变化量

Fig. 2 Change in relative abundance of the microflora in rice of different varieties

注："优势菌群"表示相对丰度较高的微生物；"稀少菌群"表示相对丰度较低的微生物；图中横向虚线表示微生物对eCO_2从负响应变为正响应的分界，也是人为划定的优势菌群和稀少菌群的分界。括号内的百分比表示微生物在aCO_2处理中的相对丰度。"优势菌群 Dominant (%)"表示相对丰度较高的微生物的相对丰度之和；"稀少菌群 Rare(%)"表示相对丰度较低的微生物的相对丰度之和

Note: "Dominant groups" mean the microbial groups high in relative abundance, while "Rare groups" the microbial groups low in relative abundance; The horizontal dotted line represents the boundary line between the negative response and positive response to eCO_2, as well as the boundary artificially set between the dominant and rare microbial groups. The percentage in "()" refers to the relative abundance of the microbes at aCO_2. "Dominant (%)" represents the sum of all the dominant groups in relative abundance, and "Rare (%)" the sum of all the rare groups in relative abundance

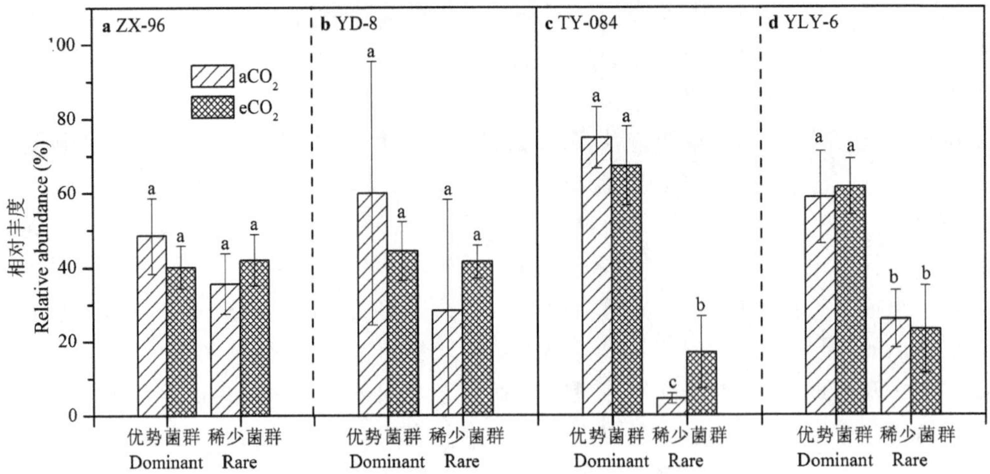

图 3 不同水稻优势菌群和稀少菌群的相对丰度

Fig. 3 Relative abundance of dominant and rare groups in rice of different varieties

注：图中优势菌群(Dominant)和稀少菌群(Rare)相对丰度是将图 2 中所划分的各个优势菌群和稀少菌群相对丰度分别相加得到的。图中不同字母表示差异显著($p<0.05$)

Note: The relative abundance of the dominant groups and of the rare groups is the sum of the relative abundances of all the individual dominant groups and of all the individual rare groups as is shown in Fig. 2. Different letters in the figure represent significant difference ($p<0.05$)

2.4 大气 CO_2 浓度升高下水稻根系内生菌的敏感微生物菌群

通过比较水稻根系内生菌的相对丰度在 FACE 圈和对照圈是否具有统计上的显著差异($p<0.05$)以及微生物相对丰度的差异大小确定对大气 CO_2 浓度升高的敏感微生物菌群。结果表明，鞘脂杆菌科的相对丰度达到了统计上的显著性差异($p<0.05$)，而水稻根系内生菌群的其他微生物均未达到显著性差异，鞘脂杆菌科被认为是最为敏感的微生物菌群。对照处理下，该微生物菌群在扬稻 8 号(YD-8)和 II 优 084(TY-084)的相对丰度分别为 0.48% 和 0.03%；而 CO_2 浓度升高处理下分别达到了 3.65%($p=0.040$)和 9.09%($p=0.009$)(图 1)，增加倍数分别达到 6.62 倍和 344 倍(表 3)。此外，丛毛单胞菌科、黄杆菌科和草酸杆菌科也可能是对大气 CO_2 比较敏感的微生物菌群。尽管大气 CO_2 浓度升高对这三类微生物相对丰度差异的影响并未达到统计显著性水平($p>0.05$)，但与对照处理相比，FACE 圈中该微生物相对丰度的增加量达到几倍至 10 倍，表现出明显的增加趋势(表 3)。

表 3 微生物菌群在科水平上的相对丰度改变倍数

Table 3 Multiple of the change in relative abundance of microflora at family level

微生物菌群(在科水平) Microflora(at family level)	籼稻 Indica variety		杂交稻 Hybrid variety	
	ZX-96	YD-8	TY-084	YLY-6
肠杆菌科 *Enterobacteriaceae*	−0.17	−0.26	−0.17	0.46
假单胞菌科 *Pseudomonadaceae*	0.42	1.08	0.05	−0.40
黄单胞菌科 *Xanthomonadaceae*	0.13	0.16	0.08	0.09

（续表）

微生物菌群（在科水平）Microflora(at family level)	籼稻 Indica variety		杂交稻 Hybrid variety	
	ZX-96	YD-8	TY-084	YLY-6
气单胞菌科 *Aeromonadaceae*	0.54	−0.07	−0.02	−0.80
鞘脂杆菌科 *Sphingobacteriaceae*	−0.94	6.62	345	−0.56
丛毛单胞菌科 *Comamonadaceae*	−0.20	0.34	6.37	0.11
黄杆菌科 *Flavobacteriaceae*	−0.22	0.69	2.03	7.55
草酸杆菌科 *Oxalobacteraceae*	−0.73	10.1	−0.90	−0.20
其他菌群 Other	0.13	0.21	−0.23	−0.01

注：相对丰度改变倍数=（微生物菌群在 eCO_2 处理中的相对丰度－微生物菌群 aCO_2 的相对丰度）/微生物在 aCO_2 的相对丰度

Note: Multiple of the change in relative abundance is calculated as (the relative abundance of the microflora at eCO_2 minus the relative abundance of the microflora at aCO_2)/ the relative abundance of the microflora at aCO_2

2.5　大气 CO_2 浓度升高对水稻根系内生菌群落结构的影响

对各样本序列进行主成分分析（Principal component analysis, PCA）评价大气 CO_2 浓度升高下微生物群落结构的变化规律。结果表明，FACE 圈样品的细菌群落趋向集中于左下象限和右上象限（图 4 实线圈出部分），对照圈样品的细菌群落趋向集中于右下象限（图 4 虚线圈出部分）。尽管 FACE 圈和对照圈水稻根系内生菌群落在 PCA 坐标体系的分布有部分重叠，但整体微生物群落分布在坐标体系的不同区域（图 4），表明大气 CO_2 浓度升高下水稻根系内生菌群落结构发生了一定的变化。

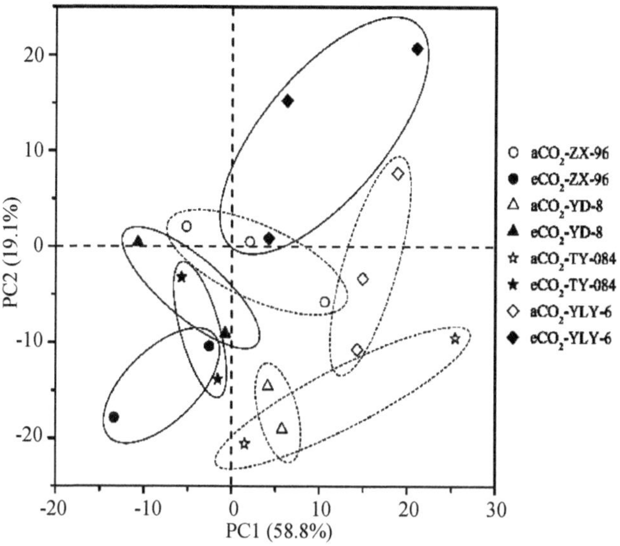

图 4　微生物群落的主成分分析

Fig. 4　Principal component analysis (PCA) of microbial communities

注：括号里面的数值表示被各个坐标轴解释的百分比

Note: The values in parentheses indicate percentage explained by each ordination axis

3 讨论

植物组织内部具有丰富的微生物资源，深度挖掘这些微生物资源，认知其微生物多样性具有重要意义。对已有的文献进行整合分析，结果表明水稻根系内生菌仅包括约 60 个属的微生物菌群，本研究则发现水稻根系内生菌群在 97%的序列相似度水平上共计有 9503 个不同的 OTUs，并可归为 10 个门、18 个纲、37 个目、62 个科、133 个属。传统微生物研究方法的局限性可能是导致这种差异的主要原因。例如，目前几乎所有的水稻根系内生菌研究均采用传统的分离培养法（Ladha *et al.*，1983；Muthukumarasamy *et al.*，2007；Engelhard *et al.*，2000；Mano *et al.*，2006，2007；Yanni *et al.*，1997；Singh，*et al.*，2010）、以及分辨率较低的 DGGE（Hardoim *et al.*，2011）和克隆文库指纹图谱法（Sun *et al.*，2008）。本研究通过新一代 454 高通量测序技术对水稻根系内生菌的整体微生物群落 16S rRNA 基因进行深度测序，所有 19 个样品共得到了 120193 条高质量序列，平均每个样品的序列数是 6326，单个样品的检测通量高于常规方法的上百倍。

本实验结果表明，γ-变形菌纲是水稻根组织内部最为优势的微生物菌群，占所有微生物 OTUs 总数的百分比高达 71.7%（表 2），这与以前的研究不一致。已有的用克隆文库方法对水稻根系内生菌的研究表明，β-变形菌纲是水稻根系内部最为优势的微生物菌群，占整个克隆文库的 27.1%（Sun *et al.*，2008）。这可能是水稻品种差异、环境条件（如水分含量、营养元素含量）的异质性、研究方法和采样时间的差异性、或者微生物之间的相互作用（Hunter *et al.*，2010）所导致的。此外，变形菌门在形态、生理及代谢方面的生物多样性也可能是造成水稻根系内生微生物群落结构差异的重要原因。例如，已有研究表明，变形菌门在碳（Badger *et al.*，2008）、氮（Galloway，1998）、磷（Longnecker *et al.*，2010）等元素生物地球化学循环过程中均发挥了重要作用。在微生物分类学科的水平，肠杆菌科是水稻根组织内部最为优势的微生物菌群，其占细菌总数量的 30.8%～59.8%（图 1）。此类微生物菌群可能在满足植物对营养元素的需求方面发挥了重要作用。已有研究表明，肠杆菌科的多个种属具有固氮功能，例如 *Enterobacter cloacae*（Ladha *et al.*，1983）、*Klebsiella planticola*（Ladha *et al.*，1983）、*Klebsiella pneumoniae*（Engelhard *et al.*，2000）等，但这些微生物的生理生态功能及其对大气 CO_2 浓度升高的响应仍需进一步的研究。

不同水稻品种根系内生菌的主要微生物群落组成基本相同，但微生物的相对丰度在不同品种间差异较大。例如，对照处理中，假单胞菌科仅占扬稻 8 号水稻品种根系内生菌整体微生物群落的 3.40%，而在扬两优 6 号品种中其相对丰度高达 27.8%，几乎相差一个数量级。同样，黄单胞菌科和气单胞菌科在不同水稻品种根系内生菌群落中的相对丰度也有明显差异，分别在 6.63%～14.5%和 1.11%～14.6%区间变化。这一现象表明，水稻品种在影响微生物群落结构方面发挥了重要作用，与前人的研究结果具有一致性。例如，Hardoim 等（Hardoim *et al.*，2011）研究表明水稻品种之间的差异在很大程度上决定了根系内生菌的群落结构，水稻品种越相近其微生物群落组成也越趋向类似，并且品种之间的差异大于土壤类型及养分的利用效率的影响。

大气 CO_2 浓度升高下，水稻品种镇籼 96、扬稻 8 号和 II 优 084 根系内生菌群落的响应模式基本一致，即大气 CO_2 升高抑制了根系内生优势菌群的相对丰度，而提高了稀少菌群的相对丰度(图 2,图 3)。这可能是因为大气 CO_2 浓度升高增加了植物的光合作用,使植物体内积累了更多的碳水化合物,体内 C 素/营养比增大,以 C 素为基础的次生代谢物质(如有机酸、酚类等碳水化合物)就会积累,然后转运到根部(陈改苹等,2005)。而对上述碳源(食物源)有“习惯性”依赖的优势微生物菌群生态位发生变化,导致数量减少,而数量上不占优势的微生物菌群极可能占据原有优势物种灭亡后的生态位,表现出种群数量的增加。因此,上述现象可能是微生物与环境条件相互作用的结果,也是环境条件对微生物长期选择的结果。已有研究表明,100 年的荒漠变为农田、樟子松地或杨树林后,数量上占优势的微生物菌群明显减少,而数量上不占优势的微生物的相对丰度明显提高(Wang *et al.*,2012)。

然而,环境干扰下微生物群落的响应规律仍不清楚,极可能与生态系统类型、微生物组成及干扰强度等因素紧密相关。例如,对于水稻品种扬两优 6 号,其根系内生菌的优势菌群对大气 CO_2 浓度升高的响应模式与其他水稻品种并不完全一致,大气 CO_2 浓度升高下优势菌群(肠杆菌科和假单胞菌科)的相对丰度增加,而稀少菌群的相对丰度降低(图 2d,图 3d)。这可能与根系内生菌各种微生物的相对丰度有关。例如,对于水稻品种镇籼 96、扬稻 8 号和 II 优 084,其根系内部最优势菌肠杆菌科在对照处理中的相对丰度是 48.4%~59.8%;而对于水稻品种扬两优 6 号,其优势菌群肠杆菌科和假单胞菌科在对照处理中的相对丰度分别是 30.8%和 27.8%,并且此两类菌群的相对丰度远低于最优势菌(肠杆菌科)在其他三种水稻品种的相对丰度,表明水稻品种扬两优 6 号优势菌群的单一性远小于其他三种水稻品种,其群落结构更为复杂多样,这可能是导致扬两优 6 号品种根系内生菌对大气 CO_2 浓度升高响应更为复杂的原因之一。这些结果表明水稻根系内生菌对大气 CO_2 升高的响应模式极可能与微生物的相对丰度及多样性组成紧密相关。可能存在一个相对丰度阈值,当相对丰度大于此阈值时,微生物表现为正响应;反之,微生物表现为负相应,但其生物学机制仍需进一步研究。

大气 CO_2 浓度升高改变了水稻根系内生菌的群落结构(图 4),与已有的土壤微生物群落结构研究结果较为一致(He *et al.*,2012)。环境微生物群落结构的变化是各种微生物对大气 CO_2 浓度升高响应不同的综合反映。例如,在水稻品种镇籼 96、扬稻 8 号和 II 优 084 中,大气 CO_2 浓度升高抑制了优势菌群的相对丰度,而促进稀少菌群的相对丰度,鞘脂杆菌科增幅高达 344 倍(表 4),可能是对 CO_2 响应最敏感的微生物菌群,而不同微生物菌群相对丰度的改变也最终导致整个微生物群落结构的改变(图 4)。

4　结论

水稻根系最为优势的内生菌群是 γ-变形菌纲的肠杆菌科,占整体微生物群落比例高达 30.8%~59.8%,大气 CO_2 浓度升高下水稻品种镇籼 96、扬稻 8 号和 II 优 084 根系的优势菌群相对丰度降低,而稀少菌群相对丰度增加。扬两优 6 号根系内生菌的响应模式与其他三种品种不完全一致,优势菌群(肠杆菌科和假单胞菌科)的相对丰度升高,而稀少菌群的相对丰度降低。微生物的相对丰度可能是影响水稻根系内生细菌对大气 CO_2 浓度升高响应的重要因素。

致 谢

感谢本课题组严陈、许静、王皖蒙、M. Saiful Alam 同学以及扬州大学周娟博士、赖上坤博士、杨连新教授在样品采集过程中提供的帮助！

参考文献

陈改苹，朱建国，程磊．2005．高 CO_2浓度下根系分泌物的研究进展．土壤，**37**(6)：602-606．

刘钢，韩勇，朱建国，等．2002．稻麦轮作 FACE 系统平台 Ⅰ．系统结构与控制．应用生态学报，**13**(10)：1253-1258．

马莹，骆永明，滕应，等．2013．内生细菌强化重金属污染土壤植物修复研究进展．土壤学报，**50**(1)：195-202．

苑学霞，林先贵，褚海燕，等．2006．大气 CO_2浓度升高对几种土壤微生物学特征的影响．中国环境科学，**26**(1)：25-29．

Badger M R, Bek E J. 2008. Multiple Rubisco forms in proteobacteria: Their functional significance in relation to CO_2 acquisition by the CBB cycle. *Journal of Experimental Botany*, **59**(7): 1525-1541.

Barraquio W L, Revilla L, Ladha J K. 1997. Isolation of endophytic diazotrophic bacteria from wetland rice. *Plant and Soil*, **194**(1/2): 15-24.

Berg G, Krechel A, Ditz M, *et al*. 2005. Endophytic and ectophytic potato-associated bacterial communities differ in structure and antagonistic function against plant pathogenic fungi. *FEMS Microbiology Ecology*, **51**(2): 215-229.

Dennis W. 1995. Endophyte: The evolution of a term, and clarification of its use and definition. *Oikos*, **73**(2): 274-276.

Dong Z, Canny M J, McCully M E, *et al*. 1994. A Nitrogen-fixing endophyte of sugarcane stems. *Plant Physiology*, **105**(4): 1139-1147.

Edgar R C. 2010. Search and clustering orders of magnitude faster than BLAST. *Bioinformatics*, **26**(19): 2460-2461.

Engelhard M, Hurek T, Reinhold-Hurek B. 2000. Preferential occurrence of diazotrophic endophytes, *Azoarcus* spp., in wild rice species and land races of *Oryza sativa* in comparison with modern races. *Environmental Microbiology*, **2**(2): 131-141.

Galloway J N. 1998. The global nitrogen cycle: Changes and consequences. *Environmental Pollution*, **102**(1): 15-24.

Garbeva P, van Overbeek L, van Vuurde J, *et al*. 2001. Analysis of endophytic bacterial communities of potato by plating and denaturing gradient gel electrophoresis (DGGE) of 16S rDNA based PCR fragments. *Microbial Ecology*, **41**(4): 369-383.

Hardoim P R, Andreote F D, Reinhold-Hurek B, *et al*. 2011. Rice root-associated bacteria: Insights into community structures across 10 cultivars. *FEMS Microbiology Ecology*, **77**(1): 154-164.

He Z L, Piceno Y, Deng Y, *et al*. 2012. The phylogenetic composition and structure of soil microbial communities shifts in response to elevated carbon dioxide. *The ISME Journal*, **6**(2): 259-272.

Hunter P J, Hand P, Pink D, *et al*. 2010. Both leaf properties and microbe-microbe interactions influence within-species variation in bacterial population diversity and structure in the lettuce (*Lactuca* species) phyllosphere. *Applied and Environmental Microbiology*, **76**(24): 8117-8125.

Ladha J K, Barraquio W L, Watanabe I. 1983. Isolation and identification of nitrogen-fixing*Enterobacter cloacae* and *Klebsiella planticola* associated with rice plants. *Canadian Journal of Microbiology*, **29**(10): 1301-1308.

Lipson D, Blair M, Barron-Gafford G, *et al*. 2006. Relationships between microbial community structure and soil processes under elevated atmospheric carbon dioxide. *Microbial Ecology*, **51**(3): 302-314.

Longnecker K, Lomas M W, van Mooy BAS. 2010. Abundance and diversity of heterotrophic bacterial cells assimilating phosphate in the subtropical North Atlantic Ocean. *Environmental Microbiology*, **12**(10): 2773-2782.

Mano H, Tanaka F, Nakamura C, *et al*. 2007. Culturable endophytic bacterial flora of the maturing leaves and roots of rice plants (*Oryza sativa*) cultivated in a paddy field. *Microbes and Environment*, **22**(2): 175-185.

Mano H, Tanaka F, Watanabe A, *et al*. 2006. Culturable surface and endophytic bacterial flora of the maturing seeds of rice plants (*Oryza sativa*) cultivated in a paddy field. *Microbes and Environments*, **21**(2): 86-100.

Muthukumarasamy R, Kang U G, Park K D, *et al*. 2007. Enumeration, isolation and identification of diazotrophs from Korean wetland rice varieties grown with long-term application of N and compost and their short-term inoculation effect on rice plants. *Journal of Applied Microbiology*, **102**(4): 981-991.

Palus J, Borneman J, Ludden P, *et al*. 1996. A diazotrophic bacterial endophyte isolated from stems of *Zea mays* L. and *Zea luxurians* Iltis and Doebley. *Plant and Soil*, **186**(1): 135-142.

Singh B K, Bardgett R D, Smith P, *et al*. 2010. Microorganisms and climate change: Terrestrial feedbacks and mitigation options. *Nature Reviews Microbiology*, **8**(11): 779-790.

Singh R, Mishra R N, Jaiswal H, *et al*. 2006. Isolation and identification of natural endophytic rhizobia from rice (*Oryza sativa* L.) through rDNA PCR-RFLP and sequence analysis. *Current Microbiology*, **52**(5): 345-349.

Sun L, Qiu F B, Zhang X X, *et al*. 2008. Endophytic bacterial diversity in rice (*Oryza sativa* L.) roots estimated by 16S rDNA sequence analysis. *Microbial Ecology*, **55**(3): 415-424.

Wang B Z, Zhang C X, Liu J L, *et al*. 2012. Microbial community changes along a land-use gradient of desert soil origin. *Pedosphere*, **22**(5): 593-603.

Wang Q, Garrity G M, Tiedje J M, *et al*. 2007. Naïve Bayesian classifier for rapid assignment of rRNA sequences into the new bacterial taxonomy. *Applied and Environmental Microbiology*, **73**(16): 5261-5267.

Yanni Y, Rizk R Y, Corich V, *et al*. 1997. Natural endophytic association between *Rhizobium leguminosarum* bv. *trifolii* and rice roots and assessment of its potential to promote rice growth. *Plant and Soil*, **194**(1/2): 99-114.

RESPONSE OF ENDOPHYTIC BACTERIA IN RICE ROOTS TO ELEVATED CO_2

REN Gaidi[1,2], ZHANG Huyong[1], LIN Xiangui[1], ZHU Jianguo[1] and JIA Zhongjun[1]

1) State Key Laboratory of Soil and Sustainable Agriculture, Institute of Soil Science, Chinese Academy of Sciences, Nanjing 210008, China;

2) University of Chinese Academy of Sciences, Beijing 100049, China

Abstract

Using the new high-throughput sequencing technique, study was carried out on responses to elevated CO_2 (eCO_2) of endophytic bacteria at the whole microbial community level in the roots of ZhenXian-96 (ZX-96), YangDao-8 (YD-8), IIYou-084 (TY-084), and YangLiangYou-6 (YLY-6), commonly used in the FACE experiment in China. Results show that the family of Enterobacteriaceae within the class of γ-proteobacteria was the highest in relative abundance, accounting for 30.8% ~ 59.8% of the whole community. In ZX-96, YD-8, and TY-084, eCO_2 would probably inhibit growth of the microbial groups that are dominant in population (dominant groups) but stimulate growth of those that are less (rare groups). For instance, in rice TY-084, Enterobacteriaceae, Pseudomonadaceae, Xanthomonadaceae, and Aeromonadaceae were the four dominant bacterial groups, of which each exceeded 14.6% in relative abundance, and their total relative abundance declined from 74.8% to 67.2% under eCO_2. On the contrary, the rare groups in the roots, consisting of Sphingobacteriaceae, Comamonadaceae, Flavobacteriaceae, and Oxalobacteraceae, increased from 4.13% to 16.9% in total. Especially the family of Sphingobacteriaceae increased by up to 344 folds, and hence is the microbial group the most sensitive to eCO_2. However, in YLY-6, the responses of endophytic bacteria differed in pattern from those in other varieties of rice. These findings indicate that relative abundance of microbes may probably be an important factor affecting the response of endophytic bacteria in the roots to elevated CO_2, which may be used as certain basis for the study on variation of structure and function of the whole microbial community in response to the global climate change.

Key words: high throughput sequencing; microbial community; endophytic bacteria; FACE (free air CO_2 enrichment)

第二部分

气候变化影响的相关研究与进展

Part Two

The Related Study and Development on the Responses to Climate Change

CHEMICAL CHARACTERIZATION AND SOURCE APPORTIONMENT OF $PM_{2.5}$ IN BEIJING: SEASONAL PERSPECTIVE①

ZHANG Renjian[1], JING Junshan[1,2], TAO Jun[3], HSU Shih-Chieh[4]②, WANG Gehui[5], CAO Junji[5], Celine Siu Lan Lee[6], ZHU Lihua[3], CHEN Zhongming[7], ZHAO Yue[7], SHEN Zhenxing[8]

1) Key Laboratory of Regional Climate-Environment Research for Temperate East Asia, Institute of Atmospheric Physics, Chinese Academy of Sciences, Beijing

2) National Meteorological Center of CMA, Beijing

3) South China Institute of Environmental Sciences, Ministry of Environmental Protection, Guangzhou

4) Research Center for Environmental Changes, Academia Sinica, Taipei

5) K LAST, SKLLQG, Institute of Earth Environment, Chinese Academy of Sciences, Xi'an

6) Institute of Earth Sciences, Academia Sinica, Taipei

7) Peking University, Beijing

8) Department of Environmental Science and Engineering, Xi'an Jiaotong University, Xi'an

Abstract

In this study, 121 daily $PM_{2.5}$ (aerosol particle with aerodynamic diameter less than 2.5 μm) samples were collected from an urban site in Beijing in four months between April 2009 and January 2010 representing the four seasons. The samples were determined for various compositions, including elements, ions, and organic/elemental carbon. Various approaches, such as chemical mass balance, positive matrix factorization (PMF), trajectory clustering, and potential source contribution function (PSCF), were employed for characterizing aerosol speciation, identifying likely sources, and apportioning contributions from each likely source. Our results have shown distinctive seasonalities for various aerosol speciations associated with $PM_{2.5}$ in Beijing. Soil dust waxes in the spring and wanes in the summer. Regarding the secondary aerosol components, inorganic and organic species may behave in different manners. The former preferentially forms in the hot and humid summer via photochemical reactions, although their precursor gases, such as SO_2 and NO_x, are emitted much more in winter. The latter seems to favorably form in the cold and dry winter. Synoptic meteorological and climate conditions can overwhelm the emission pattern in the formation of secondary aerosols. The PMF model identified six main sources: soil dust, coal combustion, biomass burning, traffic and waste

① The paper published in *Atmos. Chem. Phys.*, **13**: 7053-7074, 2013.

② The first four authors have equal contributions to the paper.

Corresponding author: schsu815@rcec.sinica.edu.tw

incineration emission, industrial pollution, and secondary inorganic aerosol. Each of these sources has an annual mean contribution of 15%, 18%, 12%, 4%, 25%, and 26%, respectively, to $PM_{2.5}$. However, the relative contributions of these identified sources significantly vary with changing seasons. The results of trajectory clustering and the PSCF method demonstrated that regional sources could be crucial contributors to PM pollution in Beijing. In conclusion, we have unraveled some complex aspects of the pollution sources and formation processes of $PM_{2.5}$ in Beijing. To our knowledge, this study is the first systematical study that comprehensively explores the chemical characterizations and source apportionments of $PM_{2.5}$ aerosol speciation in Beijing by applying multiple approaches based on a completely seasonal perspective.

Key words: $PM_{2.5}$; Beijing; Secondary aerosol; Source apportionment; PMF; PSCF; Airborne dust

1 Introduction

Particulate matter (PM) is composed of various chemical components (Seinfeld, 1989). PM profoundly affects our living environments in terms of air quality (in close relation to public health), visibility, direct and indirect radiative forcing, climate effects and ecosystems (Watson, 2003; Streets *et al.*, 2006; Andreae and Rosenfeld, 2008; Mahowald, 2011). Numerous epidemiological studies have demonstrated that long-term exposure to pronounced $PM_{2.5}$ increases morbidity and mortality (Dockery and Pope, 1994; Pope *et al.* 1995; Schwartz *et al.* 1996). Given its tiny size, fine-mode PM (i. e., $PM_{2.5}$, PM with aerodynamic diameter less than 2.5 μm) can readily penetrate the human bronchus and lungs (Pope *et al.*, 1995; Oberdorster, 2001). Through absorption and scattering of solar radiation and serving as cloud condensation nuclei, $PM_{2.5}$ extensively affects the global climate (Bardouki *et al.*, 2003), and thus, the hydrological cycle (Ramanathan and Feng, 2009). The diverse effects of $PM_{2.5}$ could be a function of its complex chemical components and composition (He *et al.*, 2009; Niwa *et al.*, 2007; Malm *et al.*, 2005; Eatough *et al.*, 2006).

Due to the rapid economic and industrial developments and urbanization in the past few decades, there is an escalating increase in energy consumption and the number of motor vehicles in China, where air pollution has become ubiquitous (Chan and Yao, 2008). According to Shao *et al.* (2006), nearly 70% of urban areas in China do not meet China's national ambient air quality standards, which are even much laxer than the air quality exposure standards/guidelines of the World Health Organization (WHO, 2005). The Beijing—Tianjin—Hebei region, the Yangtze River delta, and the Pearl River delta are of special concern because of their severe PM pollution, which can be explicitly shown by the spatial distribution of aerosol optical depth (AOD) retrieved by satellites (He *et al.*, 2009; Lee *et al.*, 2010). Three megacities that are representatives of each region, namely, Beijing, Shanghai, and Guangzhou, are the foci, because of their dense population. Coal is the primary energy source in China, and its consumption reached up to 1528 Mtce in 2005, accounting for nearly 70% of the total energy consumption (followed by petroleum at over 20%). Such quantity ranks number one in the world, representing ~37% of global consumption (Fang *et al.*, 2009; Chen *et al.*, 2010). The use of coal in China encompass from large power plants, industries to individual domestic households where coal combustion is the largest contributor of air pollution (Liu and Diamond, 2005; Chan and Yao, 2008). Given the rapid growth in vehicle numbers at a rising rate of ~20%, traffic has become a major urban pollutant emitter (He *et al.*, 2002; Fan *et al.*, 2009). Other than anthropogenic pollutants, desert and loess dust of natural origins with annual emissions of over 100 million tons also serves as important PM source in China, particularly in late winter and spring (Zhang *et al.*, 1997; Sun *et al.*, 2001).

Atmospheric pollutants in China are a complex mixtures of various sources, from gases to particulates, from natural to anthropogenic, from primary to secondary, and from local to regional and the term "Air Pollution Complex" or "Complex Atmospheric Pollution" has been emerged in the last decade (He *et al.*, 2002; Shao *et al.*, 2006; Chan and Yao, 2008; Fan *et al.*, 2009). One of the major air pollutants is PM, particularly $PM_{2.5}$, which remains a nationwide problem despite considerable efforts for its removal (Fan *et al.*, 2009). As the capital of China and a rapidly industrialized and typical urbanized city, Beijing has elicited much more attention domestically and internationally (Zhang *et al.*, 2003a; 2007; Zhou *et al.*, 2012). $PM_{2.5}$ in Beijing is abnormally elevated, often rising to more than 100 $\mu g \cdot m^{-3}$, and characterized by multiple components and sources, ranging from inorganic to organic constituents, from anthropogenic to natural origins, from primary to secondary components, and from local to long-range transported sources, and in dynamic variability with time and/or meteorological conditions and climate regimes (He *et al.*, 2001; Wang *et al.*, 2005; Duan *et al.*, 2006; Okuda *et al.*, 2011; Song *et al.*, 2012). In spite of many scientific research programs conducted by academic institutions and the political strategies implemented by the government, the state of air pollution in Beijing (and even across China) seems to improve slowly. For instance, in January 2013, Beijing (and the entire inland China) suffered from the worst $PM_{2.5}$ pollution in history, registering the highest $PM_{2.5}$ hourly concentration of 886 $\mu g \cdot m^{-3}$ (http://www.nasa.gov/multimedia/imagegallery/image_feature_2425.html). Some essential questions remain unknown, although the government has devoted itself to improving air quality and numerous studies have been conducted. Therefore, a systematically comprehensive investigation of employing multiple techniques in conjunction with chemical measurements is inevitably needed, particularly to unravel the likely contributors of $PM_{2.5}$.

Receptor models are used to quantitatively estimate the pollutant levels contributed by different sources through statistical interpretation of ambient measurement. The positive matrix factorization (PMF) developed by the Environmental Protection Agency of USA is a well adopted receptor model for source apportionment analysis. A few studies have applied

PMF to identify the likely dominant sources and apportion their respective contributions. For example, Wang *et al.* (2008) analyzed certain elements, ions, and black carbon in $PM_{2.5}$ and PM_{10} samples collected in Beijing in summer and winter (one month representative for each season) between 2001 and 2006. Based on the obtained data set, they performed PMF analyses and identified six main sources: soil dust, vehicular emission, coal combustion, secondary aerosol, industrial emission, and biomass burning. By applying the PMF model with only elemental data as input data, Yu *et al.* (2013) identified seven likely sources of $PM_{2.5}$ in Beijing, with relative contributions following the order secondary sulfur (26.5%), vehicle exhaust (17.1%), fossil fuel combustion (16.0%), road dust (12.7%), biomass burning (11.2%), soil dust (10.4%), and metal processing (6.0%). Song *et al.* (2007) analyzed a few elements, ions, and organic/elemental carbon (OC/EC) in $PM_{2.5}$ collected from multiple stations in Beijing during a short period in January and August 2004, and subjected them to PMF analyses. Six potential sources were registered: coal combustion, biomass burning, secondary sulfate, motor vehicles, secondary nitrate, and road dust, with emphases on coal combustion in winter and secondary aerosols in summer. Xie *et al.* (2008) conducted PMF analyses of PM_{10} (instead of $PM_{2.5}$) collected in Beijing in 10 days each in January, April, July, and October 2004 by using chemical data on metal elements, ions, and OC/EC as input data. Seven main sources were identified, including urban fugitive dust, crustal soil, coal combustion, secondary sulfate, secondary nitrate, biomass burning with municipal incineration, and vehicle emissions. All these studies were limited to a particular season and based on selected PM species.

To attain a better understanding of the chemical characteristics and sources of fine aerosols on a seasonal basis, we conducted a delicate investigation in Beijing. We continuously collected daily $PM_{2.5}$ samples at an urban site for four months, each of which in the respective seasons (i. e., spring, summer, autumn and winter). The samples were subjected to chemical measurements of various aerosol compositions as a whole, such as a suite of crustal and anthropogenic elements, major water-soluble ions, and OC/EC. Furthermore, we identified and apportioned the main sources to $PM_{2.5}$ by employing chemical mass closure construction and the PMF model in conjunction with trajectory cluster and potential source contribution function analyses according to the hybrid single-particle Lagrangian integrated trajectory (HYSPLIT) model. This study will elucidate the source profile of $PM_{2.5}$ in different seasons and the relative contribution from each source in the complex urban air shed in Beijing and provide vital information in formulating the future air management framework to address the current alarming level of PM pollution in China which has been affecting the air quality on a vast regional scale.

2 Methodology

2.1 Sample and Chemical Analysis

2.1.1 Sampling Site

Beijing is located on the northern edge of the North China Plain, surrounded by the Yanshan Mountains in the west, north, and northeast (Fig. 1). According to the spatial distribution of fine AOD ranging from 0.0 to 1.0 that has been retrieved from Moderate-resolution Imaging Spectrometer (MODIS) sensors on board Terra and Aqua satellites (Fig. 1), Beijing is one of the $PM_{2.5}$ hot spots in China. The four seasons are characterized by variable meteorological conditions: spring by high-speed winds and low rainfall, summer by high temperature and frequent rain usually accounting for 75% of annual rainfall, autumn by sunny days and northwest winds, and winter by cold and dry air. The population is 17.55 million. In 2009, the number of motor vehicles increased to 4.019 million, and energy consumption was equivalent to 65.73 million tons of standard coal

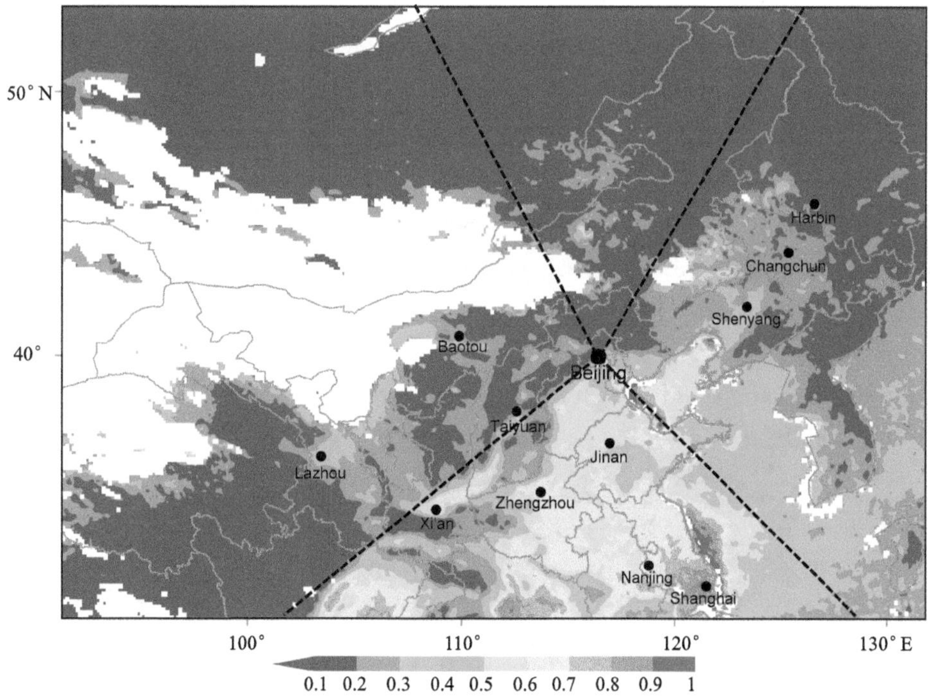

Fig. 1 Sampling location (116.30°E, 39.99°N) on a regional map superimposed with spatial distribution of annual mean fine aerosol optical depth (AOD) retrieved from MODIS satellite remote sensing in 2009. Dashed lines define four regions based on the trajectory clustering results discussed in Section 4.2 (seen in the text). Also shown are several major cities around Beijing.

(2010 Beijing Statistics Yearbook). The sampling station was set up at the roof of the Science Building in Peking University (116.30°E, 39.99°N) 26 m above the ground level. A few field experimental campaigns have been conducted at this urban site (He *et al.*, 2010; Guo *et al.*, 2012). This site is located within the educational, commercial, and residential districts, and no main pollution sources exist nearby. Thus, the observations could be typical of the general urban pollution in Beijing.

2.1.2 Sample Collection

Daily $PM_{2.5}$ samples were collected in April, July, and October 2009 and January 2010, representing spring, summer, autumn, and winter, respectively. Two collocated aerosol samplers (frm OMNI™, BGI, USA) were used to collect $PM_{2.5}$ samples from 10:00 am to 10:00 am the next day simultaneously. The two substrates used in each sampler were 47 mm quartz filter (Whatman QM/A, England) and Teflon filter (pore size $=2\ \mu m$; Whatman PTFE, England). The flow rate was set at $5\ L \cdot min^{-1}$. The quartz filters were baked at 800℃ for 3 h before use. The filter samples were stored at −18℃ until pretreatment.

2.1.3 Gravimetric Weighing

Before and after each sampling, the PTFE filters were conditioned at 22℃±1℃ in relative humidity of 35%±2% for 24 h and then weighed in a weighing room by using an electronic balance with a detection limit of 1 μg (Sartorius, Gottingen, Germany). The corresponding $PM_{2.5}$ mass concentration of each filter was equal to the weight difference before and after sampling divided by the sampled air volume.

2.1.4 Chemical Analysis of Trace Elements and Water-soluble Ions

Prior to extraction and digestion, each aerosol-laden PTFE membrane filter was cut into two equal halves with ceramic scissors. One half was subjected to Milli-Q water extraction for ionic measurement and the other half to acid digestion for elemental measurement. For the acid digestion, the polypropylene support O-ring on half of each PTFE filter sample was carefully removed with a ceramic knife from contamination. The filter samples were digested with an acid mixture (5 ml HNO_3 + 2 ml HF) by using an ultra-high throughout microwave digestion system (MARSXpress, CEM, Matthews, NC). A blank reagent and two filter blanks were prepared in each run following the same procedure used for the samples. All the acids used in this study were of ultra-pure grade (Merck, Germany). The detailed digestion method has been published elsewhere (Hsu *et al.*, 2008). Another half of all filter samples were used for extraction with 20 ml Milli-Q purified water (specific resistivity = 18.2 MΩ · cm; Millipore, Massachusetts, USA) for 1 h. The detailed extraction procedures have been described in Hsu *et al.* (2007, 2010a).

Ionic species (Na^+, NH_4^+, K^+, Mg^{2+}, Ca^{2+}, F^-, Cl^-, SO_4^{2-} and NO_3^-) in the leachate were analyzed through a Dionex model ICS-90 (foranions) and ICS-1500 (for cations) ion chromatograph equipped with a conductivity detector (ASRS-ULTRA). Trace elements in the digestion solutions, including Al, Fe, Na, Mg, K, Ca, Ba, Ti, Mn, Co, Ni, Cu, Zn, Mo, Cd, Sn, Sr, Sb, Pb, Tl, Ge, Cs, Ga, V, Cr, As, Se, and Rb, were analyzed by inductively coupled plasma-mass spectrometry (ICP-MS). Quality assurance and control of the ICP-MS was guaranteed by the analysis of a certified reference standard, NIST SRM-1648 (urban particulates). The resulting recoveries fell within ±10% of the certified values for most elements, except for Se, As, Cs, Sb, and Rb (±15%) (Hsu *et al.*, 2009; 2010a).

2.1.5 OC and EC Measurements

A punch of 0.526 cm^2 from each quartz filter was heated stepwise by a thermal/optical carbon analyzer (DRI 2001, Atmoslytic, US) in a pure helium atmosphere at 140℃(OC1), 280℃ (OC2), 480℃(OC3), and 580℃ (OC4), and then in 2% O_2/98% Heatmosphere at 580℃ (EC1), 740℃ (EC2), and 840℃ (EC3) to convert any particulate carbon on the filter to CO_2. After catalyzed by MnO_2, CO_2 was reduced to CH_4, which was then directly measured. Mass concentrations of OC and EC were obtained according to the IMPROVE protocol (Chow *et al.*, 2007), OC=OC1+OC2+OC3+OC4+OP; EC=EC1+EC2+EC3-OP, where OP is the optical pyrolyzed OC. Detailed descriptions can be found in Zhang *et al.* (2012a).

2.2 Data Analysis Methods

2.2.1 Chemical Mass Closure

In this study, we constructed chemical mass closure (CMC) on a seasonal basis by considering mineral dust, SO_4^{2-}, NO_3^-, NH_4^+, EC, particulate organic matter (POM), chloride salt (instead of sea salt; reason given below), trace element oxide (TEO), and biomass burning-derived K^+. SO_4^{2-}, NO_3^-, and NH_4^+ can be regarded as the secondary inorganic aerosols.

The aluminosilicate (i.e., soil, dust, or mineral) component is often estimated through the following formula (Malm *et al.*, 1994; Chow *et al.*, 1994), which includes Si.

$$[\text{Mineral}]=2.20\ \text{Al}+2.49\ \text{Si}+1.63\ \text{Ca}+2.42\ \text{Fe}+1.94\ \text{Ti}$$

$$[\text{Mineral}]=1.89\ \text{Al}+2.14\ \text{Si}+1.40\ \text{Ca}+1.43\ \text{Fe}$$

However, Si is volatilized as SiF_4 in the acid digestion of aerosol samples when using HF. Therefore, a few studies estimated Si from Al in the calculation of the mineral component (Hueglin *et al.*, 2005). However, once Al is used in estimating Si concentrations, it generates relatively large uncertainty in the mineral component proportion as Si/Al mass ratios could largely vary in China's dust (Yan *et al.*, 2012, and references therein). Accordingly, we adopted a straightforward method conventionally used in estimating dust aerosols from Al:

$$[\text{Mineral}]=\text{Al}/0.07$$

where 0.07 is the average Al content (7%) reported by Zhang *et al.* (2003b). A similar estimation has been applied previously (Ho *et al.*, 2006; Hsu *et al.*, 2010b).

In estimating POM, we adopted a factor of 1.6 in converting OC to POM (Viidanoja *et al.*, 2002), where as a wide range of 1.4—2.2 has been utilized in previous investigations (Turpin and Lim, 2001; Andreae *et al.*, 2008). The main determinants in selecting a conversion factor are the origin and age of the organic aerosols. The factor of 1.6 was employed in this study because the latest result shows a OM/OC ratio averaged at 1.59 ± 0.18 in $PM_{2.5}$ over China (Xing *et al.*, 2013). This factor was used for the $PM_{2.5}$ of Beijing by Dan *et al.* (2004), who also observed a similar seasonality for EC and OC and a OC/EC ratio (2—3) close to our results.

Sea salt is usually calculated as [Sea salt] $= 1.82 \times Cl^-$ or $= 2.54\ Na^+$. Given that Beijing is about 150 km away from East China's coastal oceans (i.e., Bohai Sea), sea spray-generated sea salt particles are not readily transported and are therefore insignificant to fine aerosols in Beijing. Nevertheless, dust blowing from Northern and Northwestern China is often associated with NaCl and Na_2SO_4 from salt lake sediments and saline soils (Zhang *et al.*, 2009a). On the other hand, Cl^- may be essentially contributed by coal combustion in Beijing, particularly in winter (Yao *et al.*, 2002). Thus, we considered chloride salt, instead of sea salt, as an individual component of $PM_{2.5}$ aerosols in Beijing: $[\text{Cl salt}] = [Cl^-] + [Na^+] + [\text{ss-}Mg^{2+}]$. By considering chloride depletion in sea salt particles within the marine boundary layer because of the heterogeneous reaction, Hsu *et al.* (2010a) successfully evaluated such formula.

Following Landis *et al.* (2001), we estimated the contribution of heavy metals as metal oxides by employing the following equation:

$$\begin{aligned} TEO = 1.3 \times [&0.5 \times (Sr + Ba + Mn + Co + Rb + Ni + V) + \\ &1.0 \times (Cu + Zn + Mo + Cd + Sn + Sb + Tl + Pb + As + Se + Ge + Cs + Ga)]. \end{aligned}$$

The enrichment factor (EF) of a given element (E) was calculated by using the formula $EF = (E/Al)_{Aerosol} / (E/Al)_{Crust}$ (Hsu *et al.*, 2010a), where $(E/Al)_{Aerosol}$ is the ratio of the element to the Al mass in aerosols and $(E/Al)_{Crust}$ is the ratio in the average crust (Taylor, 1964). The result of the EF is shown in Fig. S1. Elements with EFs of $\leqslant 1.0$, such as Cr and Y, were not considered, as they are of exclusive crustal origin. Elements with EFs between 1 and 5 were multiplied by a factor of 0.5, as they are possibly originated from two sources (i.e., anthropogenic and crustal sources). Elements with EFs $\geqslant 5.0$ were multiplied by unity, as they are dominated by anthropogenic origins. Furthermore, the multiplicative factor was set at 1.3 so that metal abundance could be converted to oxide abundance, similar to those used by Landis *et al.* (2001). We also considered biomass burning-derived K^+ (K_{BB}) as an individual component, although K_{BB} salt may exist in the chemical forms of KCl and K_2SO_4 (Pósfai *et al.*, 2004), where both Cl^- and SO_4^{2-} have already been considered in other components.

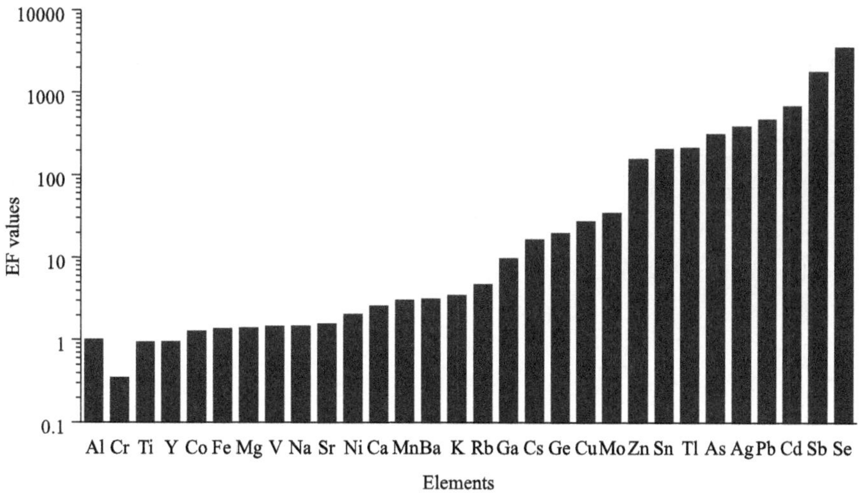

Fig. S1 Enrichment factor (EF) values varied with the selected elements. Note that the Y scale is in log

2.2.2 PMF Model

PMF is an effective source apportionment receptor model that does not require the source profile sprior to analysis and has no limitation on source numbers (Hopke, 2003; Shen *et al.*, 2010). The principles of PMF can be found elsewhere in detail (Han *et al.*, 2006; Song *et al.*, 2006, Yu *et al.*, 2013). In the present study, PMF3.0 was employed with the inclusion of 34 chemical species in the model computation: $PM_{2.5}$, Al, Fe, Na, Mg, K, Ca, Ba, Ti, Mn, Co, Ni, Cu, Zn, Mo, Cd, Sn, Sb, Pb, V, Cr, As, Se, Rb, Na^+, NH_4^+, K^+, Mg^{2+}, Ca^{2+}, Cl^-, SO_4^{2-}, NO_3^-, OC, and EC. Six physically realistic sources were identified.

2.2.3 Airmass Back Trajectory Cluster

We calculated 48h air mass back trajectories arriving at the sampling site (116.30°E, 39.99°N) during our sampling period by using the National Oceanic and Atmospheric Administration (NOAA) HYSPLIT-4 model with a 1°× 1° latitude-longitude grid and the final meteorological database. The six-hourly final archive data were generated from the National Center for Environmental Prediction's Global Data Assimilation System (GDAS) wind field reanalysis. GDAS uses a spectral medium - range forecast model. More details about the HYSPLIT model can be found at http://www.arl.noaa.gov/ready/open/hysplit4.html (NOAA Air Resources Laboratory). The model was run four times per day at starting times of 04, 10, 16, and 22 UTC (1200, 1800, 0000, and 0600 local time, respectively). The arrival level was set at 100 m above ground level. The method used in trajectory clustering was based on the GIS-based software Traj Stat (http://www.meteothinker.com/TrajStatProduct.aspx).

2.2.4 Potential Source Contribution Function

The potential source contribution function (PSCF) is a method for identifying regional sources based on the HYSPLIT model. The zone of concern is divided into $i \times j$ small equal grid cells. The PSCF value in the ijth cell is defined as m_{ij}/n_{ij}, where n_{ij} is designated as the numbers of endpoints that fall in the ijth cell and m_{ij} denotes the numbers of "polluted" trajectory endpoints in the ijth cell. In this analysis, average concentrations were treated as the "polluted" threshold (Hsu *et al.*, 2003). To better reflect the uncertainty in cells with small n_{ij} values (Polissar *et al.*, 1999), the weighting function w_{ij} was adopted:

The study domain was in the range of 30°—60°N, 75°—130°E. The resolution was 0.5°×0.5°.

3 Results

3.1 Annual Average

Table 1 provides a statistical summary of the obtained data on atmospheric concentrations for $PM_{2.5}$, Al (a tracer of aluminosilicate dust), water-soluble ions, OC, and EC during the sampling period. The annual mean $PM_{2.5}$ concentration reached 135±63 μg·m^{-3}. This mean value is nearly three times higher than that (35 μg·m^{-3}) of the interim target-1 standard for annual mean $PM_{2.5}$ recommended by the WHO. The level of $PM_{2.5}$ in Beijing is much higher than other mega-cities around the world. In comparison with that of domestic cities, $PM_{2.5}$ seems to display a spatial tendency, increasing northward and decreasing southward (Zhang *et al.*, 2012b). Such a spatial pattern may be related to the low rainfall and high dust in northern China (Qian *et al.*, 2002; 2005). According to Wang *et al.* (2008), $PM_{2.5}$ concentrations in winter were much higher than in summer in 2001 to 2002. However, such a trend seemed to be reversed in 2005 to 2006, with rather higher concentrations in summer.

Table 1 Statistical summary showing the means (with one standard deviation) and ranges of atmospheric concentrations for $PM_{2.5}$ (in unit μg·m^{-3}) and selected species (in unit ng·m^{-3}) in the entire sampling (annual) and four-season (months) periods

Species	Annual	Spring	Summer	Autumn	Winter
$PM_{2.5}$	135±63	126±59	138±48	135±55	139±86
	39—355	39—280	41—226	45—251	48—355
SO_4^{2-}	13.6±12.4	14.7±11.5	23.5±14.5	7.9±7.4	8.5±8.6
	0.9—52.8	2.3—52.8	2.5—52.0	0.9—25.7	1.3—34.4
NO_3^-	11.3±10.8	15.5±13.7	11.8±8.2	10.7±11.0	7.3±8.1
	0.3—63.8	1.3—63.8	1.8—31.5	0.3—34.7	1.6—35.5
NH_4^+	6.9±7.1	7.5±8.1	11.0±6.9	4.7±5.8	4.5±5.7
	0.1—39.1	0.6—39.1	0.5—23.9	0.1—17.7	0.3—23.3

(Continued)

Species		Annual	Spring	Summer	Autumn	Winter
Cl^-		1.42 ± 2.18	0.72 ± 0.81	0.30 ± 0.56	1.12 ± 0.98	3.52 ± 3.32
		0.03—10.34	0.04—3.74	0.03—3.06	0.09—3.71	0.19—10.34
Na^+		0.46 ± 0.55	0.31 ± 0.18	0.17 ± 0.09	0.30 ± 0.22	1.08 ± 0.80
		0.04—2.82	0.08—0.94	0.04—0.42	0.05—1.06	0.11—2.82
K^+		0.92 ± 0.75	1.08 ± 0.71	0.66 ± 0.47	1.13 ± 0.90	0.81 ± 0.77
		0.03—3.66	0.14—3.14	0.20—2.47	0.03—3.66	0.05—2.53
Mg^{2+}		0.16 ± 0.13	0.24 ± 0.20	0.07 ± 0.03	0.16 ± 0.07	0.18 ± 0.09
		0.02—1.04	0.03—1.04	0.02—0.16	0.06—0.31	0.06—0.45
Ca^{2+}		1.6 ± 1.5	2.6 ± 2.2	0.6 ± 0.3	1.7 ± 1.0	1.5 ± 0.9
		0.2—11.3	0.2—11.3	0.2—1.7	0.5—4.2	0.5—4.0
Al		1.8 ± 1.5	2.5 ± 1.7	0.7 ± 0.4	2.0 ± 1.4	2.1 ± 1.5
		0.1—6.9	0.3—6.6	0.1—2.0	0.3—6.7	0.7—6.9
OC		16.9 ± 10.0	13.7 ± 4.4	11.1 ± 1.8	17.8 ± 5.6	24.9 ± 15.6
		5.9—58.6	5.9—23.7	7.4—16.6	7.5—26.2	8.5—58.6
EC		5.0 ± 4.4	2.8 ± 1.1	4.2 ± 1.2	5.3 ± 2.8	7.5 ± 7.4
		0.6—28.1	0.6—5.8	1.5—6.8	1.3—12.1	1.2—28.1

For the ionic concentrations, SO_4^{2-} ranked the highest among the water-soluble ions analyzed, with an annual mean of 13.6±12.4 μg·m^{-3}, followed by NO_3^- (11.3±10.8 μg·m^{-3}), NH_4^+ (6.9±7.1 μg·m^{-3}), Ca^{2+} (1.6±1.4 μg·m^{-3}), Cl^- (1.4±2.2 μg·m^{-3}), K^+ (0.92±0.75 μg·m^{-3}), Na^+ (0.46±0.55 μg·m^{-3}), and Mg^{2+} (0.16±0.13 μg·m^{-3}). Such levels of mean concentrations are rather comparable with those measured in many Chinese cities such as Shanghai, Tianjin, Jinan, and Guangzhou (Yao *et al.*, 2002; Tao *et al.*, 2009; Gao *et al.*, 2011; Gu *et al.*, 2011). On average, the combination of SO_4^{2-}, NO_3^-, and NH_4^+, which could be regarded as secondary inorganic aerosols, constituted the majority (88%) of the total ionic concentrations, consistent with earlier studies (Yao *et al.*, 2002; Duan *et al.*, 2003). The annual mean concentrations of OC and EC reached up to 17.0±10.0 and 5.0±4.4 μg·m^{-3}, respectively. Such levels are close to those observed for regional sites across China (16.1±5.2 μg·m^{-3} for OC and 3.6±0.93 μg·m^{-3} for EC) by Zhang *et al.* (2008), who extensively measured carbonaceous aerosols around China; however, such concentrations are approximately half of those observed at urban sites (33.1±9.6 μg·m^{-3} for OC and 11.2±2.0 μg·m^{-3} for EC) by Zhang *et al.* (2008).

3.2 Seasonality

As illustrated in Fig. 2, the seasonalities of $PM_{2.5}$ and these primary species were characterized by distinctive features. The seasonality of $PM_{2.5}$ concentration was not very evident and typical, with nearly equal concentrations of around 140 μg·m^{-3} in summer,

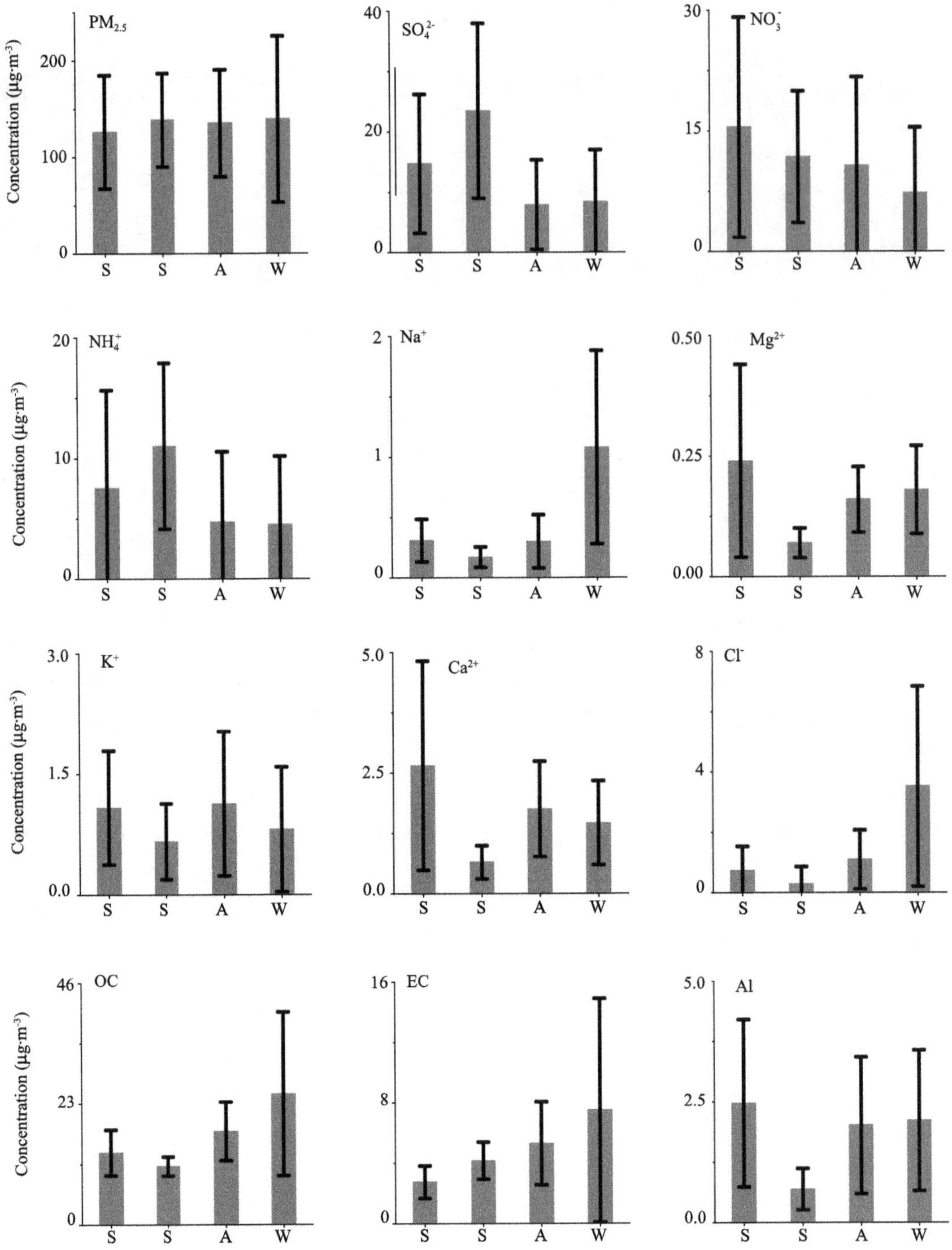

Fig. 2 Seasonal variations of $PM_{2.5}$ mass concentration and associated species, including SO_4^{2-}, NO_3^-, NH_4^+, Na^+, Mg^{2+}, K^+, Ca^{2+}, Cl^-, OC, EC, and Al concentrations. Shown here are the mean and one standard deviation for each bar.

autumn, and winter and a relative minimum (~125 $\mu g \cdot m^{-3}$) in spring. The minimum concentration typically occurs in summer because precipitation in Beijing is usually concentrated at that period (Fig. S2). However, this is not the case, because the maximum concentrations of secondary sulfate and ammonium were observed in summer, arising from strong photochemistry and accounting for a large proportion (~25%) of $PM_{2.5}$ in Beijing (Yao *et al.*, 2003). This suggestion can be supported by the fact that ambient temperatures through February to October 2009 in Beijing are higher than the climatology

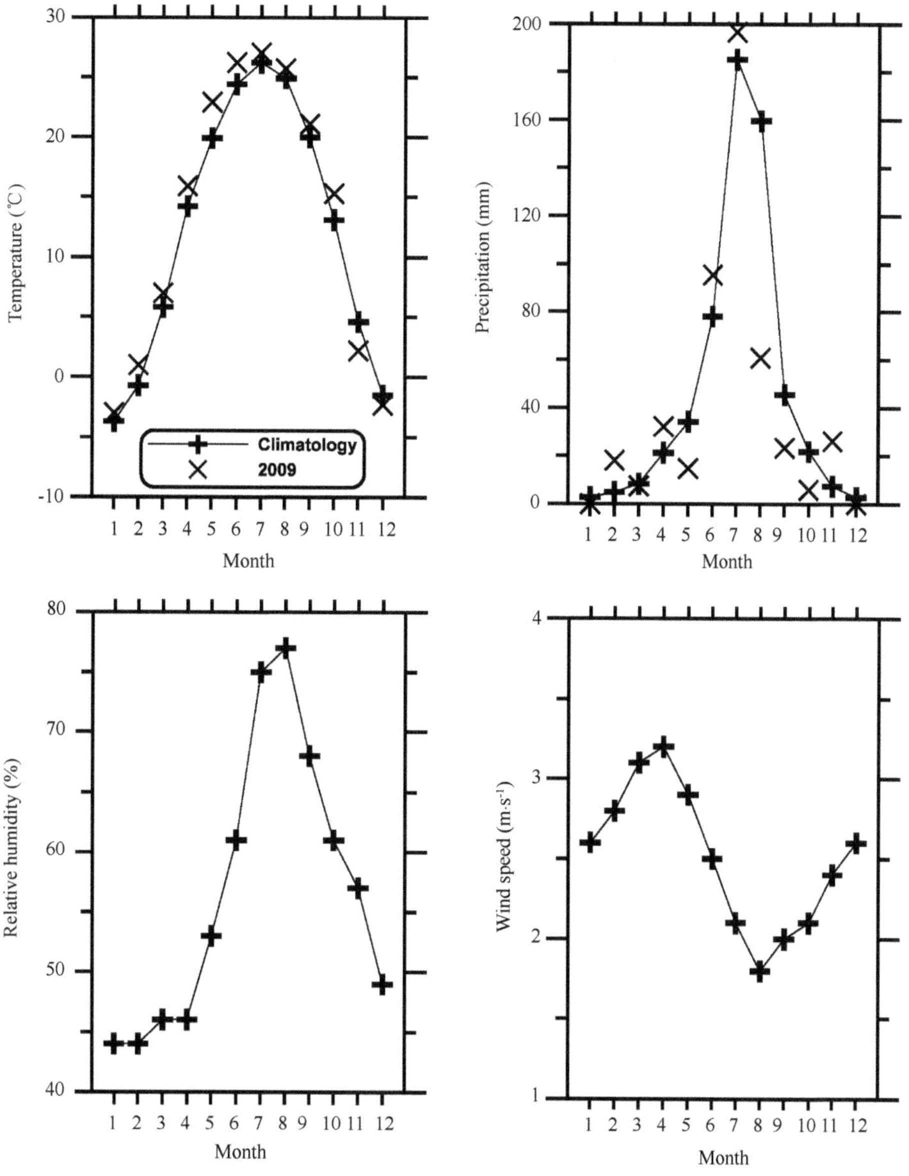

Fig. S2　The climatology of monthly mean temperature, precipitation, relative humidity, and wind speed in Beijing, which is compared with temperature and precipitation of 2009 (data source: http://cdc.bjmb.gov.cn/public.asp)

by ~1℃ or higher (Fig. S2), while precipitation in July 2009 (i. e., 197 mm) is even slightly larger than the climatology (185 mm), demonstrating that the photochemical effect might overwhelm the precipitation scavenging effect for fine aerosol pollutants.

In contrast to $PM_{2.5}$, sulfate and ammonium revealed a typical seasonality with higher concentrations in spring and summer and lower concentrations in autumn and winter, consistent with the seasonal variability of AOT (Xia *et al.*, 2006). The summertime maximum concentrations of sulfate and ammonium were 24 and 12 $\mu g \cdot m^{-3}$, respectively, which were higher than those in Beijing before 2003 (~15 and ≤10 $\mu g \cdot m^{-3}$, respectively) (He *et al.*, 2001; Duan *et al.*, 2006; Wang *et al.*, 2005) but rather comparable to those observed in the last few years (Okuda *et al.*, 2011; Song *et al.*, 2012). By contrast, the wintertime concentration of sulfate (8.5 $\mu g \cdot m^{-3}$) was significantly reduced compared with earlier literature data (He *et al.*, 2001; Hu *et al.*, 2002, Wang *et al.*, 2005). The decrease in wintertime sulfate concentration seemed to result from the effective control of SO_2 emissions over China in the recent years (Itahashi *et al.*, 2012), particularly from coal combustion (Hao *et al.*, 2005). High summertime sulfate concentration is ascribed to enhanced photochemistry during summer, and relatively high humidity accelerates the conversion rate of SO_2 to the particulate form (Yao *et al.*, 2003). However, the precursor SO_2 concentrations are much higher in winter (Fig. S3) because of higher emission at that time (Zhang *et al.*, 2009b). One might further conclude that in Beijing, photo chemistry plays a more vital role in the sulfate aerosol formation and variability than the change in precursor SO_2 emission as well as rain scavenging process. In the present study, artificial biases, particularly of nitrate and ammonium, possibly occurred during sampling because no denuder and/or back-up filter was used to trap ammonia and nitric acid (Pathak *et al.*, 2004). The maximum concentration (15.5 $\mu g \cdot m^{-3}$) of nitrate was observed in spring rather than summer, which was different from that of sulfate (Wang *et al.*, 2008). This observation may be ascribed to the volatility of ammonium nitrate, which is one of the main chemical forms of nitrate associations revealed by the ionic relationships. Thus, ammonium nitrate could evaporate at relatively high temperature. Besides, there are distinct emission sources for their respective precursor gases, SO_2 and NO_x. The minimum concentration of nitrate was observed in winter. Nitrate levels could be a function of various factors in terms of emissions, such as vehicular exhaust, coal combustion, and biomass burning, and complex chemical processes with respect to photochemistry, heterogeneous reaction, renoxification, and gas-aerosol equilibrium. The overall trend of NO_x emission in China is increasing primarily due to persistently increasing energy demand for industrial development and transportation, though control measures for NO_x emissions have been implemented in coal-fired power plants (Zhao *et al.*, 2013).

Crustally derived ions and elements, such as Mg^{2+}, Ca^{2+}, and Al, waxed in the spring and waned in the summer, followed by significant increases toward autumn and winter. Such levels of seasonal mean concentrations and seasonalities are consistent with those

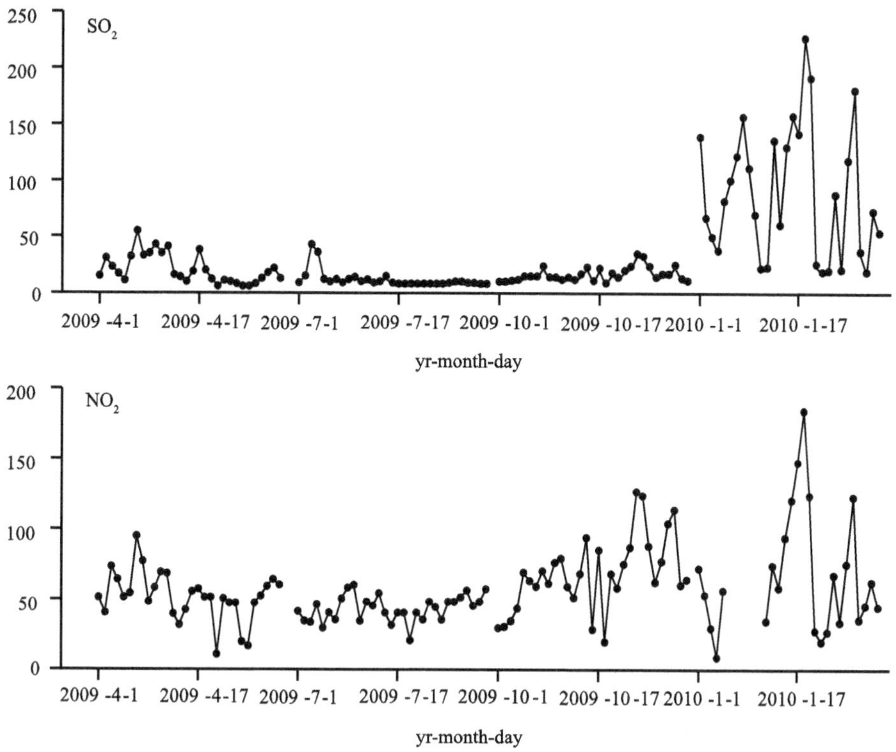

Fig. S3　Time-series of SO_2 and NO_2 daily concentrations measured at an air quality monitoring station (Wanliu), ~1.5 km west to our sampling site, during the sampling periods. Note that the concentration is in unit $\mu g \cdot m^{-3}$

observed in previous studies (Duan *et al.*, 2006; Wang *et al.*, 2005), which are related to dust storms and anthropogenic and fugitive dust. The seasonal concentrations of Na^+ and Cl^- peaked in winter, consistent with Hu *et al.* (2002) and Wang *et al.* (2005). The seasonality of Mg^{2+} was distinguishable from that of Na^+, demonstrating the difference in their dominant sources. K^+ had relatively higher concentrations in both spring and autumn than in summer and winter, which was closely associated with the agricultural burning around Beijing (Zheng *et al.*, 2005). Such seasonality was also found by some previous studies (He *et al.*, 2001; Zheng *et al.*, 2005), but distinct from other previous studies, in which winter often had the highest concentration (Duan *et al.*, 2006; Wang *et al.*, 2005). Specifically, the highest levoglucosan, which is suggested to serve as an excellent tracer of biomass burning pollutants relatively to K^+, has been measured in autumn (He *et al.*, 2006), although no spring sample was measured in that study.

Both OC and EC had similar seasonal patterns of waxing in winter and waning in spring (for EC) or summer (for OC). Zhang *et al.* (2008) observed a persistently common seasonality for both OC and EC at 18 background, regional, and urban stations in China, namely, a maximum in winter and a minimum in summer. The seasonalities may be governed by the variability in emission strengths and meteorology. For instance, lower-molecular weight semi-volatile organic compounds are mostly in gaseous phase at high temperature in summer (Yassaa *et al.*, 2001). The OC/EC mass ratio 2.0 indicates the presence of secondary organic matter (Chow *et al.*, 1996). In this work, the OC/EC ratios mostly fell within the range of 2—5, with mean ratios of 4.8, 2.5, 3.0, and 2.7 in spring, summer, autumn, and winter, respectively (Fig. 3). These figures are very similar to those observed in previous studies (Duan *et al.*, 2006; Zhang *et al.*, 2008), which suggests the relative domination of secondary organic aerosols in spring but of primary sources in other seasons (Zhang *et al.*, 2008). Another reason for the relatively higher springtime OC/EC mass ratio may be the open biomass burning source, consistent with higher K^+ in the spring (Fig. 2), as the aerosols from open biomass burning are generally characterized by elevated OC/EC ratios (Cao *et al.*, 2007).

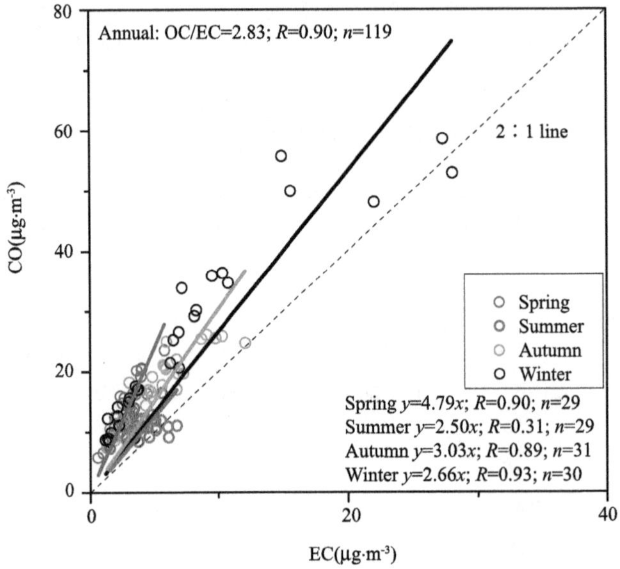

Fig. 3 Scatter plot showing the correlation between OC (y-axis) and EC (x-axis) in $PM_{2.5}$ collected from Beijing. Different symbols denote the four seasons. Linear regression equations are given in the annual and seasonal cases

3.3 Stoichiometric Analyses of Cations and Anions

Note that equivalent concentrations ($\mu eq \cdot m^{-3}$) are used throughout this section. Figure 4 shows the scatter plots of (a) Mg^{2+} vs. Na^+, (b) Mg^{2+} vs. Ca^{2+}, (c) Cl^- vs. Na^+, and (d) Cl^- vs. K^+. Figure 5 shows the scatter plots of (a) total cations vs. total anions, (b) NH_4^+ vs. SO_4^{2-}, (c) NH_4^+ vs. [SO_4^{2-} + NO_3^-], (d) [NH_4^+ + Ca^{2+}] vs. [SO_4^{2-} + NO_3^-], (e) [NH_4^+ + Ca^{2+} + Mg^{2+}] vs. [SO_4^{2-} + NO_3^-], and (f) [NH_4^+ + Ca^{2+} + Mg^{2+}] vs. [SO_4^{2-} + NO_3^- + Ex-Cl^-]. Ex-Cl^- is the excessive Cl^-, defined as the excessive amount of Cl^- relative to the amount sea salt can sustain with Na^+ as the tracer of sea salt: Ex-Cl^- = Cl^- - [Na^+] × 1.17, where 1.17 is the typical Cl^-/Na^+ equivalent ratio of average seawater (Chester, 1990). If the resulting Ex-Cl^- is negative, then no Cl^- excess exists. In other words, Cl^- is totally contributed by sea salt and is even depleted by heterogeneous reactions. Nevertheless, total Na^+ does not necessarily originate from sea salt alone, but could partially come from dust. The resultant biases are hence likely insignificant.

Figure 4a illustrates that Mg^{2+} mostly comes from non-sea salt sources, except in wintertime, because the regression slopes that represent the Mg^{2+}/Na^+ ratios (1.56, 0.70, 0.68, and 0.25 for spring, summer, autumn, and winter, respectively) are clearly deviated from the ratio (0.23) of average seawater (Chester, 1990). Instead, the dominant source of Mg^{2+} is mineral dust, mainly carbonate minerals (Li *et al.*, 2007), as reflected by the good correlations (0.97, 0.83, 0.90, and 0.80 for spring, summer, autumn, and winter, respectively) between Mg^{2+} and Ca^{2+} (Fig. 4b). Similarly, most Cl^-/Na^+ ratios (1.70, 1.21, 2.51, and 2.25 for spring, summer, autumn, and winter, respectively) in $PM_{2.5}$ are larger than the mean ratio (1.17) of seawater, except the summertime samples (Fig. 4c). This difference indicates the dominance of the non-sea salt sources, of which the most likely contributor of Cl^- is coal combustion (Yao *et al.*, 2002), particularly in winter when Cl^- and SO_2 are maximal (Figs. 2 and S2). In summer, air masses are dominated by the southerly monsoon from the Bohai Sea (as supported by the trajectories below), leading to a mean Cl^-/Na^+ ratio close to that of average seawater. Moreover, the correlations between K^+ and Cl^- largely varied with the seasons, with better correlation and higher ratios in autumn and winter and moderate correlations and lower ratios (less than unity) in summer and spring (Fig. 4d). Thus, the results suggested that K^+ was not present in chemical form KCl at high temperature, but as K_2SO_4 (Pósfai *et al.*, 2004). By contrast, low temperature in winter may favor the presence of KCl.

Furthermore, the ratio of total cation concentration to total anion concentration is averaged at near unity throughout the year (Fig. 5a), which indicates excellent charge balance in $PM_{2.5}$ and high data quality. Figure 5b shows good correlations between NH_4^+ and SO_4^{2-} for the annual data set, with ratios (represented by the slope of the linear regression line) between 1.25 and 1.77 (all higher than unity). These good correlations reveal the dominance of $(NH_4)_2SO_4$ (Ianniello *et al.*, 2011), rather than NH_4HSO_4 and the possible full neutralization of SO_4^{2-} by NH_4^+ throughout the year. We further considered the combination of NO_3^- and SO_4^{2-} in this charge balance analysis (Fig. 5c). The resulting NH_4^+ to $[SO_4^{2-} + NO_3^-]$ ratios ranged from 0.83 to 0.94 (all lower than unity), which demonstrates the presence of NH_4NO_3 in the fine-mode aerosols. Moreover, ratios lower than unity suggest that nitrate may be present in other chemical forms than NH_4NO_3. Heterogeneous reactions between NO_x (and its products, such as HNO_3 and N_2O_5) and dust carbonate are often observed in northern China (Li and Shao, 2009). Accordingly, we examined the correlations of $[NH_4^+ + Ca^{2+}]$ *vs.* $[SO_4^{2-} + NO_3^-]$ (Fig. 5d) and of $[NH_4^+ + Ca^{2+} + Mg^{2+}]$ *vs.* $[SO_4^{2-} + NO_3^-]$ (Fig. 5e), given that the

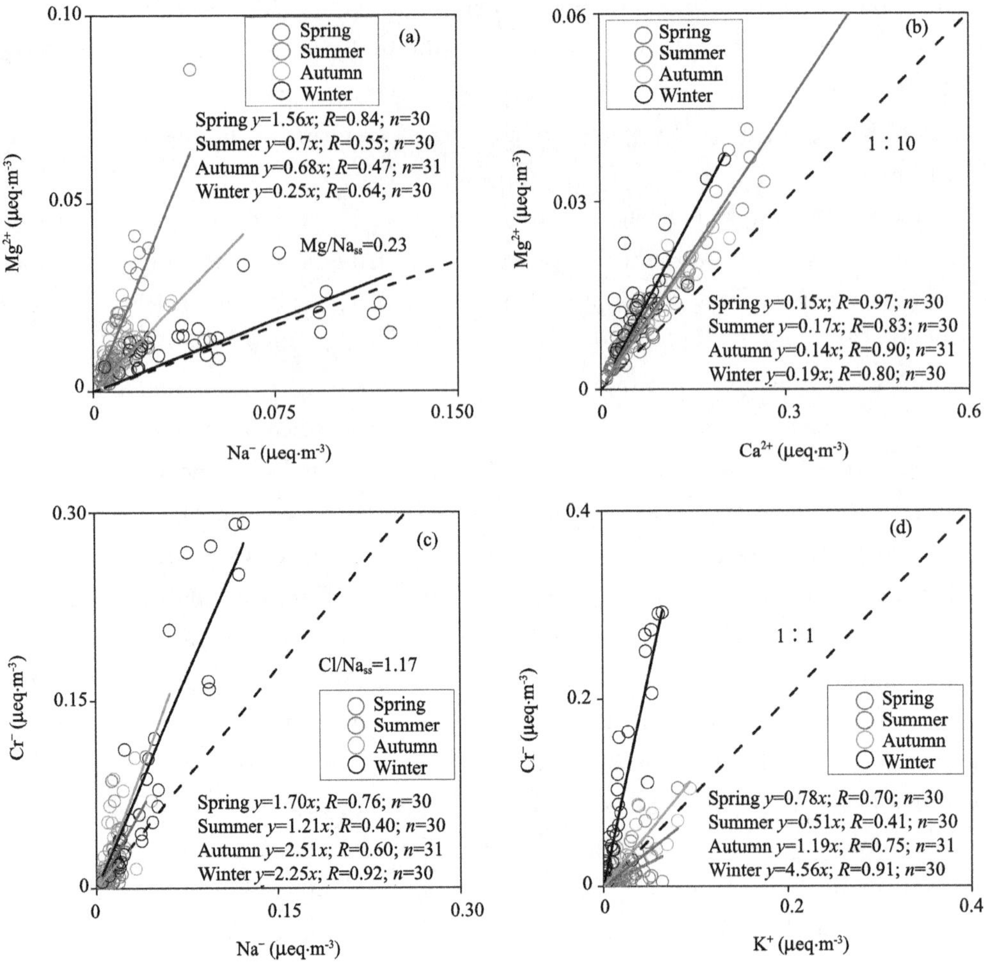

Fig. 4 Correlations between certain cations and anions: (a) Mg^{2+} versus Na^+, (b) Mg^{2+} versus Ca^{2+}, (c) Cl^- versus Na^+, and Cl^- versus K^+. Four seasons are considered and represented by different color symbols. Linear regression equations are also given. In Figures (a) and (c), the dashed lines indicate the Mg/Na and Cl/Na equivalent concentration ratios in seawater (sea salt), respectively

good correlations between Mg^{2+} and Ca^{2+} suggest the possible existence of water-soluble Mg in reacted carbonate dust. These ions are strongly correlated throughout the year, with high coefficients (all 0.99 or higher) and slopes of regression lines around unity. These correlations indicate that nitrate is partly present in $Ca(NO_3)_2$ and $Mg(NO_3)_2$, not just in NH_4NO_3. We assumed that fine-sized sulfate is exclusively associated with ammonium and that Na^+ is present only in the associated NaCl. However, these assumptions may not always be true because Na_2SO_4 is observed in dust particles from dried lakes in northern China (Zhang *et al.*, 2009a) and NaCl could react with nitric acid to form $NaNO_3$ via heterogeneous reaction (Hsu *et al.*, 2007). Therefore, based on the aforementioned equivalent interrelationships and assumptions, we quantitatively estimated that the former two chemical forms ($Ca(NO_3)_2$ and $Mg(NO_3)_2$) represent ~20% of the total nitrate and

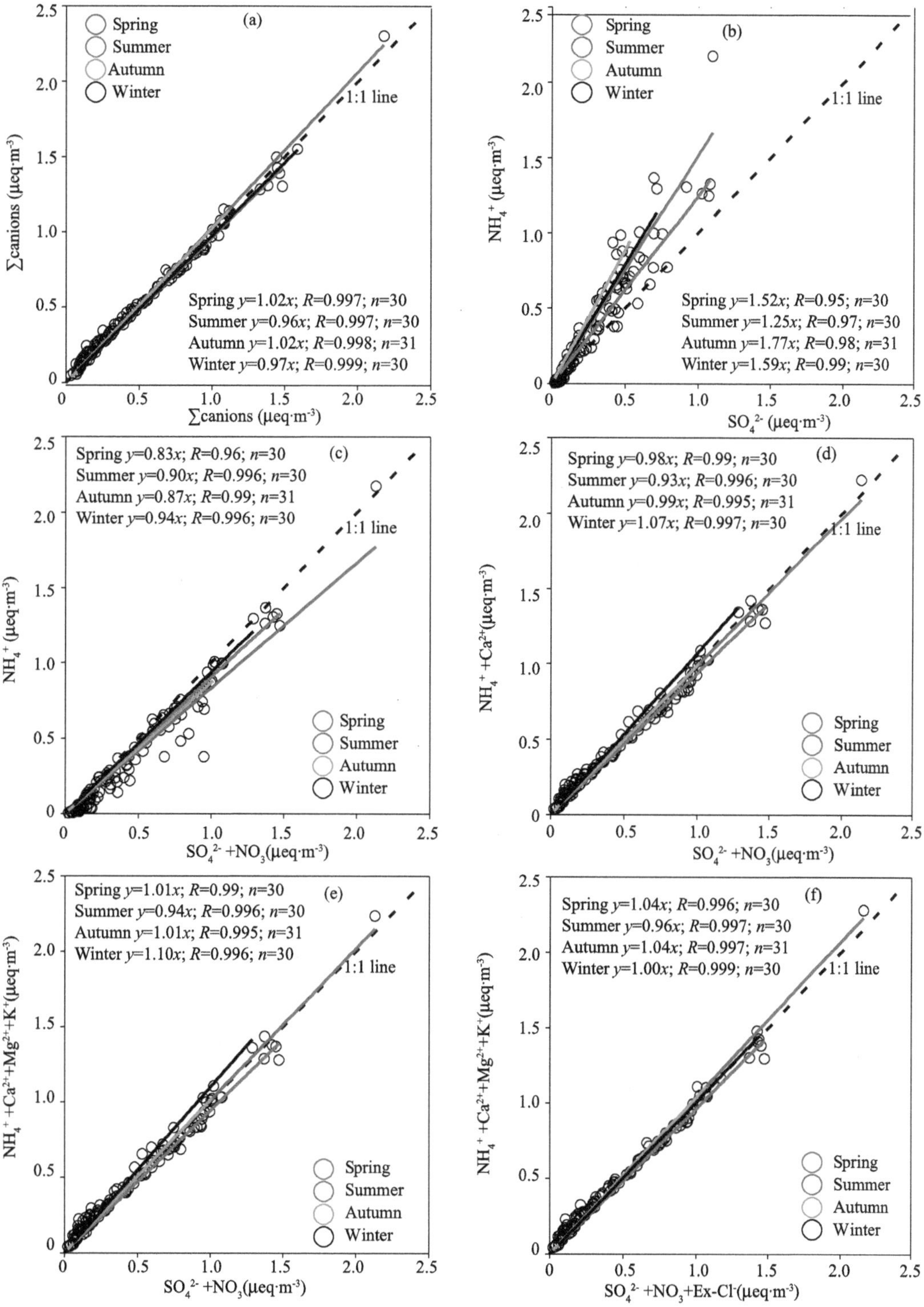

Fig. 5　Same as Fig. 4, but for (a) total cations versus total anions, (b) NH_4^+ versus SO_4^{2-}, (c) NH_4^+ versus [SO_4^{2-} + NO_3^-], (d) [NH_4^+ + Ca^{2+}] versus [SO_4^{2-} + NO_3^-], (e) [NH_4^+ + Ca^{2+} + Mg^{2+}] versus [SO_4^{2-} + NO_3^-], and (f) [NH_4^+ + Ca^{2+} + Mg^{2+}] versus [SO_4^{2-} + NO_3^- + Ex-Cl^-]. The detailed definition of Ex-Cl^- (Excessive-Cl^-) can be found in the text. The 1 : 1 lines are given for comparison

that NH_4NO_3 is the dominant association accounting for the remaining ~80%. The elevated Cl^-/Na^+ ratio (>1.17) shows that excessive Cl^- seemed to be attributed to coal combustion rather than sea salt particles from dried salt lake sediment. The addition of excessive Cl^- (Fig. 5f) insignificantly changed the correlations of positive and negative charges. Nevertheless, we noted that in wintertime, the equivalent ratio improved from 1.10 to 1.00, which indicates the presence of chloride salts such as KCl, $CaCl_2$, and $MgCl_2$ other than NaCl at low ambient temperature (Ianniello *et al.*, 2011). KCl may have originated from biomass burning, and $CaCl_2$ and $MgCl_2$ could have been formed through heterogeneous reactions between the dust carbonate and HCl emitted from coal combustion.

3.4 Chemical Mass Closure

By employing the methods in Section 2.6, we constructed the CMC of $PM_{2.5}$ in Beijing on a seasonal and annual basis. The reconstructed $PM_{2.5}$ mass concentrations were compared with the gravimetric $PM_{2.5}$ mass concentrations, as shown in Fig. 6, which shows a good correlation with one another in each season and throughout the year. However, the ratios seasonally changed, with higher ratios of 0.82 in spring and 0.75 in winter and lower ratios of 0.59 in summer and 0.68 in autumn. The proportions of all specific components in $PM_{2.5}$ together with the unidentified constituents as a whole are schematically illustrated by five pie charts for the four seasonal and annual cases (Fig. 7).

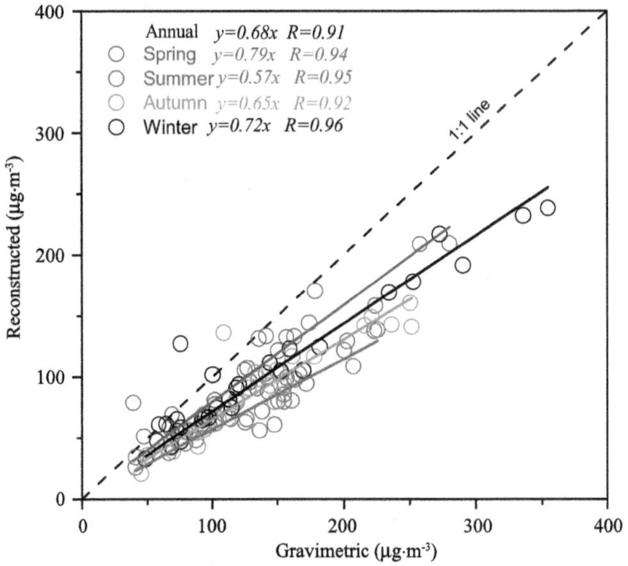

Fig. 6 Scatter plot showing the correlations between the $PM_{2.5}$ mass concentrations reconstructed from the chemical mass balance method and obtained from gravimetric measurement, which are presented on a seasonal basis. Linear regression lines are shown with equations for the four-season period, along with the annual case

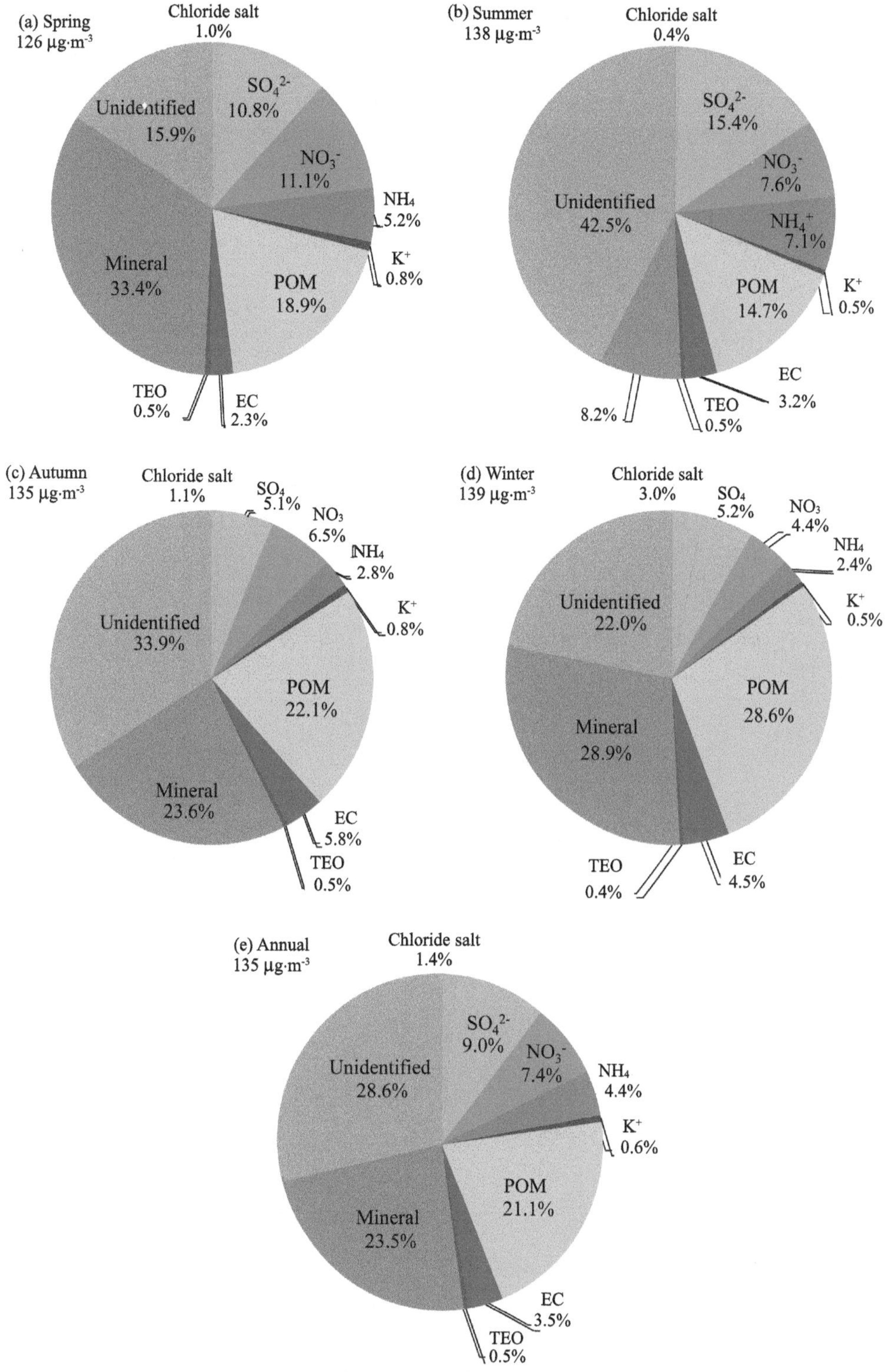

Fig. 7　Pie-charts showing the constructed chemical mass closures for $PM_{2.5}$ in Beijing: (a) spring; (b) summer; (c) autumn; (d) winter; and (e) annual. The components include mineral dust, secondary aerosol ions (sulfate, nitrate, and ammonium), POM, EC, trace element oxides (TEO), chloride salt, and biomass burning-derived potassium. Other than the identified components, unidentified fractions comprise a significant portion of $PM_{2.5}$

Overall, the major components are secondary inorganic aerosols (combination of sulfate, nitrate, and ammonium), mineral dust, and POM, which account for each ~20%, albeit with seasonal variations. The minor components include EC, chloride salt, potassium salt, and TEO, each of which represents less than 5%. Specifically, the proportions of mineral dust are maximal (33.4%) in spring, minimal (only 8.2%) in summer, and intermediate (23.6 and 28.9%) in the other two seasons, consistent with the tendency of seasonal Al concentrations. The totals of secondary inorganic species (SO_4^{2-}, NO_3^-, and NH_4^+) have the largest proportion (27% to 30%) in spring and summer and a minimal percentage (<15%) in autumn and winter. However, sulfate peaks were noted in summer (15.4%), where as nitrate peaks were observed in spring (11.1%). Ammonium decreased from around 5% to 7% in spring and summer to half (2% to 3%) in autumn and winter. The POM fractions largely varied as follows: summer (14.7%) < spring (18.9%) < autumn (22.1%) < winter (28.6%). EC and chloride salt exhibited the largest proportions (5.2% and 3.3%, respectively) in winter. Potassium salt and TEO had slightly higher proportions in spring and autumn than in summer and winter.

Both the primary and secondary components of $PM_{2.5}$ in Beijing are equally important, albeit with seasonal variability, which is typical of PM pollution in China (Shao *et al.*, 2006). In general, given that the seasonal variability in $PM_{2.5}$ mass concentrations is relatively small, temporal trends in the proportions of each component of $PM_{2.5}$ resemble the atmospheric concentrations of their corresponding chemical species. The likely factors for such seasonalities are partially addressed in Section 3.2 and discussed in detail in the following two sections.

On average, the unidentified components reached 28.6% of the total $PM_{2.5}$. They also showed seasonal variability, with the smallest (15.9%) in spring when dust was prevalent, and the largest (42.5%) in summer when secondary inorganic aerosol formation was favorable. Such high uncertainties in the CMCs were caused by the water absorption of water-soluble components in the weighing environment, though relative humidity was controlled (Speer *et al.*, 1997; Tsai and Kuo, 2005). The absorption likely led to positive biases in $PM_{2.5}$ concentrations. Alternatively, such uncertainties may be partly due to the volatilization of NH_4NO_3 and organic matter, particularly in summer and autumn during the storage of the weighted samples prior to extraction, which may have resulted in negative biases in the specific components. Another likely reason for the non-match of the reconstructed and gravimetric mass concentrations is the varying factors used in transferring a given analyzed species (e.g., OC and Al) to a certain component (e.g., POM and mineral soils) (Rees *et al.*, 2004; Hsu *et al.*, 2010a; Yan *et al.*, 2012). For example, a few studies adopted a factor of only 1.4 for converting OC content to organic matter (Duan *et al.*, 2006; Song *et al.*, 2007; Guinot *et al.*, 2007). Another study obtained a much higher POM/OC mean ratio over China (Xing *et al.*, 2013) of up to 1.92±0.39 based on a mass balance method. If we adopt this higher ratio, the unidentified percentage would be reduced by 3%. In the present study, the EF_{crust} of Ca averages at 2.6, which shows its enrichment relative to average crust composition. In Beijing, fine-sized Ca-rich dust is partly attributed to construction activities. Therefore, we may underestimate carbonate abundance in the mineral component estimated from Al concentration alone (Guinot *et al.*, 2007). This may have resulted in the underestimation (~2%) of the total mass reconstructed.

4 Discussion

4.1 Source Identification and Apportionment

By utilizing the PMF model with the obtained full data set as input data, we identified six main sources: mineral dust, biomass burning, coal combustion, traffic emissions plus waste incineration, industrial pollution, and secondary inorganic aerosol. Table 2 summarizes the source apportionment results of the relative contributions from each identified source to the $PM_{2.5}$ on both seasonal and annual bases in Beijing. These sources have average contributions of 15%, 18%, 12%, 4%, 25%, and 26%, respectively (Table 2). Figures 8 and 9 show the modeled source profiles and the time series of modeled concentrations for each identified main source, respectively. Again, the relative dominance of each identified source largely varies with changing seasons, which is roughly consistent with the CMC results. For mineral dust, only one of the six sources mostly dominated by nonvolatile substance, its proportions (e.g., annual mean ~20%) and relative order in the four seasons are consistent with the CMC results, with the highest contribution in spring, the lowest contribution in summer, and intermediate contribution in autumn and winter. This consistency indirectly verifies the reliability of the PMF results. The other five sources all appear to be related to high-temperature activities and/or photochemical processes and involved with volatile species. We then compared the contribution percentages of the secondary inorganic aerosol (SIA) with the CMC results as this source is also identified in the CMC analyses. Apparently, the percentages of SIA in the four seasons differ from those obtained by CMC in terms of the values (e.g., 6% to 54% versus 14% to 33% for SIA), although the seasonal trends are quite similar. Thus, CMC method only offers chemical characterization instead of source apportionment. The PMF model provides real information on sources of aerosol speciation.

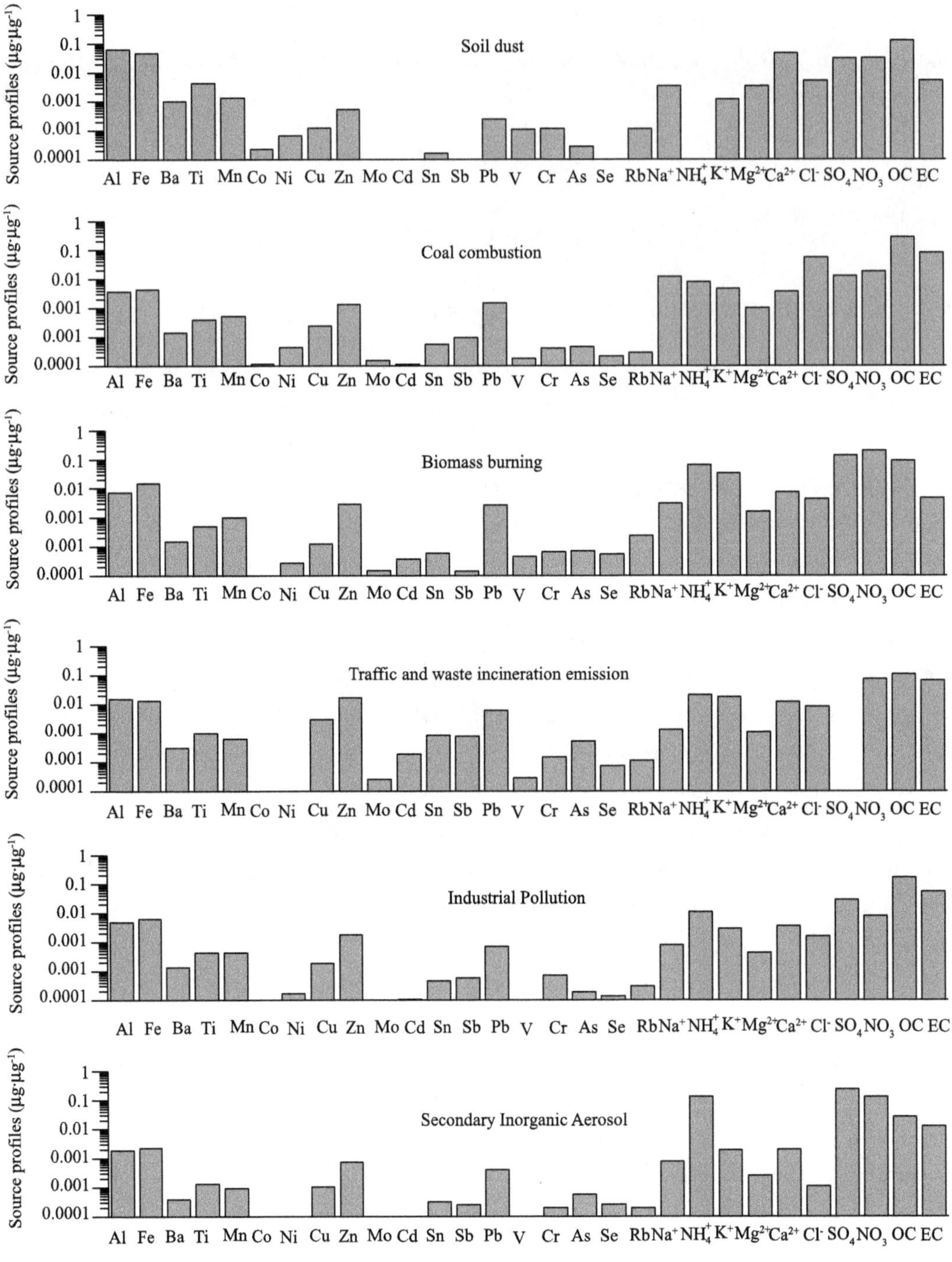

Fig. 8 Profiles of six sources identified from the PMF model, including soil dust, coal combustion, biomass burning, traffic and waste incineration emission, industrial pollution, and secondary inorganic aerosol (from the upper to the lower panels)

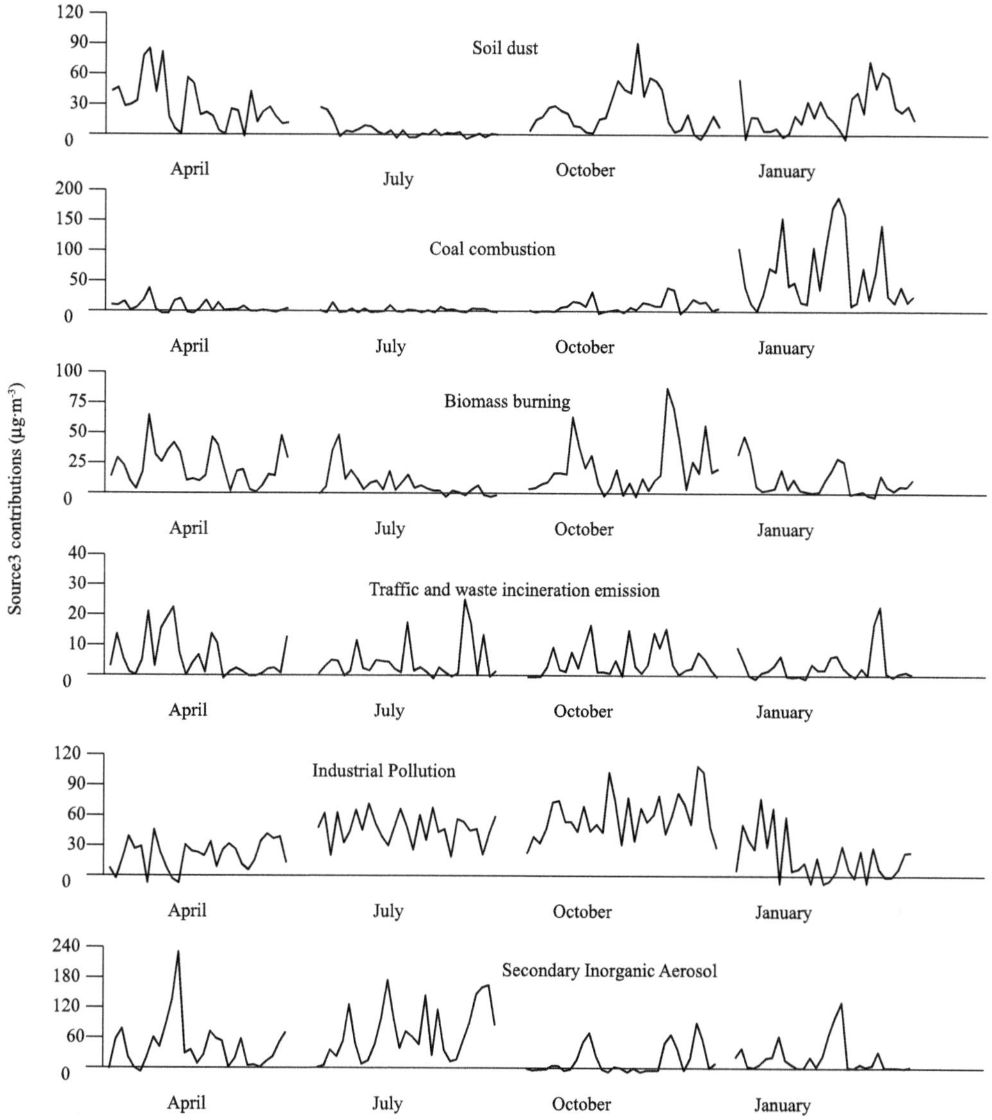

Fig. 9　Time-series of daily contributions from each identified source, including soil dust, coal combustion, biomass burning, traffic and waste incineration emission, industrial pollution, and secondary inorganic aerosol (from the upper to the lower panels) during the study period between April 2009 and January 2010

Table 2　Relative contributions from six identified sources of $PM_{2.5}$ in Beijing within the one-year and four-season periods

Source	Spring	Summer	Autumn	Winter	Annual
Soil dust(%)	23	3	18	16	15
Coal combustion(%)	5	1	7	57	18
Biomass burning(%)	19	6	17	7	12
Traffic and waste incineration emission(%)	5	4	4	2	4
Industrial pollution(%)	14	32	42	12	25
SIA(%)	34	54	13	6	26

Moreover, we evaluated the distinctive characteristics of each modeled source profile according to well-accepted knowledge. The first component, mineral dust, is typically characterized by high crustal elements, such as Al, Ca (Ca^{2+}), Fe, Mg, K, and Ti. Ca content is higher than Al, which indicates Ca-rich dust. OC content is close to ~12%, which suggests resuspended dust because of the presence of high levels of OC (Watson and Chow, 2001). Thus, this source possibly mixes desert/loess dust, anthropogenic construction dust, fugitive dust, and resuspended road dust. Construction activities are prevalent in the urban cities of China, and no effective measures for dust control are implemented. Therefore, calcium is used as an indicator element for construction dust in Beijing (Zhang and Iwasaka, 1999). The second source, coal combustion, is characterized by elevated Cl^- associated with high Na^+, OC, and EC. Extraordinarily high Cl^- associated with fine aerosols in winter is a distinctive feature in Beijing and even around inland China, which is ascribed to coal combustion (Yao *et al.*, 2002; Wang *et al.*, 2008). Coal combustion is the predominant source of fine aerosols over China (Yao *et al.*, 2009), which has resulted in severe air pollution problem not only locally, but also regionally and globally; for instance, it alone contributes more than 10% (268 Mg) of the global anthropogenic mercury emission (2319 Mt) annually (Pirrone *et al.*, 2010). Besides, coal fly ash could be one of the main contributors of aerosol Pb in China as they contain abundant Pb (Zhang *et al.*, 2009c). Sodium has also been found to be enriched in fine particulates from coal combustion (Takuwa *et al.*, 2006). Wang *et al.* (2008) and Zhang *et al.* (2009a) attributed the observed high Na and Cl to the presence of Na_2SO_4 and NaCl that maybe originated from dried lake salt sediment in Inner Mongolia, a non-local dust source (Xie and Liu, 2007). However, as discussed, salt lake aerosols alone cannot account for such strikingly high Cl^- in winter, which suggests that coal combustion is the most likely dominant source. Different investigations have obtained significantly different contributions of coal combustion to $PM_{2.5}$ in Beijing, which range from 7% to 19%. Yao *et al.* (2009) concluded that the likely fraction ranges between 15% and 20%. Nevertheless, previous studies have not considered that other main identified sources, such as secondary inorganic and organic aerosols, have contributions from coal combustion.

The third source, biomass burning, is characterized by high K (K^+), which is an excellent tracer of biomass-burning aerosols (Cachier and Ducret, 1991; Watson and Chow, 1998), and by rich Rb, OC, and NO_3^-. Biomass burning has higher contributions in spring and autumn than in summer and winter, consistent with cultivation in spring and harvest in autumn (Duan *et al.*, 2004). The fourth source is a mixed source of traffic and metropolitan incineration emissions, which is characterized by high NO_3^-, EC, Cu, Zn, Cd, Pb, Mo, Sb, and Sn. These aerosol species are all enriched in vehicular and/or waste incineration emissions (Lee *et al.*, 1999; Alastuey *et al.*, 2006; Birmili *et al.*, 2006; Marani *et al.*, 2003; Dall'Osto *et al.*, 2012; Tian *et al.*, 2012). For instance, Wåhlin *et al.* (2006) observed that traffic-generated aerosol particles are rich in Cu, Zn, Mo, and

Sb. Christian *et al.* (2010) analyzed the aerosol particles emitted from garbage burning, which are rich in Zn, Cd, Sb, and Sn. Leaded gasoline was phased out in 1997 in Beijing and in 2000 in the rest of China. Coal burning was then suggested as the most important source of Pb aerosols in China (Mukai *et al.*, 2001). However, Widory *et al.* (2010) argued that in Beijing, metal-refining plants are the dominant sources of aerosol Pb, followed by thermal power stations and other coal combustion sources.

The fifth source is industrial pollution, which is characterized by high contents of OC, EC, Zn, Mn, and Cr. This source may also be involved with secondary organic aerosols. Coal is the primary energy source commonly used in industries in China. Both coal combustion and vehicle emissions are the main sources of primary OC (Zhang *et al.*, 2007; Cao *et al.*, 2011). However, Zhang *et al.* (2008) estimated that secondary OC represents more than half of the measured OC at regional sites (~67%) and urban sites (~57%), which is higher than those reported by Cao *et al.* (2007)(i.e., 30% to 53%). Therefore, industrial pollution could act as a vital source of carbonaceous aerosols, which seems to be widely ignored. Furthermore, given that Zn and Cr contents are high, this source may be relevant to smelters and metallurgical industries (Dall'Osto *et al.*, 2012). The sixth source is relevant to secondary inorganic aerosols, which are typically characterized by remarkable SO_4^{2-}, NO_3^-, and NH_4^+. Certain identified sources, such as biomass burning, coal combustion, vehicle exhausts, and waste incineration, can also contribute to secondary inorganic and organic aerosols through the emission of their precursor gases.

Figure 9 illustrates the time series of daily concentrations contributed by each identified source. To examine if the results are reasonable, we compared the modeled seasonalities of each source with the observed seasonalities of the specific chemical species that could represent respective contribution sources (Fig. 2). For instance, we compared the maximal and minimal contributions of mineral dust in spring and summer, respectively, which are consistent with the seasonality of aerosol Al. Dust storms are essentially responsible for springtime dust aerosols, whereas in autumn and winter, fugitive dust from construction and the resuspension of street dust are the main contributors. Obviously, the reconstructed time series of daily concentrations from coal combustion reveals a pronounced wintertime maximum, consistent with those of aerosol Cl^- (Fig. 2) and even gaseous SO_2 (Fig. S2). Moreover, the time series of biomass burning contributions show relatively higher concentrations in spring and autumn and lower concentrations in summer and winter, consistent with the seasonality of K^+. For traffic and waste-burning emissions, the resulting time series do not reveal evident seasonality, corresponding with the seasonalities of nitrate and some trace metals, such as Pb, Cu, Sb, and Cd (Fig. S4). Industrial pollution has higher contributions in summer and autumn, possibly corresponding with the seasonalities of Zn and Cr. However, such seasonality is inconsistent with that of OC, with a wintertime maximum, because OC may be from various sources, including the former five sources identified. Coal combustion has the largest contribution in winter, and low temperature in winter facilitates the formation of secondary organic aerosols. SIA has higher contributions in summer and spring, mirroring the seasonalities of sulfate, nitrate, and ammonium. This result is definitely related to the photochemistry that accounts for SIA formation. The formed SIA species may not appear in their original emission sources (i. e., coal combustion, biomass burning, traffic exhausts, waste incineration, and industrial pollution), but in the SIA component. Based on the PMF results and chemical data in January and August 2004, Song *et al*. (2007) found that the most predominant sources of $PM_{2.5}$ are coal combustion in winter and secondary aerosols in summer, along with other significant sources, such as motor vehicle emissions, road dust, and biomass burning. The PMF-modeled results seem to be promising because the corresponding time series of each source's contribution are very consistent with the observations.

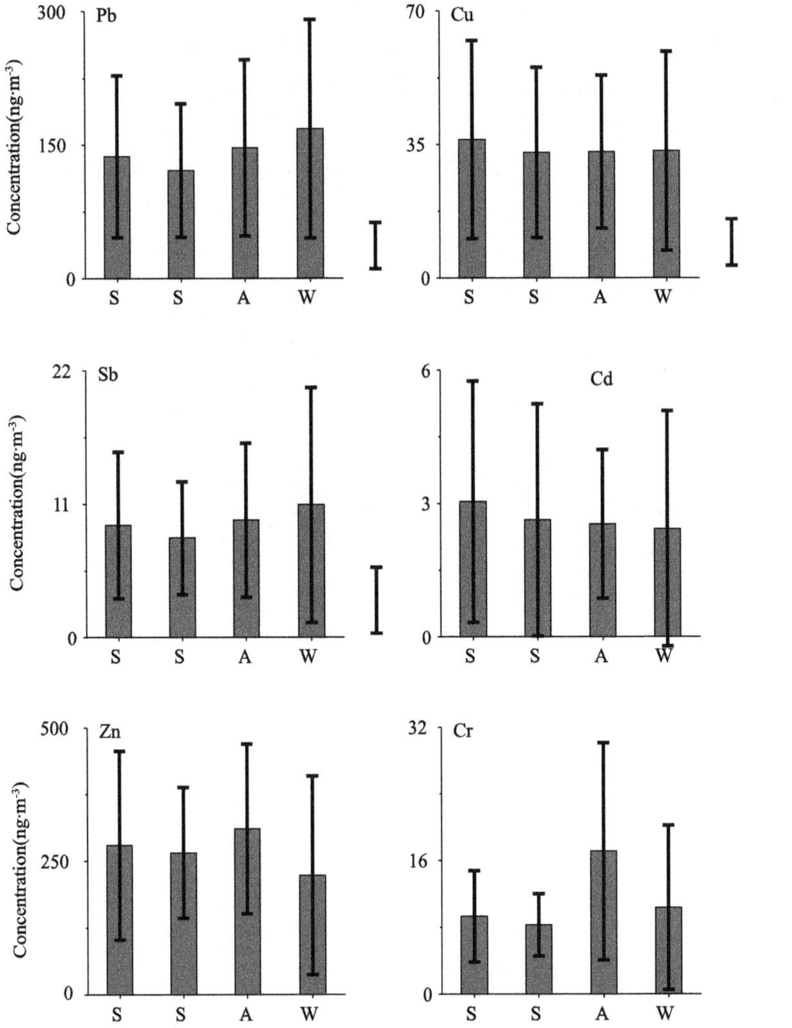

Fig. S4 Seasonal patterns of trace elements including Pb, Cu, Sb, Cd, Zn, and Cr. The concentration unit is in ng • m^{-3}

4.2 Regional Sources Deduced from Trajectory and PSCF Analyses

The regional sources and transport of air pollutants exert a profound impact on local air quality in Beijing (e. g., Wang *et al.*, 2004). To address this issue, both trajectory clustering and PSCF methods were employed. The 48h back trajectories starting at 100 m from Beijing were computed by using the HYSPLIT model of NOAA (http://www.arl.noaa.gov/ready.html). Four clusters were made (Fig. 10): northwestern (including western, NW), northern (N), eastern (from northeastern to southeastern, E), and southern (S) directions. The NW cluster was further differentiated into two types, namely, fast (NW_f) and slow (NW_s), according to the motion speed ($\leqslant 7\ m \cdot s^{-1}$ for NW_s and $>7\ m \cdot s^{-1}$ for NW_f,) and distance of air parcels. The classification is consistent with the spatial distribution of fine AOD retrieved by remote sensing (Fig. 1).

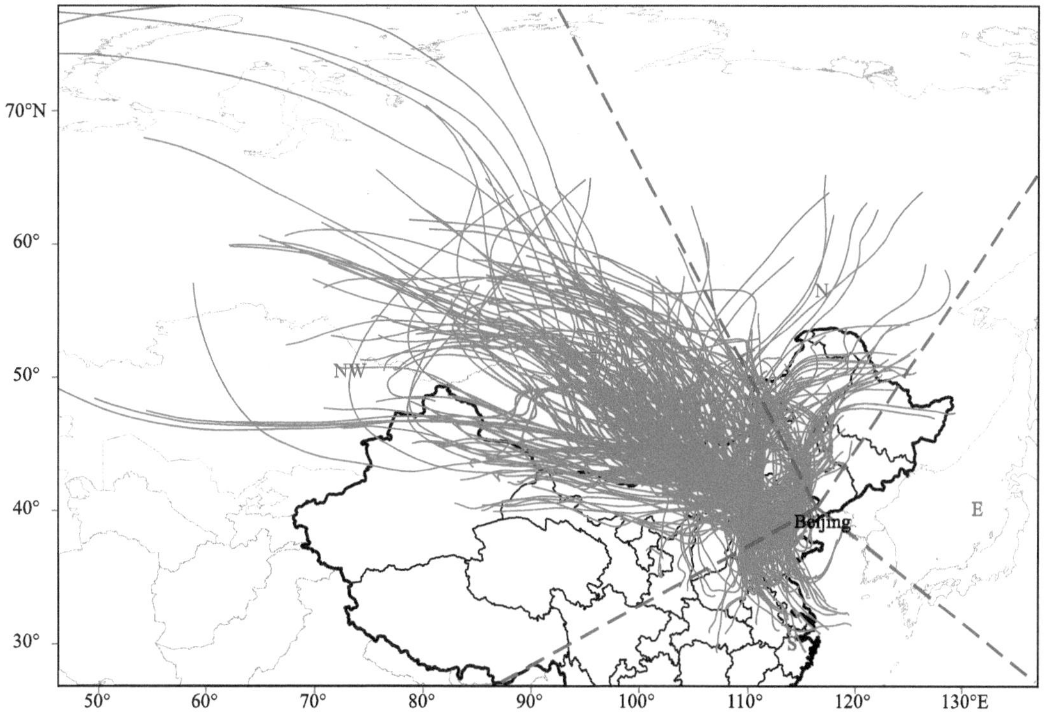

Fig. 10　Analytical results of the 48h air-mass back trajectories at 100 m elevation during the sampling periods, which were run four times per day. Four regions were defined based on the trajectory clustering results, namely, NW, N, E, and S regions

Table 3 summarizes the percentages of each trajectory cluster in the total on annual and seasonal bases and the corresponding mean concentrations of $PM_{2.5}$ and various aerosol species associated within each trajectory cluster. Annually, the trajectory clusters are dominated by both NW and S, accounting for 44% and 34%, respectively. The E and N clusters represent the rest (15% and 7%, respectively). However, the variability is large and season-dependent. For instance, the predominant clusters are N (30%) and S (44%) in spring, S (73%) in summer, NW_f (50%) in autumn, and NW_f (88%) in winter. The resulting mean concentrations of main aerosol species seasonally vary with certain types of air masses. In winter, a few $PM_{2.5}$ pollution cases (only 12% of the wintertime trajectories) with mean concentration as high as 209 $\mu g \cdot m^{-3}$ are associated within the E trajectories that passed over Hebei and Liaoning Provinces, where heavy industries are concentrated in certain cities (e.g., Tianjin, Tangshan, Dalian, Shenyang). However, in spring, summer, and autumn, high $PM_{2.5}$ (> 150 $\mu g \cdot m^{-3}$) is preferentially associated with the S trajectory cluster. Overall, the general patterns agree with the spatial distribution of the MODIS-retrieved fine AOD around Beijing (Fig. 1).

Furthermore, we applied an alternative approach called PSCF to explore the likely regional sources and transport pathways of various $PM_{2.5}$-associated speciations, such as sulfate, nitrate, ammonium, OC, EC, and mineral dust in Beijing, as illustrated in Fig. 11. A few main features were found: (a) sulfate, nitrate, and ammonium have similar spatial patterns, with higher values in the east to the south, covering Tianjin, Shijiazhuang, and Zhengzhou; (b) both OC and EC show similar spatial distribution, with higher values in the northwest, the south, and the northeast, covering the border of Hebei and Shanxi Provinces, Inner Mongolia, the border of the Hebei, Shanxi, and Henan Provinces, and the area from Tianjin to Shenyang; (c) the higher value for mineral aerosols is localized in the northwest and the south; and (d) for these six aerosol speciations, the southern area appears to be a common hot spot. The overall PSCF results are rather consistent with the spatial distributions of fine AOD (Fig. 1) and their respective corresponding species' emissions such as SO_2, NO_x, NH_3, OC, and EC in China (Zhang *et al.*, 2009b, Fu *et al.*, 2012; Huang *et al.*, 2012). In regard of dust, our PSCF result has shown that the dust transported to Beijing is primarily originated from northern/northwestern China (Wang *et al.*, 2004) while Taklimakan Desert is a very important dust source in China (Laurent *et al.*, 2005). The statistics obtained from the trajectory clustering (Table 3) shows that the southern air masses bring high levels of secondary inorganic and carbonaceous aerosols and the northwestern air masses are enriched in mineral dust and carbonaceous aerosols. Sun *et al.* (2006) and Street *et al.* (2007) found that the S sector has much higher secondary species, such as sulfate, nitrate, and ammonium. During haze-fog events in Beijing, chemical constituents of secondary inorganic aerosols are also much higher when the winds blow from the south. Such high amounts of secondary fine-sized aerosols in southern air parcels may be related to high humidity (water vapor) and enhanced heterogeneous reaction in clouds/fog, aside from strong photochemistry. The association of high dust with the NW trajectories is consistent with Wang *et al.* (2004) and Yu *et al.* (2011).

Table 3　Mean concentrations (in unit μg/m^3) of $PM_{2.5}$ and selected aerosol speciations in the identified trajectory clusters within the one-year (annual) and four-season periods. Also given are the percentages of each trajectory cluster classified in the one-year and four-season periods. For details on trajectory clustering, please refer to the text

Air-mass type	Annual					Spring				Summer				Autumn			Winter	
	NW_s	NW_f	N	E	S	NW_s	N	E	S	NW_s	N	E	S	NW_f	E	S	NW_f	E
Percent (%)	11.9	32.4	7.4	14.6	33.7	8.7	29.6	17.4	44.3	9.6	8.7	8.7	73.0	49.6	26.8	23.6	88.2	11.8
$PM_{2.5}$	148	111	87	110	172	145	70	108	167	108	110	63	155	113	144	173	131	209
Sulfate	10.2	5	6.3	10.9	25.4	11.9	4.8	13.6	23.1	11.1	9.6	7.6	28.3	4.8	10.9	11.2	7	19.5
Nitrate	11.2	5.1	4.3	10.2	19.2	14	4.8	14.4	24.4	4.5	4	3.5	14.3	6.4	13.4	16.9	6.3	14.9
Ammonium	5.4	2.3	2.3	5.7	13.2	4.7	1.7	8	12.2	4.8	3.7	3.4	13.4	2.4	6.5	7.6	3.7	10.9
OC	22.2	18.1	10.8	13.8	16.4	15.5	10	11.5	16.7	12.4	12.2	9.4	11	15.6	18.7	21.6	23.8	33.6
EC	6.5	4.9	2.5	3.7	5.4	3.7	1.9	2.2	3.4	3.5	3.6	2.6	4.5	4	5.9	7.5	7	11.4
Mineral	31.3	36.5	26	16.4	20.8	64.8	32.8	15.3	39.1	15	21	4.8	8.5	33.3	19.4	30.4	31.3	21.8

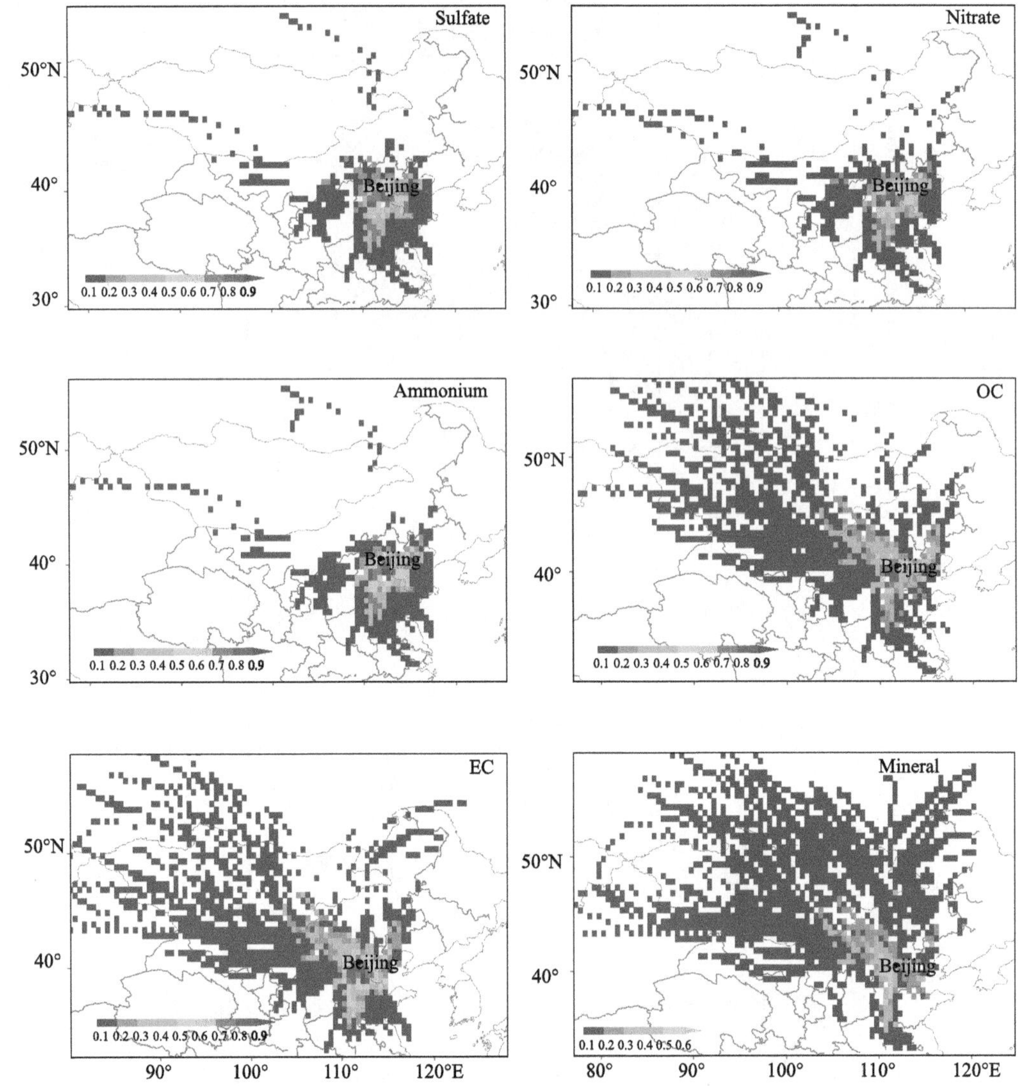

Fig. 11 The PSCF maps for sulfate, nitrate, ammonium, OC, EC, and mineral

4.3 Implications for Atmospheric Chemistry, PM Control Measures, and Climate

Rigorous efforts exerted for air pollutant governance prior to the 2008 Beijing Summer Olympics, such as changing the energy source structure, reducing local dust emissions, controlling vehicle exhaust emissions, and relocating major industrial emitters, have achieved air quality improvement during the games. Effective control of coarse PM pollution seemed possible and the main urban air pollutants became finer PM ($PM_{2.5}$). The annual mean concentration of $PM_{2.5}$ in Beijing is nearly three times and over an order of magnitude higher than the annual exposure level (35 $\mu g \cdot m^{-3}$) and air quality guideline (only 10 $\mu g \cdot m^{-3}$) recommended by the WHO, which indicates that tremendous efforts of multi-pollutant alleviation measures and air quality management policies are still needed for PM abatement (Zhang *et al.*, 2012b). The PM pollution level in Beijing is governed by the emission sources involved with natural and anthropogenic origins and particulate and gaseous phases and by synoptic meteorological conditions and atmospheric circulation systems. With sulfate/SO_2 as an example, our results demonstrate that meteorological conditions could be a crucial factor for determining fine aerosol levels and formation other than their own and/or precursor emission strengths. Accordingly, this study offers insights into the likely impact on atmospheric chemistry because of the changing climate. The increasingly warm climate predicted (IPCC, 2007) will enhance photochemistry in summer, which may offset the mitigation measures in China to some extent. Moreover, the decreasing wind speed forecasted in China (Chen *et al.*, 2012) favors air quality degradation because of the likely reduction in ventilation efficiency (Zhang *et al.*, 2007; Song *et al.*, 2008). Complex PM pollution in terms of chemical and physical properties such as multiple sources (natural versus anthropogenic), mixing states (internal versus external), various chemical composition, size spectrum, and hygroscopicity, which are closely related to optical and direct/indirect radiative properties, complicate the modeling assessment of aerosol effects on the climate (cooling versus warming) in China and in the region.

If the contributions from the three main sources (coal combustion, industrial pollution, and SIA) are combined, fossil fuel burning-related emissions may dominate $PM_{2.5}$ pollution in Beijing, representing two thirds (~68%) of $PM_{2.5}$. Rapid industrial development in provinces around Beijing, including Liaoning, Hebei, Shandong, Shanxi, and Henan, has exacerbated regional air quality because of massive quantities of air pollutant emissions, resulting in cross-border transport. Better understanding of the pollution characteristics of $PM_{2.5}$ in Beijing, particularly after the 2008 Summer Olympics, in terms of chemical composition and sources of $PM_{2.5}$ from a regional and seasonal perspective, is urgently needed. Relevant air pollution control measures should be implemented locally, regionally, and nationally in China. Such measures would improve pollution abatement, public health, and climatic modeling capacity.

5 Summary

The levels of daily $PM_{2.5}$ concentrations are still elevated in Beijing throughout the year, with an annual mean of up to $135 \pm 63\ \mu g \cdot m^{-3}$, which is several times higher than the 24h exposure standard and guideline recommended by the WHO. Seasonality is not very evident, although the highest occurred in winter and the lowest in spring. Distinctive seasonalities occurred for various aerosol species. Sulfate and ammonium peaked in summer, mainly because of photochemistry, whereas the maximum of nitrate was observed in spring, which is attributed to the heterogeneous reaction between dust carbonate and nitric acid (the product of NOx) along with the photochemical process. Crustally derived species, such as Ca^{2+}, Mg^{2+}, and Al, showed typical seasonality, waxing in spring and waning in summer. The maximum of OC, BC, and Cl^- was observed in winter, and their minimum in summer or spring, which suggests a relation with fossil fuel (mainly coal) combustion processes. The correlations between total cation and anion equivalent concentrations were excellent with a mean ratio around unity, which indicates good charge balance. The mean NH_4^+/SO_4^{2-} equivalent concentration ratios were higher than unity throughout the year, which reveals that NH_4^+ could fully neutralize sulfuric acid. The correlations between NH_4^+ and $[SO_4^{2-}+NO_3^-]$ suggest that ammonium could partially be present as NH_4NO_3. Furthermore, the correlations between $[NH_4^+ + Ca^{2+}]$ and $[SO_4^{2-}+NO_3^-]$ demonstrate the presence of $Ca(NO_3)_2$ and NH_4NO_3, which was due to the heterogeneous reaction between dust carbonate and nitric acid. A few chloride salts may form, particularly in winter when low temperature is favorable for their occurrence.

CMCs were successfully constructed on a seasonal basis, although with an unidentified fraction averaged at 28.6%. The major aerosol speciation considered in this study included mineral dust, POM, and SIA (the combination of sulfate, nitrate, and ammonium), and the minor components were EC, chloride salts, trace element oxides, and biomass burning-derived potassium salts. Their respective fractions in $PM_{2.5}$ largely varied with the season, generally corresponding to the seasonalities of their concentrations. The higher unidentified fraction, particularly in summer, may be partly due to the volatilization of volatile or semi-volatile species, such as ammonium nitrate and organic species. This observation highlights the requirement for the immediate analysis of aerosol samples after collection. Through the PMF model, six potential sources were identified: mineral dust, coal combustion, biomass burning, traffic and waste incineration emission, industrial pollution, and SIA. Similarly, the relative contributions of these sources to $PM_{2.5}$ in Beijing greatly varied with the changing seasons, which proves the "Complex Air Pollution" in Beijing. On the average, the annual mean contributions were 15%, 18%, 12%, 4%, 25%, and 26%, respectively. The highest contributions occurred in spring for soil dust (23%) and traffic and waste incineration emission (5%), in winter for coal combustion (57%), in autumn for industrial pollution (42%), in summer for SIA (54%), and in both spring and autumn for biomass burning (19% and 17%, respectively). Based on the trajectory clustering and PSCF method results, regional sources and transport pathways of various aerosol speciation associated with $PM_{2.5}$ in Beijing were explored. The southern region may be an important source of all major aerosol components. The eastern region is significant for anthropogenic species such as SIA and carbonaceous aerosols. The northwestern region could be the common source of naturally derived mineral dust and anthropogenically derived carbonaceous aerosols. This observation indicates that air pollution control measures must be simultaneously implemented in the surrounding provinces.

The results of our studies clearly suggest that chemical constituents and sources of $PM_{2.5}$ can largely vary with seasons, which are characterized by variable meteorology and diverse air pollution sources. Source apportionment results do not precisely reflect the large temporal contribution variations from various sources when PMF analyses using data from the full four seasons are not considered. In addition, apart from the emission strengths of primary aerosols and gaseous precursors, the dynamically variable synoptic weather conditions and circulation patterns also have a crucial role in the anomalies of PM (both fine and coarse size) concentrations.

Acknowledgement

This work is supported by the National Natural Science Foundation of China (No. 41175131), the National Basic Research Program of China(2010CB428503), and the Ministry of Science and Technology of China(2010DFA22770).

References

Alastuey A, Querol X, Plana F, *et al*. 2006. Identification and chemical characterization of industrial particulate matter sources in southwest Spain. *J. Air & Waste Manage. Assoc.*, **56**:993-1006.

Andreae M O, Rosenfeld D. 2008. Aerosol-cloud-precipitation interactions. part 1. The nature and sources of cloud-active aerosols. *Earth-Sci. Rev.*, **89**:13-41.

Bardouki H, Liakakou H, Economou C. 2003. Chemical composition of size resolved atmospheric aerosols in the eastern Mediterranean during summer and winter. *Atmos. Environ.*, **37**:195-208.

Beijing EPB (Beijing Municipal Environmental Protection Bureau). 2005. Beijing communiqué on environmental quality, 2001—2005, www.bjee.org.cn, Viewed 24 July 2006.

Birmili W, Allen A G, Bary F, *et al*. 2006. Trace metal concentrations and water solubility in size-fractionated atmospheric particles and influence of road traffic. *Environ. Sci. Technol.*, **40**:1144-1153.

Cachier H, Ducret J. 1991. Influence of biomass burning on equatorial African rains. *Nature*, 352, 228-230.

Cao J J, Chow J C, Tao J, *et al*. 2011. Stable carbon isotopes in aerosols from Chinese cities: Influence of fossil fuels. *Atmos. Environ.*, 1359-1363.

Cao J J, Lee S C, Chow J C, *et al*. 2007. Spatial and seasonal distributions of carbonaceous aerosols over China. *J. Geophys. Res.*, **112**, D22S11. doi:10.1029/2006JD008205.

Chan C K, Yao X H. 2008. Air pollution in mega cities in China. *Atmos. Environ.*, **42**: 1-42.

Chen L, Pryor S C, Li D L. 2012. Assessing the performance of Intergovernmental Panel on Climate Change AR5 climate models in simulating and projecting wind speeds over China. *J. Geophys. Res.*, **117**, D24102, doi:10.1029/2012JD017533.

Chen W Y, Xu R N. 2010. Clean coal technology development in China. *Energy Policy*, **38**:2123-2130.

Chester R. 1990. *Marine Geochemistry*, Cambridge University Press, London, 698.

Chow J C, Watson J G, Chen L W, *et al*. 2007. The IMPROVE—A temperature protocol for thermal/optical carbon analysis: Maintaining consistency with a long-term database. *J. Air & Waste Manage. Assoc.*, **42**,1014-1023.

Chow J C, Watson J G, Fujita E M, *et al*. 1994. Temporal and spatial variation of $PM_{2.5}$ and PM_{10} aerosol in the Southern California air quality study. *Atmos. Environ.*, **28**:2061-2080.

Chow J C, Watson J G, Lu Z, *et al*. 1996. Magliano, Descriptive analysisof $PM_{2.5}$ and PM_{10} at regionally representative locations during SJVAQS/AUSPEX. *Atmos. Environ.*, **30**:2079-2112.

Christian T J, Yokelson R J, Cárdenas, *et al*. 2010. Trace gas and particle emissions from domestic and industrial biofuel use and garbage burning in central Mexico. *Atmos. Chem. Phys.*, **10**:565-584.

Dall'Osto M, Querol X, Amato F, *et al*. 2012. Hourly elemental concentrations in $PM_{2.5}$ aerosols sampled simultaneously at urban background and road site. *Atmos. Chem. Phys. Discuss.*, **12**:20135-20180.

Dan M, Zhuang G S, Li X X, *et al*. 2004. The characteristics of carbonaceous species and their sources in $PM_{2.5}$ in Beijing. *Atmos. Environ.*, **38**:3443-3452.

Dockery D W, Pope C A. 1994. Acute respiratory effects of particulate airpollution. *Annu. Rev. Publ. Health*, **15**:107-132.

Duan F K, He K B, Ma Y L, *et al*. 2006. Concentration and chemical characteristics of $PM_{2.5}$ in Beijing, China: 2001—2002. *Sci. Total Environ.*, **355**: 264-275.

Duan F K, Liu X, Yu T, *et al*. 2004. Identification and estimate of biomass burning contribution to the urban aerosol organic carbon concentrations in Beijing. *Atmos. Environ.*, **38**:1275-1282.

Duan F K, Liu X D, He K B, *et al*. 2003. Atmospheric aerosol concentration level and chemical characteristic of its water soluble ionic species in wintertime in Beijing, China. *J. Environ. Monit.*, **4**: 569-573.

Eatough D J, Cui W X, Hull J, *et al*. 2006. Fine particulate chemical composition and light extinction at Meadview, AZ. *J. Air & Waste Manage. Assoc.*, **56**:1694-1706.

Fang M, Chan C K, Yao X H. 2009. Managing air quality in a rapidly developing nation: China. *Atmos. Environ.*, **43**:79-86.

Fu T M, Cao J J, Zhang X Y, *et al*. 2012. Carbonaceous aerosols in China: top-down constraints on primarysources and estimation of secondary contribution. *Atmos. Chem. Phys.*, **12**:2725-2746.

Gao X, Yang L, Cheng S, *et al*. 2011. Semi-continuous measurement of water-soluble ions in $PM_{2.5}$ in Jinan, China: temporal variations and source apportionments. *Atmos. Environ.*, **45**:6048-6056.

Gu J X, Bai Z P, Li W F, *et al*. 2011. Chemical composition of $PM_{2.5}$ during winter in Tianjin, China. *Particuology*, **9**:215-221.

Guinot B, Cachier H, Oikonomou K. 2007. Geochemical perspectives from a new aerosol chemical mass closure. *Atmos. Chem. Phys.*, **7**:1657-1670.

Guo S, Hu M, Guo Q F, *et al*. 2012. Primary sources and secondary formation of organic aerosols in Beijing, China. *Environ. Sci. Technol.*, **46**:9846-9853.

Han J S, Moon K J, Lee S J, *et al*. 2006. Size-resolved source apportionment of ambient particles by positive matrix factorization at Gosan background site in East Asia. *Atmos. Chem. Phys.*, **6**:211-223.

Hao J M, Wang L T, Li L, *et al*. 2005. Air pollutants contribution and control strategies of energy-use related sources in Beijing. *Science in China Ser.*, D. **48**(SII), 138-146.

He L Y, Hu M, Huang X F, *et al*. 2006. Seasonal pollution characteristics of organic compounds in atmospheric fine particles in Beijing. *Sci. Total Environ.*, **359**: 167- 176.

He K B, Huo H, Zhang Q. 2002. Urban Air Pollution in China: Current status, characteristics, and progress. *Annu. Rev. Energy Environ.*, **27**:397-431.

He K B, Yang F M, Ma Y L, *et al*. 2001. The Characteristics of $PM_{2.5}$ in Beijing, China. *Atmos. Environ.*, **35**:4959-4970.

He S Z, Chen Z M, Zhang X, *et al*. 2010. Measurement of atmospheric hydrogen peroxide and organic peroxides in Beijing before and during the 2008 Olympic Games: Chemical and physical factors influencing their concentrations. *J. Geophys. Res.*, **115**, D17307, doi:10. 1029/2009JD013544.

He X, Li C C, Lau A K H, *et al*. 2009. An intensive study of aerosol optical properties in Beijing urban area. *Atmos. Chem. Phys.*, **9**:8903-8915.

Ho K F, Lee S C, Cao J J, *et al.* 2006. Seasonal variations and mass closure analysis of particulate matter in Hong Kong. *Sci. Total Environ.*, **355**:276-287.

Hopke P K. 2003. Recent developments in receptor modeling. *J. Chemometrics*, **17**:255-265.

Hsu S C, Liu S C, Huang Y T, *et al.* 2009. Long-range southeastward transport of Asian biosmoke pollution: signature detected by aerosol potassium in Northern Taiwan. *J. Geophys. Res.*, **114**: D14301.

Hsu S C, Liu S C, Huang Y T, *et al.* 2008. A criterion for identifying Asian dust events based on Al concentration data collected from Northern Taiwan between 2002 and early 2007. *J. Geophys. Res.*, **113**:D18306, doi:10.1029/2007JD009574.

Hsu S C, Liu S C, Kao S J, *et al.* 2007. Water soluble species in the marine aerosol from the Northern South China sea: High chloride depletion related to air pollution. *J. Geophys. Res.*, **112**, doi: 10.1029/2007JD008844.

Hsu S C, Liu S C, Tsai F, *et al.* 2010a. High wintertime particulate matter pollution over an offshore island (Kinmen) off southeastern China: An overview. *J. Geophys. Res.*, **115**, D17309, doi: 10.1029/2009JD013641.

Hsu S C, Liu S C, Arimoto R, *et al.* 2010b. Effects of acidic processing, transport history, and dust and sea salt loadings on the dissolution of iron from Asian dust. *J. Geophys. Res.*, **115**, D19313, doi: 10.1029/2009JD013442.

Hsu Y K, Holsen T M, Hopke P K. 2003. Comparison of hybrid receptor models to locate PCB sources in Chicago. *Atmos. Environ.*, **37**:545-562.

Hu M, He L Y, Zhang Y H, *et al.* 2002. Seasonal variation of ionic species in fine particles at Qingdao, China. *Atmos. Environ.*, **36**:5853-5859.

Huang X, Song Y, Li M, *et al.* 2012. A high-resolution ammonia emission inventory in China. *Global Biogeochem. Cycles*, **26**, GB1030, doi:10.1029/2011GB004161.

Hueglin C, Gehrig R, Baltensperger U, *et al.* 2005. Chemical characterisation of $PM_{2.5}$, PM_{10} and coarse particles at urban, near-city and rural sites in Switzerland. *Atmos. Environ.*, **39**:637-651.

Ianniello A, Spataro F, Esposito G, *et al.* 2011. Chemical characteristics of inorganic ammonium salts in $PM_{2.5}$ in the atmosphere of Beijing (China). *Atmos. Chem. Phys.*, **11**:10803-10822.

IPCC. 2007. Climate change 2007—Synthesis Report.

Itahashi S, Uno I, Yumimoto K, *et al.* 2012. Interannual variation in the fine-mode MODIS aerosol optical depth and its relationship to the changes in sulfur dioxide emissions in China between 2000 and 2010. *Atmos. Chem. Phys.*, **12**:2631-2640.

Landis M S, Norris G A, Williams R W, *et al.* 2001. Personal exposures to $PM_{2.5}$ mass and trace elements in Baltimore, MD, USA. *Atmos. Environ.*, **35**:6511-6524.

Laurent B, Marticorena B, Bergametti G, *et al.* 2005. Simulation of the mineral dust emission frequencies from desert areas of China and Mongolia using an aerodynamic roughness length map derived from the POLDER//ADEOS 1 surface products. *J. Geophys. Res.*, **110**: D18S04, doi: 10.1029/2004JD005013.

Lee E, Chan C K, Paatero P. 1999. Application of positive matrix factorization in source apportionment of particulate pollutants in Hong Kong. *Atmos. Environ.*, **33**(19):3201-3212.

Lee K H, Li Z Q, Cribb M C, *et al.* 2010. Aerosol optical depth measurements in eastern China and a new calibration method. *J. Geophys. Res.*, **115**: D00K11, doi:10.1029/2009JD012812.

Li G, Chen J, Chen Y, *et al*. Dolomite as a tracer for the source regions of Asian dust. *J. Geophys. Res.*, **112**,D17201.

Li W J, Shao L Y. 2009. Observation of nitrate coatings on atmospheric mineral dust particles. *Atmos. Chem. Phys.*, **9**:1863-1871.

Liu J, Diamond J. 2005. China's environment in a globalizing world. *Nature*, **435**:1179-1186.

Mahowald N. 2011. Aerosol indirect effect on biogeochemical cycles and climate. *Science*, **334**:794-796.

Malm W C, Sisler J F, Huffman D, *et al*. 1994. Spatial and seasonal trends in particle concentration and optical extinction in theUnited States. *J. Geophys. Res.*, **99**: D1, 1347-1370.

Malm, W C, Day D E, Carrico C,*et al*. 2005. Intercomparison and closure calculations using measurements of aerosol species and optical properties during the Yosemite aerosol characterization study. *J. Geophys. Res.*, **110**: D14302.

Marani D, Braguglia C M, Mininni G,*et al*. 2003. Behaviour of Cd, Cr, Mn, Ni, Pb, and Zn in sewage sludge incineration by fluidised bed furnace. *Waste Manage.*, **23**:117-124.

Mukai H, Tanaka A, Fujii T, *et al*. 2001. Regional characteristics of sulphur and lead isotope ratios in the atmosphere at several Chinese urban sites. *Environ. Sci. Technol.*, **35**:1064-1071.

Niwa Y, Hiura Y, Murayama T, *et al*. 2007. Nano-sized carbon black exposure exacerbates atherosclerosis in ldl-receptor knockout mice. *Circ. J.*, **71**:1157-1161.

Oberdörster G. 2001. Pulmonary effects of inhaled ultrafine particles. *Int. Arch. Occup. Environ. Health*, **74**(1):1-8.

Okuda T, Matsuura S, Yamaguchi D, *et al*. 2011. The impact of the pollution control measures for the 2008 Beijing Olympic Games on the chemical composition of aerosols. *Atmos. Environ.*, **45**:2789-2794.

Pathak R K,Yao X H, Lau A K H, *et al*. 2004. Sampling artifacts of acidity and ionic species in $PM_{2.5}$. *Sci. Total Environ.*, **38**:254-259.

Pirrone N, Cinnirella S, Feng X, *et al*. 2010. Global mercury emissions to the atmosphere from anthropogenic and natural sources. *Atmos. Chem. Phys.*, **10**:5951-5964.

Polissar A V, Hopke P K, Paatero P,*et al*. 1999. The aerosol at Barrow, Alaska: Long-term trends and source locations. *Atmos. Environ.*, **33**: 2441-2458.

Pope C A, Thun M J, Namboodiri M M, *et al*. 1995. Particulate air pollution as a predictor of mortality in a prospective study of U. S. adults. *Amer. J. of Resp. Critical Care Med.*, **151**: 669-674.

Pósfai M, Gelencsér A, Simonics R,*et al*. 2004. Atmospheric tar balls: Particles from biomass and biofuel burning. *J. Geophys. Res.*, **109**, D06213, doi:10. 1029/2003JD004169.

Qian W, Lin X. 2005. Regional trends in recent precipitation indices in China. *Meteorol. Atmos. Phys.*, **90**:193-207.

Qian W H, Quan L S, Shi S Y. 2002. Variations of the dust storm in china and its climatic control. *J. Climate*, **15**:1216-1229.

Ramanathan V,Feng Y. 2009. Air pollution, greenhouse gases and climate change: global and regional perspectives. *Atmos. Environ.*, **43**:37-50.

Rees S L, Robinson A L, Khlystov A, *et al*. 2004. Mass balance closure and the Federal Reference Method for $PM_{2.5}$ in Pittsburgh, Pennsylvania. *Atmos. Environ.*, **38**:3305-3318.

Schwartz J, Dockery D W, Neas L M. 1996. Is daily mortality associated specifically with fine particles? *J. Air & Waste Manage. Assoc.*, **46**: 927-939.

Seinfeld J H. 1989. Urban air pollution: state of the science. *Science*,**243**:745-752.

Shao M, Tang X Y, Zhang Y H,*et al*. 2006. City clusters in China: air and surface water pollution. *Front.*

Ecol. Environ., **4**(7):353-361.

Shen Z X, Cao J J, Arimoto R, *et al.* 2010. Chemical characteristics of fine particles (PM_1) from Xi'an, China. *Aerosol Sci. Technol*, **44**:461-472.

Song S J, Wu Y, Jiang J K, *et al.* 2012. Chemical characteristics of size-resolved $PM_{2.5}$ at a roadside environment in Beijing, China. *Environ. Pollut.*, **161**: 215-221.

Song Y, Zhang Y H, Xie S D, *et al.* 2006. Source apportionment of $PM_{2.5}$ in Beijing by positive matrix factorization. *Atmos. Environ.*, **40**: 1526-1537.

Song Y, Tang X Y, Xie S D, *et al.* 2007. Source apportionment of $PM_{2.5}$ in Beijing in 2004. *J. Hazard. Mater.*, **146**: 124-130.

Song Y, Miao W J, Liu B, *et al.* 2008. Identifying anthropogenic and natural influences on extreme pollution of respirable suspended particulates in Beijing using backward trajectory analysis. *J. Hazard. Mater.*, **154**: 459-468.

Speer R E, Barnes H M, Brown R. 1997. An instrument for measuring the liquid water content of aerosols. *Aerosol Sci. and Technol.*, **27**:50-61.

Streets D G, Wu Y, Chin M. 2006. Two-decadal aerosol trends as a likely explanation of the global dimming/brightening transition. *Geophys. Res. Lett.*, **33**: L15806, doi:10.1029/2006GL026471.

Streets D G, Fu J S, Jang C J, *et al.* 2007. Air quality during the 2008 Beijing Olympic Games. *Atmos. Environ.*, **41**:480-492.

Sun J, Zhang M, Liu T. 2001. Spatial and temporal characteristics of dust storms in China and its surrounding regions, 1960—1999: Relations to source area and climate. *J. Geophys. Res.*, **106**: 10325-10333.

Sun Y, Zhuang G, Tang A, *et al.* 2006. Chemical characteristics of $PM_{2.5}$ and PM_{10} inhaze-Fog episodes in Beijing. *Environ. Sci. Technol.*, **40**: 3148-3155.

Takuwa T, Mkilaha I S N, Naruse I. 2006. Mechanisms of fine particulates formation with alkali metal compounds during coal combustion. *Fuel*, **85**: 671-678.

Tao J, Ho K F, Chen L, *et al.* 2009. Effect of chemical composition of $PM_{2.5}$ on visibilityin Guangzhou, China, 2007 spring. *Particuology*, **7**: 68-75.

Taylor S R. 1964. Trace element abundances and the chondritic Earth model. *Geochim. Cosmochim. Acta*, **28**:1989-1998.

Tian H, Gao J, Lu L, *et al.* 2012. Temporal trends and spatial variation characteristics of hazardousair pollutant emission inventory from municipal solid waste incineration in China. *Environ. Sci. Technol.*, **46**: 10364-10371.

Tsai Y I, Kuo S C. 2005. $PM_{2.5}$ aerosol water content and chemical composition in a metropolitan and a coastal area in southern Taiwan. *Atmos. Environ.*, **39**: 4827-4839.

Turpin B J, Lim H J. 2001. Species contributions to $PM_{2.5}$ mass concentrations: revisiting common assumptions for estimating organic mass. *Aerosol Sci. Technol.*, **35**: 602-610.

Viidanoja J, Sillanpaa M, Laakia J, *et al.* 2002. Organic and black carbon in $PM_{2.5}$ and PM_{10}: 1 year of data from an urban site in Helsinki, Finland. *Atmos. Environ.*, **36**: 3183-3193.

Wåhlin P, Berkowicz R, Palmgren F. 2006. Characterisation of traffic-generated particulate matter in Copenhagen. *Atmos. Environ.*, **40**:2151-2159.

Wang H L, Zhuang Y H, Wang Y, *et al.* 208. Long-term monitoring and source apportionment of $PM_{2.5}$/PM_{10} in Beijing, China. *J. Environ. Sci.*, **20**:1323-1327.

Wang Y Q, Zhang X Y, Arimoto R, *et al.* 2004. The transport pathways and sources of PM_{10} pollution in

Beijing during spring 2001, 2002 and 2003. *Geophys. Res. Lett.*, **31**: L14110, doi: 10.1029/2004GL019732.

Wang Y, Zhuang G S, Tang A H, *et al.* 2005. The ion chemistry and the source of $PM_{2.5}$ aerosol in Beijing. *Atmos. Environ.*, **39**: 3771-3784.

Watson J G, Chow J C. 1998. CMB8 Applications and Validation Protocol for $PM_{2.5}$ and VOCs, Desert Research Institute, No. 1808. 2D1.

Watson J G, Chow J C. 2001. $PM_{2.5}$ chemical source profiles for vehicular exhaust, vegetation burning, geological materials and coal burning in Northwestern Colorado during 1995. *Chemosphere*, **43**: 1141-1151.

Watson J G. 2003. Visibility: science and regulation. *J. Air & Waste Manage. Assoc.*, **52**: 628-713.

WHO. 2005. World Health Organization Air Quality Guidelines Global Update. E87950.

Widory D, Liu X D, Dong S P. 2010. Isotopes as tracers of sources of lead and strontium in aerosols (TSP & $PM_{2.5}$) in Beijing. *Atmos. Environ.*, **44**: 3679-3687.

Xia X A, Chen H B, Wang P C, *et al.* 2006. Variation of column-integrated aerosol properties in a Chinese urban region. *J. Geophys. Res.*, **111**: D05204, doi:10.1029/2005JD006203.

Xie S D, Liu Z. 2007. Source apportionment of inhalable particles in Beijing. In: Proceeding of research on the formation and control technology of particles from combustion sources, 179-195.

Xie S D, Liu Z, Chen T, *et al.* 2008. Spatiotemporal variations of ambient PM_{10} source contributions in Beijing in 2004 using positive matrix factorization. *Atmos. Chem. Phys.*, **8**: 2701-2716.

Xing L, Fu T M, Cao J J, *et al.* 2013. Seasonal and spatial variability of the organic matter-to-organic carbon mass ratios in Chinese urban organic aerosols and a first report of high correlations between aerosol oxalic acid and zinc. *Atmos. Chem. Phys. Discuss.*, **13**: 1247-1277.

Yan P, Zhang R J, Huan N, *et al.* 2012. Characteristics of aerosols and mass closure study at two WMO GAW regional background stations in eastern China. *Atmos. Environ.*, **60**: 121-131.

Yao Q, Li S Q, Xu H W, *et al.* 2009. Studies on formation and control of combustion particulate matter in China: a review. *Energy*, **34**:1296-1309, doi:10.1016/j.energy.2009.03.013.

Yao X H, Chan C K, Fang M, *et al.* 2002. The Water-soluble ionic composition of $PM_{2.5}$ in Shanghai and Beijing, China. *Atmos. Environ.*, **36**: 4223-4234。

Yao X, Lau A P S, Fang M, *et al.* 2003. Size distribution and formation of ionic species in atmospheric particulate pollutants in Beijing, China. *Atmos. Environ.*, **37**(21):2991-3000.

Yassaa N, Meklati B Y, Cecinato A, *et al.* 2001. Organic aerosols in urban and waste landfill of Algiers metropolitan area:occurrence and sources. *Environ. Sci. Technol.*, **35**(2):306-311.

Yu L D, Wang G F, Zhang R J, *et al.* 2013. Characterization and source apportionment of $PM_{2.5}$ in an urban environment in Beijing. *Aerosol Air Qual. Res.*, **13**:574-583.

Yu Y, Schleicher N, Norra S, *et al.* 2011. Dynamics and origin of $PM_{2.5}$ during a three-year sampling period in Beijing, China. *J. Environ. Monit.*, **13**:334-346.

Zhang D, Iwasaka Y. 1999. Nitrate and sulfate in individual Asian dust-storm particles in Beijing, China in spring of 1995 and 1996. *Atmos. Environ.*, **33**:3213-3223.

Zhang Q, He K, Huo H. 2012b. Cleaning China's air. *Nature*, **484**: 161-162.

Zhang R J, Wang M X, Zhang X Y, *et al.* 2003a. Analysis on the chemical and physical properties of particles in a dust storm in spring in Beijing. *Powder Technol.*, **137**(1): 77-82.

Zhang R J, Cao J J, Lee S C, *et al.* 2007. Carbonaceous aerosols in PM_{10} and pollution gases in winter in Beijing. *J. Environ. Sci.*, **19**(5):564-571.

Zhang R J, Tao J, Ho K F, *et al*. 2012a. Characterization of atmospheric organic carbon and elemental carbon of $PM_{2.5}$ in a typical semi-arid area of Northeastern China. *Aerosol Air Qual. Res.*, **12**:792-802, doi: 10.4209/aaqr.2011.07.0110.

Zhang X Y, Arimoto R, An Z S. 1997. Dust emission from Chinese desert sources linked to variations in atmospheric circulation. *J. Geophys. Res.*, **102**: 28041-28047.

Zhang X Y, Gong S L, Shen Z X, *et al*. 2003b. Characterization of soil dust aerosol in China and its transport anddistribution during 2001 ACE-Asia: 1. Network observations. *J. Geophys. Res.*, **108**: NO. D9, 4261, doi:10.1029/2002JD002632.

Zhang X Y, Wang Y Q, Zhang X C, *et al*. 2008. Carbonaceous aerosol composition over various regions of China during 2006. *J. Geophys. Res.*, **113**:D14111, doi:10.1029/2007JD009525.

Zhang X Y, Zhuang G S, Yuan H, *et al*. 2009a. Aerosol particles from dried salt-lakes and saline soils carried ondust storms over Beijing. *Terr. Atmos. Ocean. Sci.*, **20**(4): 619-628.

Zhang Q, Streets D G, Carmichael G R, *et al*. 2009b. Asian emissions in 2006 for the NASA INTEX-B mission. *Atmos. Chem. Phys.*, **9**: 5131-5153.

Zhang Y P, Wang X F, Chen H, *et al*. 2009c. Source apportionment of lead-containing aerosol particles in Shanghai using single particle mass spectrometry. *Chemosphere*, **74**: 501-507.

Zhao B, Wang S X, Xu J Y, *et al*. 2013. NO_x emissions in China: historical trendsand future perspectives. *Atmos. Chem. Phys. Discuss.*, **13**:16047-16112.

Zheng X Y, Liu X D, Zhao F H, *et al*. 2005. Seasonal characteristics of biomass burning contribution to Beijing aerosol. *Science in China* Ser. B, **48**(5): 481-488.

Zhou J M, Zhang R J, Cao J J, *et al*. 2012. Carbonaceous and ionic components of atmospheric fine particles in Beijing and their impact on atmospheric visibility. *Aerosol Air Qual. Res.*, **12**:492-502.

A STUDY OF DUST RADIATIVE FEEDBACK ON DUST CYCLE AND METEOROLOGY OVER EAST ASIA BY A COUPLED REGIONAL CLIMATE-AEROSOL MODEL①

HAN Zhiwei[1), LI Jiawei[1), GUO Weidong[2), ZHE Xiong[1), WU Zhang[3)

[1) Key Laboratory of Regional Climate-Environment for Temperate East Asia (RCE-TEA), Institute of Atmospheric Physics (IAP), Chinese Academy of Sciences (CAS), Beijing 100029, China;

[2) School of the Environment, Nanjing University, Nanjing 210093, China

[3) College of Atmospheric Sciences, Lanzhou University, Lanzhou 730000, China

Abstract

An online coupled regional climate-chemistry-aerosol model was utilized to investigate the dust direct radiative feedbacks on dust deflation, transport and meteorological elements in March 2010, when a severe dust storm originated from the Gobi desert near the China-Mongolia border swept across most areas of East Asia during the period of 19 — 22 March. The predicted meteorology and aerosol concentration agree generally well with observations, and it clearly shows that the predictions of both dust concentration and meteorological elements with dust radiative feedback are closer to observation than that without feedback, suggesting the superiority of on-line coupled model in chemical and climate predictions. The direct radiative forcing by dust aerosol caused significant reductions in ground temperature and wind speed in dust deflation region, with maximums up to −7℃ and −4.0 $m \cdot s^{-1}$, respectively. The reduced wind speed and increased atmospheric stability resulted in smaller dust emission and weakened vertical diffusion, and consequently less dust aerosol in upper levels transported further downwind. While the shortwave radiative forcing dominated over longwave forcing in the daytime, leading to decreases of surface air temperature and wind speed, in the nighttime, the warming effect of longwave forcing dominated, causing increases of air temperature and wind speed up to 1℃ and 1 $m \cdot s^{-1}$, respectively, in the dust deflation region. The variation of meteorology during the dust storm period at a rural site (Yuzhong) downwind of the Gobi desert exhibited an evident decrease in surface air temperature due to the dust radiative forcing. In terms of monthly mean, the dust radiative forcing caused a surface cooling of −0.6 — −1.0℃ over wide areas from west China to northern parts of east China and the Korean peninsula, with maximums in the middle reaches of the Yellow River and portions of northeast China. Concurrently, precipitation decreased by 0.1—0.6 $mm \cdot d^{-1}$ in the middle reaches of the Yangtze River and large areas of north China, and alternating bands of increasing and decreasing precipitation (~1 $mm \cdot d^{-1}$) occurred in south China.

Key words: Mineral dust, meteorological elements, radiative effects, cooling, warming, dust-climate interaction.

---

① The paper published in *Atmospheric Environment*. **68**:54-63. 2013

1 Introduction

Dust aerosol is one of the most important aerosol components because it plays important roles in human health, air quality and climate change (Overpeck *et al.*, 1996; Tegen *et al.*, 1996; Alpert *et al.*, 1998). Dust can affect climate by scattering and absorbing solar / thermal radiation (direct effect) (Miller and Tegen, 1998) and by altering cloud property and lifetime (indirect effect) (Sassen, 2002), or by heating the atmosphere/evaporating cloud (semi-direct effect) (Helmert *et al.*, 2007). The radiative and climatic effects of dust aerosol are still uncertain due to limited observational evidence of its properties and complex nonlinear interplay between dust and climate. The fourth assessment report from the Intergovernmental Panel on Climate Change (IPCC, 2007) indicates that the global annual mean radiative forcing (RF) of mineral dust is estimated as $-0.1\ W \cdot m^{-2}$, with an uncertainty ranging from -0.3 to $+0.1\ W \cdot m^{-2}$, indicating the dust RF estimated from current methods can be either positive or negative. The climatic effect of dust aerosol is even lesser known and needs more studies.

Asia is one of the continents emitting large amounts of dust aerosols. There are two major dust sources in East Asia, one is the Taklimakan desert in west China and one is the Gobi desert across south Mongolia and north China. Dust storms often occur under certain meteorological conditions in springtime and have been observed to exert significant impacts on environment and climate in this region. In the past two decades, there have been numerous studies on the properties, sources, transport and direct radiative forcings of Asian dust aerosols (Zhang *et al.*, 2003; Gong *et al.*, 2003; Han *et al.*, 2004; Park *et al.*, 2005; Shao and Dong, 2006; Han *et al.*, 2012), but the effect of dust radiative forcing on climate is rarely explored. So far, only a few studies investigated the climatic impact of East Asian dust with regional climate or/weather models, such as Ahn *et al.* (2007) simulated the effect of direct radiative forcing of Asian dust on meteorological fields during a dust event on 18—23 March 2002 with a coupled MM5/Dust model. Wang *et al.* (2010) numerically studied the direct radiative feedback of dust aerosols on dust storm in April 2006 by using a mesoscale dust storm forecasting model. But the above two studies did not consider dust longwave radiative forcing, which has been found to be climatic important in previous global modeling and observation studies (Cautenet *et al.*, 1991; Ackerman and Chung, 1992; Zhang and Christopher, 2003; Xia and Zong, 2009; Brindley and Russell, 2009; Yue *et al.*, 2010). Zhang *et al.* (2009) used RegCM3 coupling with a desert dust aerosol model to simulate the dust radiative feedback on dust load and surface air temperature over East Asia for a multi-year period (1997—2006).

This paper utilized an online coupled regional climate-chemistry-aerosol model (namely RIEMS-Chemaero) to investigate the direct radiative feedback of dust aerosol on dust cycle and meteorological elements over East Asia, with focus on a severe dust storm on 19—22

March 2010, which originated from the Gobi desert near the China-Mongolia border. This study is an extension of Han *et al.* (2012), in which we described in detail the RIEMS-Chemaero, conducted model validation against ground observations of PM_{10} concentration and aerosol optical depth (AOD) measured at AERONET sites and retrieved from MODIS (MODerate-resolution Imaging Spectroradiometer), and simulated the direct aerosol radiative forcings (RF) including shortwave (SW), longwave (LW) and net RF during the dust storm period and in March 2010.

This paper is organized as follows: we first briefly describe RIEMS-Chemaero and input parameters, then compare the predicted PM_{10} concentration and meteorological elements with observations. Next we numerically investigate the effect of the dust direct radiative forcings (including both shortwave, longwave and net) on meteorological elements near the surface and in the troposphere and how the meteorological changes further affect dust deflation, transport and concentration. We also analyze the variation of surface meteorology during the dust storm period together with ground observations at a rural site of north China. Finally, we make statistical analysis of model predictions with observations and present monthly mean changes in surface air temperature and precipitation induced by the dust direct radiative forcings.

2 Model Description and Experiments

RIEMS(Regional Integrated Environmental Model System) was developed based on the dynamic structure of the fifth-generation Pennsylvania State University-NCAR Mesoscale Model (MM5; Grell *et al.*, 1995), applying a nonhydrostatic, sigma vertical and lambert horizontal coordinate. It contains a number of parameterizations or modules to represent major physical processes, such as land surface, planetary boundary layer, cumulus convection etc. RIEMS has ever been used to predict East Asian climate and to investigatethe impact of human-induced land cover change on East Asia monsoon (Fu, 2003). RIEMS is also one of the key models in the Regional Climate Model Intercomparison Project (RMIP) for Asia, details on structure and physics of RIEMS refer to Fu *et al.* (2005).

Major aerosol-related processes have recently been incorporated into RIEMS (Han, 2010; Han *et al.*, 2012), which includes emission, transport, diffusion, deposition and multi-phase aerosol chemistry with CB-IV (Carbon Bond Version 4) gas chemistry mechanism and ISORROPIA ("equilibrium" in Greek) thermodynamic chemistry. Dust deflation is treated by an empirical model (Han *et al.*, 2004), depending on several controlling factors, such as threshold friction velocity and relative humidity and reduction factor of vegetation. The total dust emission flux is apportioned to each size bin based on observation of size-resolved vertical dust flux from Chinese deserts during the Aerosol Characterization Experiment—Asia. The schemes for dry deposition and below-cloud scavenging of dust aerosol are described in detail in Han *et al.* (2004).

We used Mie code to calculate dust optical properties (mass extinction coefficient, single scattering albedo, and asymmetry factor) based on refractive index of mineral dust retrieved from dust samples of Chinese deserts (Wang *et al.*, 2004). A modified radiation package based on the radiation package of the NCAR Community Climate Model, version CCM3 (Kiehl *et al.*, 1996) is used to calculate aerosol perturbation to radiation transfer. The solar spectrum is divided into eighteen spectral intervals, spanning the range of 0.2 to 5.0 μm and a delta-Eddington approximation is used to calculate solar fluxes. For longwave radiation, dust transmission is assumed to follow an exponential decay law as described in Carlson and Benjamin(1980). More details on the treatments of aerosol optical properties and radiation calculation refer to Han *et al.* (2012).

The reanalysis data, four times a day with $1°\times1°$ resolution are derived from National Centers for Environmental Prediction (NCEP) to provide initial and lateral boundary conditions for RIEMS (wind, air temperature, water vapor, geopotential height, sea-level pressure, sea-surface temperature etc.), with a horizontal resolutionof 60 km, and 16 levels from the surface to 100 hPa. The simulation period is from 00 UTC 24 February to 18 UTC 31 March 2010, with the first 5 days as initialization. The study domain covers most parts of East Asia (75°—145°E, 15°—55°N).

To explore the dust radiative effect/feedback, we conducted three numerical experiments. Case 0 (base case) and Case 1 include all processes of dust aerosol without and with radiative feedback, respectively. Case 2 is the same as Case 1 except Case 2 includes only shortwave radiative feedback. The difference between Case 1 and Case 0 represents the changes in dust emission and meteorological elements induced by the net dust radiative forcing, whereas the difference between Case 0 and Case 2 reflects the sole effect of dust solar radiative forcing.

3 Results

3.1 Model Evaluation

Xiong *et al.* (2006) used ground observations to evaluate the ability of RIEMS to predict Asian climate during 1988—1998. They found the seasonal mean temperature biases were within 1－2℃ and the seasonal precipitation biases ranged from －12% to －50%, with better predictions in winter than in summer. Wu and Han (2011) reported that RIEMS predicted cloud cover generally agreed with results from ECMWF and MODIS in terms of spatial distribution and seasonal variation, but the predictions from RIEMS and ECMWF were both lower than that from MODIS, RIEMS and ECWMF exhibited similar abilities in predicting cloud cover.

Figure 1 presents the comparison of RIEMS predictions (from Case 1) to NCEP reanalysis data (just providing initial and boundary conditions to RIEMS) and precipitation from TRMM (Tropical Rainfall Measuring Mission) product 3B43 (http://disc.sci.gsfc.nasa.gov/precipitation/tovas/), which are monthly based with 0.25° grid resolution.

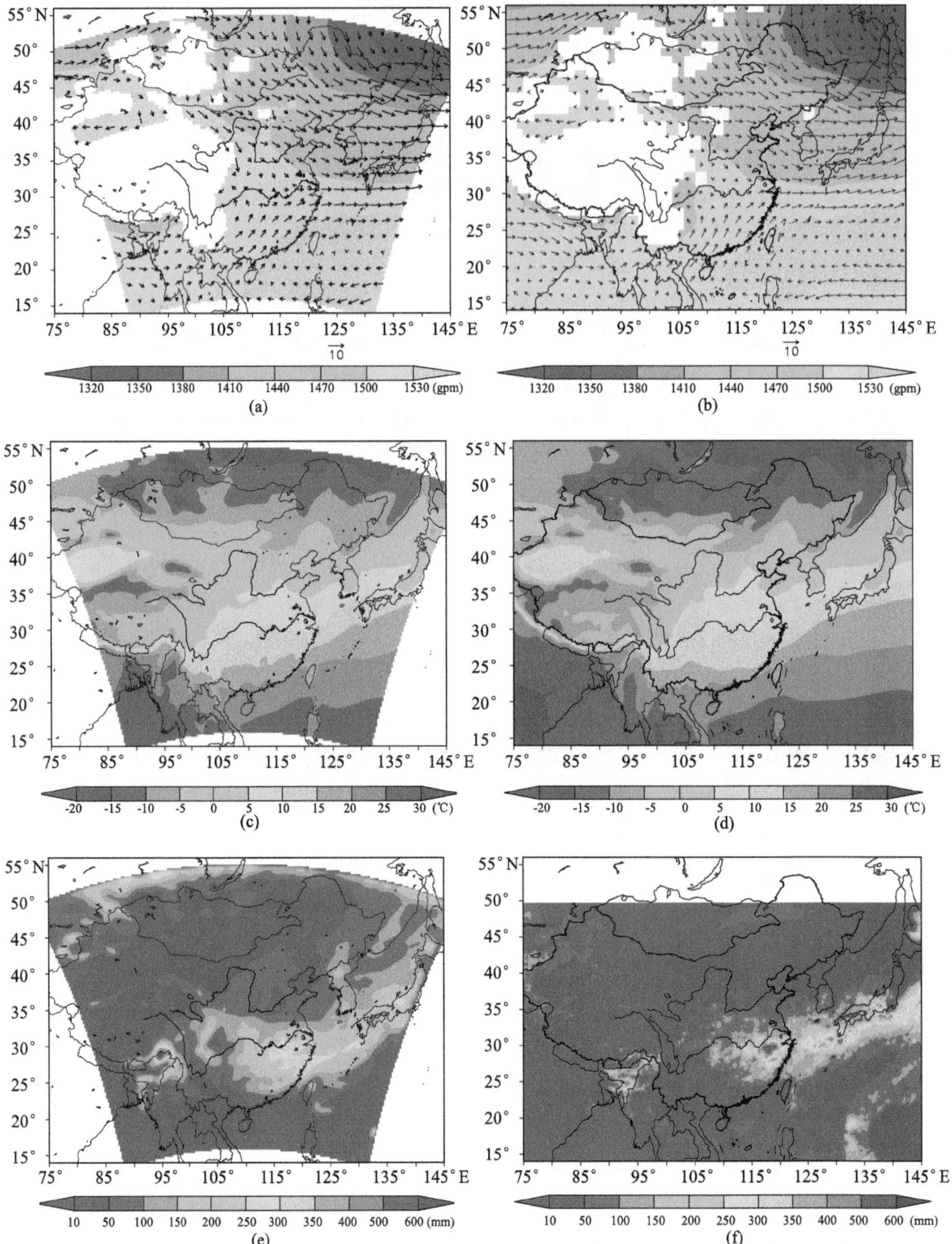

Fig. 1 Monthly mean geopotential height and wind at 850 hPa from (a) model prediction, (b) NCEP reanalysis. Air temperature at 2 m from (c) model prediction, (d) NCEP reanalysis. Monthly accumulated precipitation (mm) from (e) model prediction, (f) TRMM.

It shows that RIEMS predicts reasonably well the distribution pattern of a deep low trough over northeast Asia and a subtropical high situated over the west Pacific, which results in northwesterly winds in north China and southwesterly winds in south China. The model predicted subtropical high extends somewhat westward compared with the reanalysis data. The prediction of air temperature at 2 m (T2m) is in a good agreement with the NCEP data in the domain except a portion of the Taklimakan desert and parts of the Huabei Plain. Both the model prediction and the TRMM data exhibit a distinct distribution feature of precipitation, extending from southeast China, to East China Sea and to the sea just south of Japan, with maximums in the lower reaches of the Yangtze River. The model predicted domain averaged precipitation is 48. 8 mm (per grid), about 20% higher than the satellite retrieval (40. 7mm), but the predicted maximum in parts of southeast China is about 30% ~ 40% lower than the satellite retrieval. Precipitation is difficult to be accurately predicted by current weather/climate models, which could be attributed to uncertainties in parameterizations of a series of physical processes. The discrepancy could be also related to the different grid resolution between RIEMS and the satellite retrieval and potential inaccuracy in the satellite data.

Giorgi *et al*. (2002) used a regional climate model (RegCM) to predict Asia climate, and found the model tended to overpredict summer monsoon rainfall over east China (by about 15%~30%), where as in cold season, both positive and negative bias within ±25% of observations are found. Zhang *et al*. (2010) evaluated WRF/Chem prediction of meteorology over the continental U. S. in January and July 2001, precipitations were overpredicted in both months with normalized mean biases of 21. 2% ~ 32. 3% and are comparable to those from the fifth-generation mesoscale model (MM5) applications. Overall, RIEMS is capable in predicting major meteorological features over East Asia, and the model skill appears to be in a similar level to those of current regional weather/climate models.

Figure 2 shows the model predicted (from Case 0 and Case 1) and the observed daily mean PM_{10} concentrations at Lanzhou (the capital city of Gansu Province), which is about 600 km southeast of the Gobi desert and at Beijing, which is about 1200 km downwind of the Gobi desert. The observed PM_{10} concentration at Lanzhou is derived from Air Pollution Index (API) reported by the Ministry of Environmental Protection of China, the method for transforming API to PM_{10} concentration refers to Li *et al*. (2011). The observation at Beijing is from a rural monitoring site (39. 8°N, 116. 47°E), about 12 km southeast toward Beijing downtown, with the sampling height of 31. 3 m above sea level. The predicted PM_{10} concentration just includes dust aerosol.

On 19 March, a strong dust storm outbroke in the Gobi desert of north China, and then transported eastward and southeastward. It reached as far as the Pearl River Delta of south China on 22 March. Figure 2 clearly shows that RIEMS-Chemaero captures the arrival of the dust storm on 20 March at Lanzhou and Beijing. It is noteworthy that the

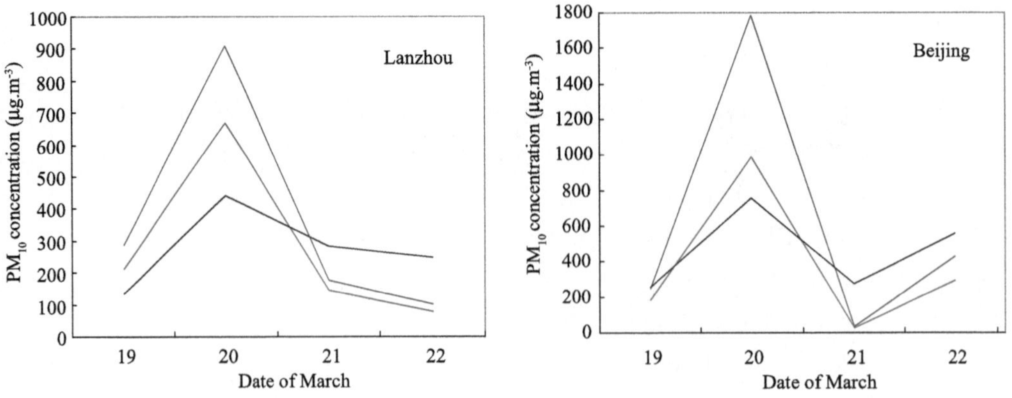

Fig. 2 Model predicted daily mean PM_{10} concentrations from Case 1 (red) and Case 0 (blue) and observation (black) at Lanzhou and Beijing (unit: $\mu g \cdot m^{-3}$)

peak values from Case 0 (909 $\mu g \cdot m^{-3}$ and 1783 $\mu g \cdot m^{-3}$ at Lanzhou and Beijing, respectively, without dust radiative feedback) are higher than that from Case 1 (670 $\mu g \cdot m^{-3}$, 987 $\mu g \cdot m^{-3}$) and observations (442 $\mu g \cdot m^{-3}$, 754 $\mu g \cdot m^{-3}$), and the predictions from Case 1 are apparently closer to observations. The underprediction of PM_{10} in non-dust days (22 March) at the two sites is mainly due to the exclusion of anthropogenic aerosols, which is out the scope of this study.

3.2 The Dust Radiative Feedbacks during the Dust Storm Period

3.2.1 Interaction between Dust and Meteorology in the Gobi Desert

In the Gobi desert where the dust storm originated, dust radiative forcing exerted large impacts on meteorological elements. Figure 3 exhibits the net dust direct radiative forcing (sum of shortwave and longwave, from Case 1 with feedback activated) at the surface and at TOA under clear-sky condition at 13:00 LST on 19 March when a strong dust deflation occurred in the Gobi desert near the China-Mongolia border. AOD at this time exceeded 5.0 and the instantaneous dust induced net radiative forcings at the surface and at TOA were up to $-480\ W \cdot m^{-2}$ (of which SW RF~$-570\ W \cdot m^{-2}$, LW RF ~ $+90\ W \cdot m^{-2}$) and $+80\ W \cdot m^{-2}$ (of which SW RF~$+65\ W \cdot m^{-2}$, LW RF~$+15\ W \cdot m^{-2}$), respectively, indicating a radiative cooling at the surface and a radiative energy gain in the column below TOA.

Figure 4 presents the changes in ground temperature and surface wind speed caused by the dust net radiative forcing, which are derived from the difference between Case 0 and Case 1 (Case 1−Case 0). It clearly shows the large decreases both in ground (and surface air) temperature and wind speed, with the values up to −7℃ (−4℃ for T2m, figure not shown) and $-4.0\ m \cdot s^{-1}$, respectively in the dust deflation area. Wang *et al*. (2010) simulated a maximum decrease in surface air temperature exceeding 5℃ in the Gobi desert during a typical dust storm on 16−18 April 2006; but they could overestimate the extent of

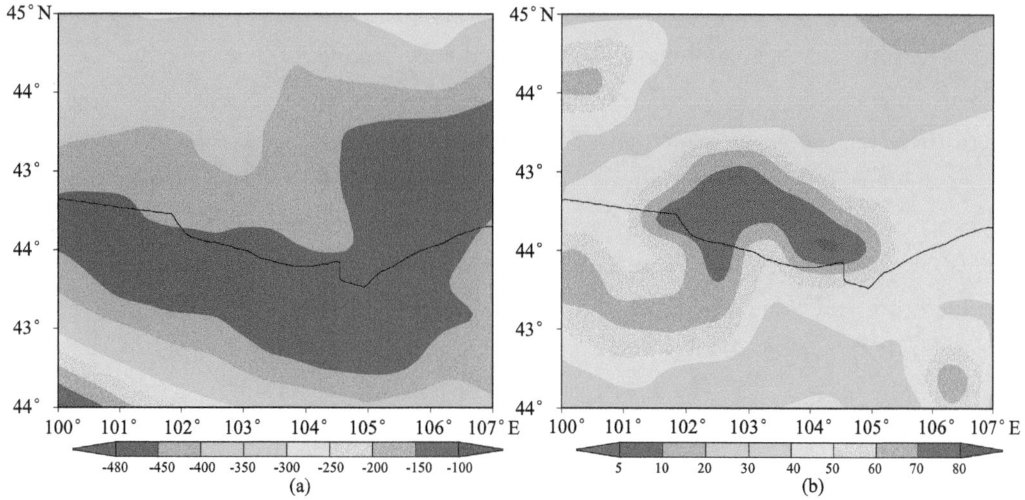

Fig. 3　The model predicted net dust radiative forcings (a) at the surface, (b) at TOA under clear-sky condition at 13:00 LST on 19 March (unit: W · m^{-2})

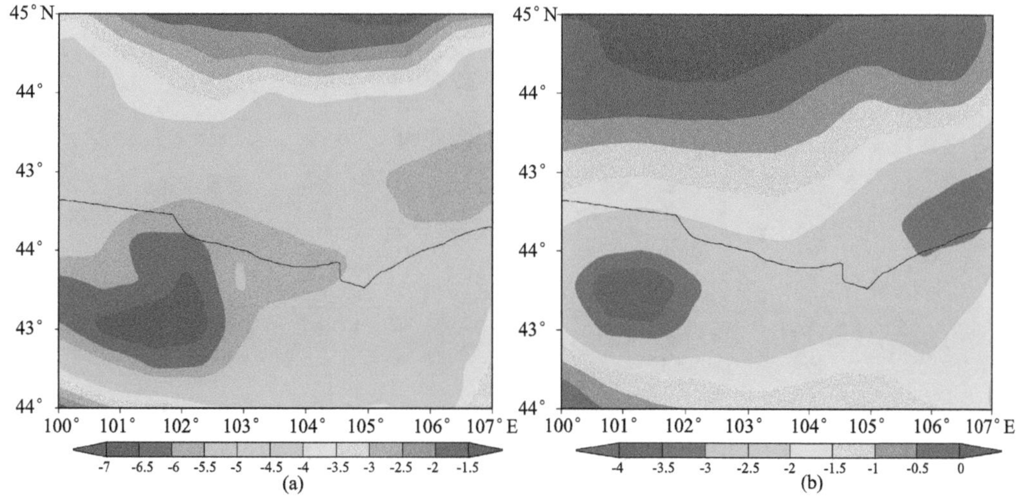

Fig. 4　Changes in (a) ground temperature (℃), and (b) surface wind speed (m · s^{-1}) due to the net dust forcing in the Gobi desert at 13:00 LST on 19 March

surface cooling because longwave radiative forcing and warming effect by dust aerosols were not considered in their study. The change in surface water vapor mixing ratio was relatively smaller than temperature and wind speed and had little effect on dust deflation. On one hand, dust absorbs and scatters sunlight, leading to a reduction in incident solar radiation acting toward cooling at the surface; on the other hand, dust absorbs and emits thermal radiation, exerting a warming effect. In the daytime, the cooling effect by the dust solar radiative forcing dominated over the warming effect by the longwave forcing, resulting in a net decrease in ground temperature. As a result of surface cooling, the surface sensible heat flux and turbulent energy in the planetary boundary layer (PBL) decreased, leading to a reduction of downward transport of momentum from upper layer above PBL to the surface, and consequently wind speed reduction near the surface.

When the ground surface was cooling, the middle troposphere of dust layer was heated by both the solar and longwave radiative forcings, causing a strong gradient of temperature between the surface and the middle troposphere. Figure 5 shows the vertical profiles of changes in air temperature and wind speed above ground at a location (101. 5°E, 41°N) of the Gobi desert (Case 1 − Case 0). It clearly shows that the dust forcing decreased air temperature and wind speed of the lower troposphere (below 4 km) by ~ 1. 5℃ and ~2. 8 m • s^{-1}, respectively, increased air temperature and wind speed of the middle troposphere by up to 1. 4 ℃ and 2. 6 m • s^{-1} at altitudes of ~3—4 km, respectively, where as in the upper troposphere, the meteorological elements changed little.

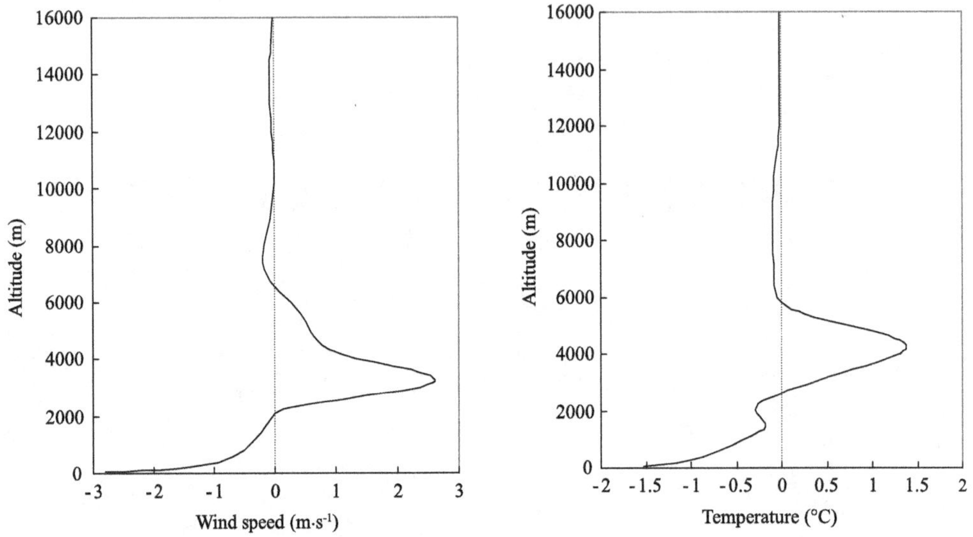

Fig. 5 The vertical profile (altitude above ground) of changes in (a) wind speed, (b) air temperature, at a location (101. 5°E, 41°N) of the Gobi desert at 13:00 on 19 March

In response to the surface cooling and wind speed decreasing, on one hand, dust deflation reduced, leading to a smaller dust emission in the source region, on the other hand, the cooling at the surface and warming at upper layer strengthens the atmospheric stability, causing less dust aerosol being uplifted into higher levels and less dust aerosol transported further downwind. This explains the lower dust concentration in Case 1 than that in Case 0 at Lanzhou and Beijing (Figure 2). Figure 6 shows the differences between Case 0 and Case 1 (Case 1—Case 0) in dust emission flux, surface dust concentration and AOD. It's striking that the online simulation with dust radiative feedback (Case 1) consistently yields lower dust emission, dust concentration, AOD and thus radiative forcings (not shown) than that without feedback (Case 0), with the differences in the emission flux, dust concentration and AOD up to 2400 μg • m^{-2} • s^{-1}, 14 mg • m^{-3} and 2. 4, respectively in the areas of dust deflation.

The above results for Asian dust radiative effect are consistent with previous global model investigations by Perlwitz *et al.* (2001) and Miller *et al.* (2004), in which they found the radiative forcing by dust generally causes a decrease of surface wind speed, dust emission and atmospheric dust load.

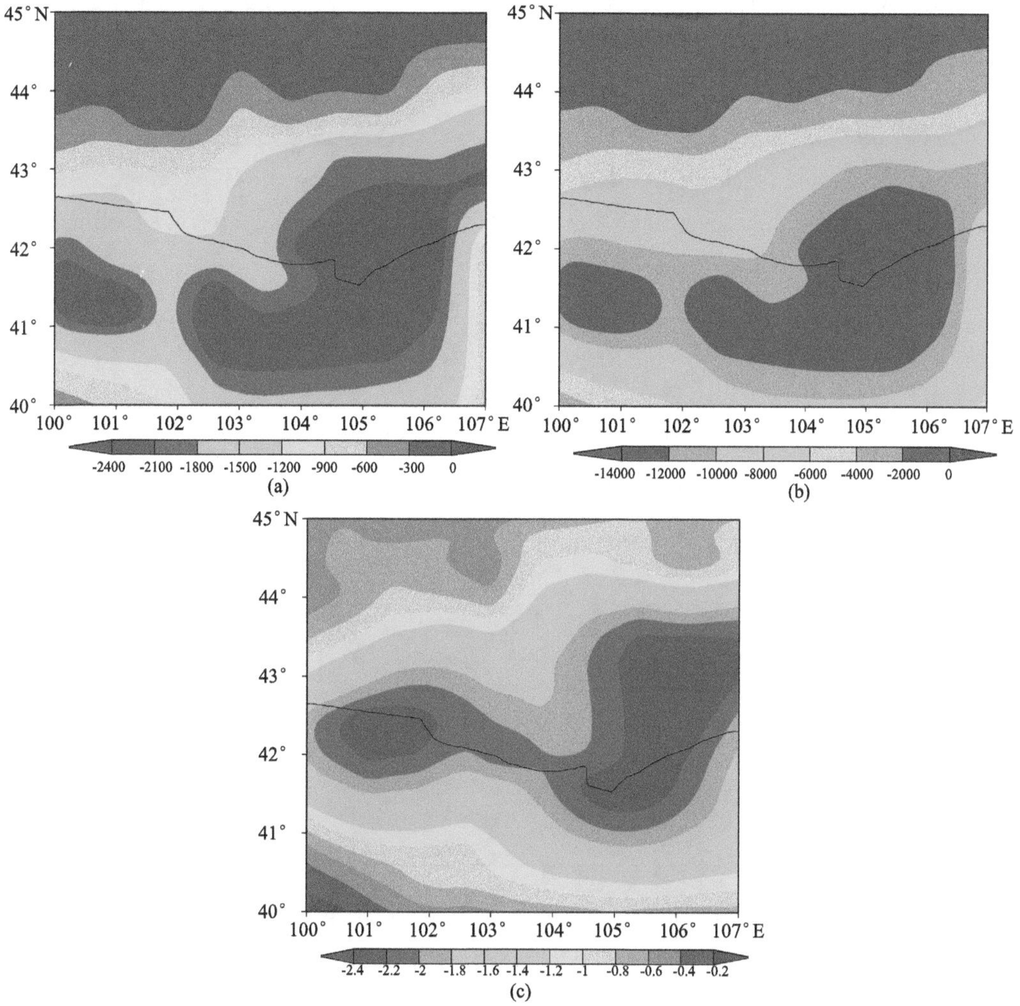

Fig. 6　Changes in (a) dust emission flux ($\mu g \cdot m^{-2} \cdot s^{-1}$) (b) dust concentration of the 1st model layer ($\mu g \cdot m^{-3}$) and (c) AOD at 13:00 LST on 19 March

3.2.2　Diurnal Variation of the Dust SW and LW Radiative Effect in the Gobi Desert

The contrast between Case 1 − Case 0 and Case 2 − Case 0 (not shown) reveals the important effect of dust longwave radiative forcing on meteorological elements. Dust solar radiative forcing led to larger decreases (than that from the dust net radiative forcing, Case 1 − Case 0) in ground temperature and surface wind speed by up to −8℃ and $-6\ m \cdot s^{-1}$ (Figure not shown), respectively, and a smaller dust emission and radiative forcings in the dust deflation region. Figure 7 illustrates the relative magnitude and effect of SW and LW dust radiative forcings at a surface location (101.5°E, 41°N) of the Gobi desert on 19

March. It shows that the SW forcing can reach $-570\ W \cdot m^{-2}$ in the afternoon, about 5 times the maximum LW forcing, but in the nighttime, the SW forcing was zero, whereas the LW forcing was in a range of $\sim 60\ W \cdot m^{-2}$. The decreasing LW forcing from afternoon to early morning next day was due to the weakening dust storm. It's noteworthy that the SW forcing dominated over the LW forcing in the daytime, leading to apparent decreases in both the surface air temperature and wind speed (about 2℃ and $5\ m \cdot s^{-1}$), but in the nighttime, the warming effect of LW forcing dominated, causing increases of air temperature and wind speed up to 1℃ and $1\ m \cdot s^{-1}$, respectively. This demonstrates that the dust LW radiative forcing can mitigate the cooling effect by the SW forcing, thus play an important role in dust cycle and aerosol climatic effect, which is in agreement with previous studies (Heinold *et al.*, 2008; Hansell *et al.*, 2010).

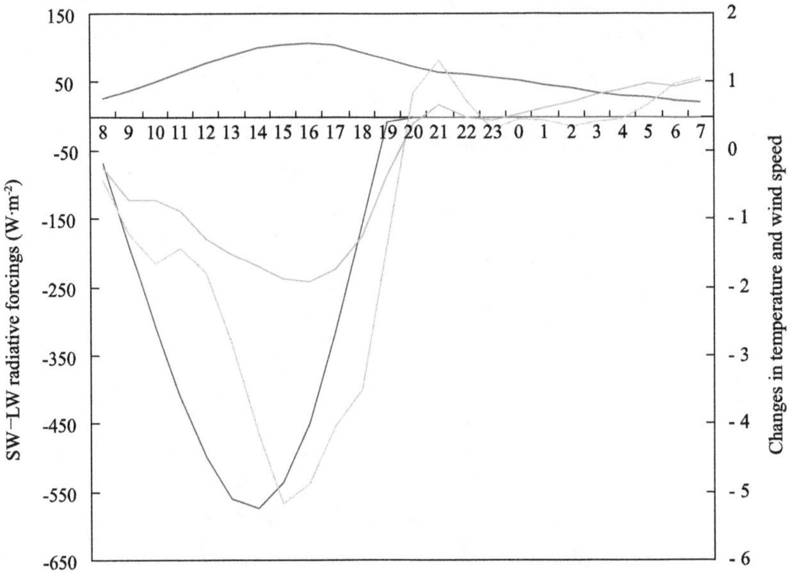

Fig. 7 Diurnal variation of the surface SW (blue) and LW(red) radiative forcings, and of the dust induced changes in near surface air temperature (℃) (green) and wind speed ($m \cdot s^{-1}$) (yellow) at a point (101.5°E, 41°N) of the Gobi desert during the period of 00:00—23:00 UTC on 19 March (x—axis denotes local time on 19—20 March)

3.2.3 The Radiative Effect of the Dust Storm on Meteorological Elements at Yuzhong Site

Section 3.2.2 investigated the dust radiative effect in the Gobi desert, this section discusses the dust effect in downwind regions. Figure 8 shows the model predicted and observed T2m and wind speed at 10 m (WS10) at a rural site Yuzhong (35°57′N, 104°08′E, an altitude of 1966m above sea level), which is the location of the Semi-Arid Climate and Environment Observatory of Lanzhou University (SACOL) and situated 48km southeast of the Lanzhou city, the topography around the site is characterized by Loess Plateau. Details about SACOL refer to Huang *et al.* (2008) and the website (http://climate.lzu.edu.cn/english/index.asp). This dust storm was originated from the Gobi desert on 19 March, reached this site in the evening of 19 March, and peaked at about 06:00 on 20 March, and then the strength of the dust storm decreased on 21 March.

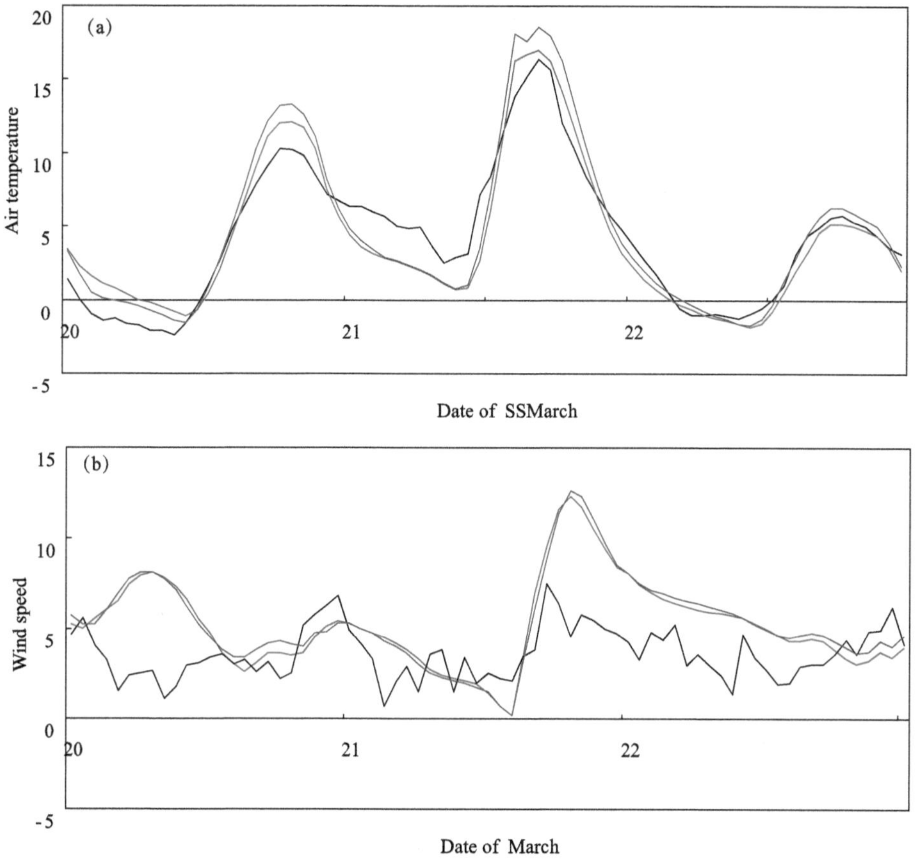

Fig. 8 The model predicted (a) near surface air temperature (℃) and (b) wind speed ($m \cdot s^{-1}$) at Yuzhong site during the dust event (black line denotes observation, red line denotes prediction from Case 1, and blue line denotes prediction from Case 0)

The predicted T2m variation agrees with observation quite well, showing a decrease of T2m on 20 March due to the dust direct radiative forcing during the dust storm, an increase of T2m on 21 March due to the weakened dust storm (Figure 2) and an abrupt T2m decrease on 22 March, which could result from the invasion of cold air masses.

It is noteworthy that the predicted daily maximum T2m in Case 1 (considering dust radiative feedback) is lower than that in Case 0 on both 20 and 21 March, which is mainly due to the reduction of solar radiation reaching the surface. The predictions in Case 1 (maximum T2m are 12.0℃, 17℃ on 20 and 21 March, respectively) are apparently closer to observation (10.5℃, 16.4℃) than that in Case 0 (13.4℃, 18.5℃). On 22 March, the difference between Case 0 and Case 1 is small in the non-dust day and both close to observation.

The model generally predicts the variation trend of wind speed, but the prediction is not as good as that for air temperature. This is because wind is also perturbed by local inhomogeneous topography and geography, which could not be accurately represented in the regional model with a grid resolution of 60 km. It shows the model overpredicts the maximum WS10 by about 60% in the late afternoon on March 21 and the prediction in the morning of 20 March is also higher than observation, which could be associated with the unrealistic representation of morning boundary layer evolution. Han *et al.* (2008) found that all the PBL schemes in MM5 overpredict PBL height in early morning, which may lead to stronger downward transport of momentum from upper layers and consequently larger wind speed near the surface, which could partly explain the overprediction of wind speed at Yuzhong site.

Figure 8 also shows that the difference in wind speed between Case 0 and Case 1 is much smaller than air temperature; this is because surface air temperature is affected by land-air energy exchange, which responds rapidly to dust radiative forcing wherever dust is present, whereas wind is mainly driven by large scale pressure gradient. Even so, it can be distinguished that the wind speed in Case 1 is generally smaller than that in Case 0, consistent with the situation in the Gobi desert.

3.2.4 Statistical Analysis of Model Predictions from the Two Cases with Observations

To evaluate the overall dust radiative feedback on model prediction, we compared the predicted T2m from case 0 and case 1 with NCEP reanalysis data (from assimilation of various observations with quality control), which just provides initial and lateral boundary conditions for RIEMS-Chemaero. The T2m (at 00:00, 06:00, 12:00, 18:00 UTC, 4 times per day) is extracted from NCEP reanalysis dataset and bilinearly interpolated to RIEMS projection for comparison. The total pairs of data are 807488 for the whole domain for March 2010 (88×74×4 times/day×31days, 88 and 74 are numbers of model grid in x-axis and y-axis, respectively)

Some traditional statistical measures, so-called, correlation coefficient(R), mean bias error (MBE), root mean square error (RMSE), normalized mean bias (NMB) are calculated and analyzed to help evaluate and interpret model predictions. Because this study focused on the continent including the Gobi desert and its downwind areas, we equally divided east China into two regions (the medium line is nearly along the Yangtze River), one is north China (100°—123°E, 32.5°—45°N), another one is south China (100°—123°E, 20°—32.5°N). Table 1 presents the statistics for the predicted and NCEP T2m for the three specific regions for March 2010.

Table 1 shows that for the entire domain, the model predicted T2m in Case 0 generally exhibits a slight warm bias of 0.18℃ (with NMB of 3%), while considering dust radiative feedback in the model, the mean bias reduced to 0.08℃ (NMB of 1%), and this tendency is more evident in north China, where the model predicted a warm bias of 0.38℃ (with NMB up to 97%) in Case 0, whereas in Case 1, the mean bias reduced to near zero. The small difference in T2m between Case 0 and Case 1 indicates that the dust radiative effect in south China is apparently weaker than that in north China due to the decrease in atmospheric dust load along the dust transport pathway, and the overall model predictions in the two cases are close.

Table 1 Statistics for T2m in specific regions for March 2010

Regions	Whole domain		North China		South China	
T2m−obs. (℃)	6.71		0.38		14.5	
Cases	Case 0	Case 1	Case 0	Case 1	Case 0	Case 1
T2m −model (℃)	6.88	6.79	0.75	0.37	14.4	14.3
MBE	0.17	0.08	0.37	−0.01	−0.15	−0.20
R	0.97	0.97	0.91	0.91	0.89	0.89
RMSE	3.51	3.50	3.30	3.29	3.60	3.60
NMB	0.03	0.01	0.97	−0.02	−0.01	−0.01

Figure 1e and 1f have shown the model prediction from Case 1 and TRMM retrieval for monthly accumulated precipitation in March 2010, which are generally consistent in distribution. TRMM data are bilinearly interpolated to RIEMS projection and the total pairs of data are 6512 (88×74) for statistical analysis. The domain averaged precipitation in Case 0 are 49.7mm, 39.2 and 91.3 mm for the whole domain, north China and south China, respectively, whereas in Case 1, the corresponding values are 48.8mm, 36.0mm and 89.2 mm, and the satellite retrievals are 40.7mm, 25.7mm and 70.0mm for the three regions, respectively. This indicates that the model generally overpredicted precipitation by 20% ~ 22% in east Asia, but the dust radiative effect (Case 1) tends to reduce precipitation and to decrease the model predicted precipitation biases in all the three regions, with the largest changes of MBE(NMB) in north China, from 13.5mm (53%) in Case 0 to 10.3mm (40%) in Case 1, and the biases for the whole domain change from 9.0 mm (22%) in Case 0 to 8.1 mm (20%) in Case 1.

The above statistics is consistent with the analysis at SACOL site during the dust storm period, demonstrating the inclusion of dust feedback effect is capable of reducing biases in climate prediction and improve the reasonability of climate model.

It should be mentioned that the causes of biases in climate model prediction could result from a variety of uncertainties in parameterizations of physical processes, such as land surface, cloud microphysics, boundary layer turbulence, as well as aerosol-climate interaction. Introducing aerosol processes in climate model can't solve all existing problems in climate prediction (overprediction of precipitation by 20% even considering aerosol's effect), but this study reveals the importance and advantage to consider aerosol effect in climate model prediction.

3.3 Monthly Mean Direct Climatic Effect of Dust Aerosol

Figure 9 shows the monthly mean changes in surface air temperature and precipitation induced by the net dust direct radiative forcing in the study domain. Dashed white line indicates the change is statistically significant at a 90% confidence level according to a two-tailed t-test. It is apparent that the dust radiative forcing mainly caused a surface cooling over wide areas from west China to the northern parts of east China and the Korean peninsula, with a maximum cooling of −0.6℃ to −1.0℃ in the middle reaches of the Yellow River and portions of northeast China. The regions of cooling with statistical significance included a part of the Taklimakan desert and large areas of east China north of the Yangtze River. The frequent dust deflation or passage made surface cooling more effective in these regions. The cooling in the above regions resulted from a combined effect of the dust shortwave and longwave forcings, which affected the surface temperature effectively in the daytime and nighttime, respectively. However, the increase of surface air temperature in the nighttime can not offset the large air temperature decrease in the daytime, leading to an overall decrease in air temperature. The surface temperature change was also a result of the competing effects of the direct radiative forcing and the semi-direct effect of dust aerosol. Han *et al.* (2012) found that the lower and middle cloud cover reduced in large portions of the middle and lower reaches of the Yellow River and Yangtze River, which were attributed to a combined effect of the dust direct radiative forcing (ground evaporation decrease and atmospheric stability increase due to surface cooling) and the semi-direct effect (heating of the troposphere of dust layer and cloud evaporation), resulting in more solar radiation reaching the ground surface, and consequently surface temperature increase. Perlwitz and Miller (2010) recently found an increase in low cloud

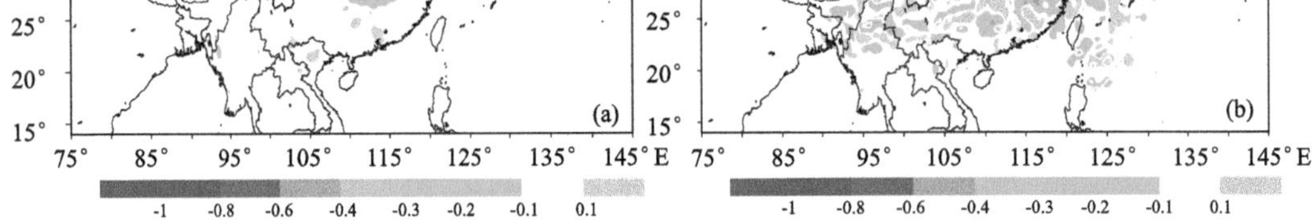

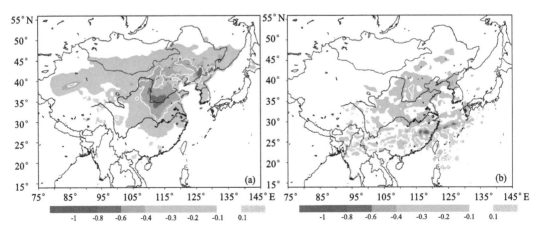

Fig. 9 Predicted dust-induced monthly mean changes in (a) surface air temperature (℃) and (b) accumulated precipitation (mm · d^{-1}). The white dashed line indicates the regions of statistically significant change at 90% confidence level.

cover by the semi-direct effect, due to an enhanced convergence of moisture driven by dust radiative heating, indicating the link between aerosols and clouds is more varied and complex.

The overall decreasing tendency in surface air temperature over the majority of the study domain indicates that the surface cooing by the dust direct radiative forcing dominated over the surface warming by the reduction of cloud cover. Figure 9a also shows the surface air temperature increased in a part of south China (but no statistically significant), which could be attributed to a decrease in cloud cover. The variation of surface air temperature in different regions depended on the relative magnitude of direct radiative forcing and semi-direct effect in association with the distribution of dust aerosol and cloud.

Figure 9b clearly shows that in the middle reaches of the Yangtze River and large areas of north China, precipitation decreased by 0.1～0.6 mm • d^{-1}, and the decrease in a portion of the middle reaches of the Yellow River was statistically significant. In portions of southeast China, precipitation decreased by up to 1 mm/d, but not statistically significant. The maximum change in precipitation in southeast China appears to be larger than that in north China, this is due to the much larger precipitation in these regions (see Figure 1e, f). It is clearly seen that in accordance with the change in surface air temperature, the dust-induced change in precipitation was more consistent (decrease) in north China than that in southeast China, where the alternating bands of increasing and decreasing precipitation occurred, which is due to the complex dust perturbation to dynamics in the moisture-rich condition and the resulting water vapor/cloud redistribution.

Zhang *et al.* (2009) conducted a 10-year model simulation from 1997 to 2006 for spring season and found the net radiative forcing by East Asian dust causes a surface air temperature decrease, with values in the range of −0.5℃ to −1.0℃ over deserts, and −0.1℃ to −0.25℃ in downwind regions, which is generally in line with our findings (−0.1℃ to −1.0℃), but our results exhibited larger air temperature decrease in areas downwind of the Gobi desert (−0.6℃ to −1.0℃ in wide areas of north China). Their simulation shows dust feedback just unsystematically causes a small decrease in precipitation, whereas in this study, precipitation decreased by 0.1～0.6 mm • d^{-1} in the middle reaches of the Yellow River with statistical significance. This could be associated with the stronger radiative effects during the severe dust storm on 19—22 March.

4 Summary and Conclusions

An online coupled regional climate-chemistry-aerosol model was utilized to investigate the dust direct radiative feedback on dust deflation, transport and meteorology in March 2010, with focus on a severe dust storm during 19—22 originated from the Gobi desert near the China-Mongolia border. The comparison shows the model is able to predict reasonably well the major distribution features of meteorological fields in the study domain. The comparison at Lanzhou, Beijing and Yuzhong site demonstrates the model predicts well the variation of PM_{10} concentration and meteorological elements during the dust storm period, although the wind speed is systematically overpredicted by the model at Yuzhong site. The predictions of both concentration and meteorology with dust radiative feedback are closer to observations than that without feedback, which could be due to the better representations of energy transfer, dust cycle and meteorology when incorporating dust feedback in the model. A caveat to the above results is the empirical dust model used in this study, which is scaled by calibration factors and can't provide an explicit physical representation of dust deflation process.

The instantaneous net direct radiative forcings by dust at the surface and at TOA under clear-sky condition at the time of dust deflation on 19 March in the Gobi desert were predicted to be up to $-480\ W \cdot m^{-2}$ and $+80\ W \cdot m^{-2}$, respectively, indicating a surface cooling and a slight TOA warming effect.

The dust radiative forcing caused large decreases both in ground temperature and wind speed, with maximums up to -7℃ and $-4.0\ m \cdot s^{-1}$, respectively in the dust deflation region. The dust radiative forcing decreased air temperature and wind speed of the lower troposphere by ~1.5℃ and ~$2.8\ m \cdot s^{-1}$, respectively, increased air temperature and wind speed of the middle troposphere by up to 1.4℃ and $2.6\ m \cdot s^{-1}$, respectively, and changed little the meteorological elements in the upper troposphere.

The online prediction considering dust radiative feedback yield lower dust emission flux, dust concentration, and consequently smaller AOD and radiative forcings than offline prediction, with the maximum differences in dust emission, PM_{10} concentration and AOD up to $2400\ \mu g \cdot m^{-2} \cdot s^{-1}$, $14\ mg \cdot m^{-3}$ and 2.4, respectively in the Gobi desert.

The sole dust solar radiative forcing led to larger decreases in ground temperature, surface wind speed and dust emission than that by the net forcing in the Gobi desert, indicating the longwave radiative forcing can partly offset shortwave forcing. The SW dust forcing dominated over the LW dust forcing in the daytime, whereas in the nighttime, the warming effect of the LW dust forcing dominated, causing increases of air temperature and wind speed up to 1℃ and $1\ m \cdot s^{-1}$, respectively.

The prediction and observation at Yuzhong observatory both exhibited a clear decrease of daytime surface air temperature on 20 March due to the arrival of dust storm.

It is noteworthy that the dust forcing generally caused a surface cooling in west China, northern parts of east China and the Korean peninsula, with maximums of -0.6℃ to -1.0℃ in the middle reaches of the Yellow River and portions of northeast China. The change in the surface air temperature in the domain is a combined result of shortwave and longwave forcings and it also depends on the relative importance of the dust direct radiative forcing and the semi-direct effect.

Due to the dust radiative forcing, precipitation decreased by $0.1 \sim 0.6\ mm \cdot d^{-1}$ in the middle reaches of the Yangtze River and large areas of north China, decreased or increased by ~1 mm · d^{-1}, in portions of southeast China. The complex precipitation response to the dust radiative forcing in south China is associated with the dust perturbation to dynamics in moist condition and the resulting water vapor/cloud redistribution.

Both comparison at the station and statistical analysis exhibit that the inclusion of dust feedback effect is capable of reducing prediction biases and thus improve climate model ability. This suggests the necessity to develop online coupled chemistry-aerosol-climate model.

This study did not consider indirect effects, which involves a series of complex processes of dust aging, mixing, hygroscopic growth, cloud microphysics, etc. and could exerts a large impact on radiation and climate. Dust storm occurs throughout the whole year in west China, and the intensity of Asian dust storm could be quite different between years, thus we plan to explore the dust indirect effects and the long-term variations of dust climatic effect in the future.

Acknowledgement

This study was supported by the "Strategic Priority Research Program" of the Chinese Academy of Sciences (Grant No. XDA05100502), the National 973 Project of China (No. 2010CB950804), the R&D Special Fund for Public Welfare Industry (Meteorology) (No. GYHY200906020), and the Ministry of Science and Technology of China (No. 2010DFA22770). The authors would like to thank SACOL for providing high quality meteorological data and the technical team for processing TRMM data via the Giovanni online system (http://giovanni. gsfc. nasa. gov/). We are grateful to the anonymous reviewers for their valuable comments and suggestions for improving the paper.

References

Ackerman S A, Chung H. 1992. Radiative effects of airborne dust on regional energy budgets at the top of the atmosphere, *J. Appl. Meteorol.* **31**:223-233.

Ahn H J, Park S U, Chang L S. 2007. Effect of Direct Radiative Forcing of Asian Dust on the Meteorological Fields in East Asia during an Asian Dust Event Period. *Journal of Applied Meteorology and Climatology*. **46**:1655-1681.

Alpert P, Kaufman Y J, Shay-El Y, *et al*. 1998. Quantification of dust-forced heating of the lower troposphere, *Nature*, **395**:367- 370, doi:10. 1038/26456.

Brindley H A, Russell J E. 2009. An assessment of Saharan dust loading and the corresponding cloud-free longwave direct radiative effect from geostationary satellite observations, *J. Geophys. Res.*, **114**, D23201, doi:10. 1029/2008JD011635.

Carlson T N, Benjamin S G. 1980. Radiative heating rates for Saharan dust Saharan dust. *Journal of the Atmospheric Sciences*. **37**(1):193-213.

Cautenet G, Legrand M, Cautenet S, *et al*. 1991. Thermal impact of Saharan dust over land. Part I: Simulation, *J. Appl. Meteorol.*, **31**:166-180.

Fu C B. 2003. Potential impacts of human-induced land cover change on East Asia monsoon. *Global and Planetary Change*. **37**:219-229.

Fu C, Wang S, Xiong Z, *et al*. 2005. Regional climate model intercomparison project for Asia. *Bulletin of the American Meteorological Society*. **86**:257-266.

Giorgi F, Bi X, Qian Y. 2002. Direct radiative forcing and regional climatic effectsof anthropogenic aerosols over East Asia: a regional coupled climate-chemistry/aerosol model study. *Journal of Geophysical Research*, **107**(D20), 4439. doi:10.1029/2001JD001066.

Gong S L, Zhang X Y, Zhao T L, *et al*. 2003. Characterization of soil dust aerosol in China and its transport and distribution during 2001 ACE-Asia: 2. Model simulation and validation. *Journal of Geophysical Research*. **108**, 4262. doi:10.1029/2002JD002633.

Grell G A, Dudhia J, Stauffer D R. 1995. A description of the Fifth-Generation Penn State/NCAR Mesoscale Model (MM5). NCAR Technical Note, NCAR/TN-398+STR, 117 pp.

Hansell R A, Tsay S C, Ji Q, *et al*. 2010. An assessment of the surface longwave direct radiative effect of airborne Saharan dust during the NAMMA field campaign, *J. Atmos. Sci.*, **67**(4):1048-1065.

Han Zhiwei. 2010. Direct radiative effect of aerosols over East Asia with a Regional Coupled Climate/Chemistry Model. *Meteorologische Zeitschrift*, **19**(3):287-298.

Han Zhiwei, Ueda Hiromasa, An Junling. 2008. Evaluation and intercomparison of meteorological predictions by five MM5-PBL parameterizations in combination with three land-surface models. *Atmospheric Environment*. **42**(2):233-249.

Han Z W, Li J W, Xia X A, *et al*. 2012. Investigation of direct radiative effects of aerosols in dust storm season over East Asia with an online coupled regional climate-chemistry-aerosol model. *Atmospheric Environment*, **54**:688-699.

Han Z W, Ueda H, Matsuda K, *et al*. 2004. Model study on particle size segregation and deposition during Asian dust events in March 2002. *Journal of Geophysical Research*. **109**, D19205. doi: 10.1029/2004JD004920.

Heinold B, Tegen I, Schepanski K, *et al*. 2008. Dust radiative feedback on Saharan boundary layer dynamics and dust mobilization, *Geophys. Res. Lett.*, **35**, L20817, doi: 10.1029/2008GL035319.

Helmert J, Heinold B, Tegen I, *et al*. 2007. On the direct and semidirect effects of Saharan dust over Europe: A modeling study. *Journal of Geophysical Research*, **112**, D13208, doi: 10.1029/2006JD007444.

Huang J P, Zhang Wu, Zuo Jinqing, *et al*. 2008. An Overview of the Semi-arid Climate and Environment Research Observatory over the Loess Plateau. *Advances in Atmospheric Sciences*, **25**(6):906-921.

IPCC. 2007. In: Solomon S, *et al*. (Eds.), Climate Change 2007: The Physical Science Basis, Contribution of Working Group I to the Fourth Assessment Report of the Intergovernmental Panel on Climate Change. Cambridge University Press, Cambridge and New York, p. 996.

Kiehl J T, Hack J J, Bonan G B, *et al*. 1996. Description of the NCAR Community Climate Model (CCM3). NCAR Technical Note, NCAR/TN-420+STR, 152 pp.

Li J W, Han Z W, Zhang R J. 2011. Model study of atmospheric particulates during dust storm period in March 2010 overEast Asia. *Atmospheric Environment*. **45**: 3954-3964.

Miller R L, Tegen I. 1998. Climate response to soil dust aerosols, *J. Clim.*, **11**(12), 3247 - 3267, doi: 10.1175/1520-0442.

Miller R L, Perlwitz J, Tegen I. 2004. Feedback upon dust emissionby dust radiative forcing through the planetary boundary layer, *Journal of Geophysical Research*, **109**, D24209, doi: 10.1029/2004JD004912.

Overpeck J, Rind D, Lacis A, *et al*. 1996. Possible role of dustinduced regional warming in abrupt climate change during the last glacial period. *Nature*, **384**(6608), 447-449, doi:10.1038/384447a0.

Park S U, Chang L S, Lee E H. 2005. Direct radiative forcing due to aerosols in East Asia during a

Hwangsa (Asian dust) event observed on 19—23 March 2002 in Korea. *Atmospheric Environment*. **39**: 2593-2606.

Perlwitz J, Miller R L. 2010. Cloud cover increase with increasing aerosol absorptivity—A counter-example to the conventional semi-direct aerosol effect. *J. Geophys. Res.*, **115**, D08203, doi: 10.1029/2009JD012637.

Perlwitz J, Tegen I, Miller R L. 2001. Interactive soil dust aerosol model in the GISS GCM1. Sensitivity of the soil dust cycle to radiative properties of soil dust aerosols, *Journal of Geophysical Research*, **106**, D16, 18,167-18,192.

Sassen K. 2002. Indirect climate forcing over the western US from Asian dust storms, *Geophys. Res. Lett.*, **29**(10), 1465, doi:10.1029/2001GL014051.

Shao Y, Dong C H. 2006. A review on East Asian dust storm climate, modeling and monitoring. *Global and Planetary Change*. **52**:1-22.

Tegen I, Lacis A A, Fung I. 1996. The influence on climate forcing of mineral aerosols from disturbed soils, *Nature*, **380**:419-422.

Wang H, Shi G Y, Teruo A, *et al*. 2004. Radiative forcing due to dust aerosol over east Asia-north Pacific region during spring 2001. *Chinese Science Bulletin*. **49**(20):2212-2219.

Wang H, Zhang X Y, Gong S L, *et al*. 2010. Radiative feedback of dust aerosols on the East Asian dust storms. *Journal of Geophysical Research*, **115**, D23214, doi:10.1029/2009JD013430.

Wu Pengping, Zhi Weihan. 2011. Indirect Radiative and Climatic Effects of Sulfate and Organic Carbon Aerosols over East Asia Investigated by RIEMS. *Atmospheric and Oceanic Science Letters*. **4**(1):7-11.

Xia X, Zong X. 2009. Shortwave versus longwave direct radiative forcing by Taklimakan dust aerosols, *Geophys. Res. Lett*, **36**, L07803, doi:10.1029/2009GL037237.

Xiong Z, Fu C B, Zhang Q. 2006. On the ability of the regional climate model RIEMS to simulate the present climate over Asia. *Advances in Atmospheric Sciences*. **23**(5):784-791.

Yue Xu, Wang H J, Liao H, *et al*. 2010. Simulation of dust aerosol radiative feedback using the GMOD: 2. Dust- climate interactions. *Journal of Geophysical Research*. **115**, D04201, doi: 10.1029/2009JD012063.

Zhang D F, Zakey A S, Gao X J, *et al*. 2009. Simulation of dust aerosol and its regional feedbacks over East Asia using a regional climate model. *Atmospheric Chemistry and Physics*. **9**:1095-1110.

Zhang J, Christopher S A. 2003. Longwave radiative forcing of Saharan dust aerosols estimated from MODIS, MISR, and CERES observations on Terra. *Geophys. Res. Lett.*, **30**(23), 2188, doi: 10.1029/2003GL018479.

Zhang X Y, Gong S L, Shen Z X, *et al*. 2003. Characterization of soil dust aerosol in China and its transport and distribution during 2001 ACE-ASIA: Network observations. *Journal of Geophysical Research*. **108** (D9), 4261. doi:10.1029/2002JD002632.

Zhang Yang, Wen X Y, Jang C J. 2010. Simulating chemistry-aerosol-cloud-radiation- climate feedbacks overthe continental U. S. using the online-coupled Weather Research Forecasting Model with chemistry (WRF/Chem). *Atmospheric Environment*, **44**:3568-3582.

EFFECT OF ATMOSPHERIC CO_2 ENRICHMENT ON SOIL RESPIRATION IN WINTER WHEAT GROWING SEASONS OF A RICE-WHEAT ROTATION SYSTEM①②

SUN Huifeng[1),2)], ZHU Jianguo[1)]③, XIE Zubin[1)], LIU Gang[1),3)] and TANG Haoye[1)]

[1)] State Key Laboratory of Soil and Sustainable Agriculture, Institute of Soil Science, Chinese Academy of Sciences, Nanjing 210008 (China)

[2)] University of Chinese Academy of Sciences, Beijing 100049 (China)

[3)] Jiangsu Biochemical Engineering Center, Nanjing 210008 (China)

Abstract

Studies on the effect of elevated CO_2 on C dynamics in cultivated croplands are critical to a better understanding of the C cycling in response to climate change in agroecosystems. To evaluate the effects of elevated CO_2 and different N fertilizer application levels on soil respiration, winter wheat (*Triticum aestivum* L. cv. Yangmai 14) plants were exposed to either ambient CO_2 or elevated CO_2 (ambient $[CO_2] + 200\ \mu mol \cdot mol^{-1}$), under N fertilizer application levels of 112.5 and 225 kg $N \cdot hm^{-2}$ (as low N and normal N subtreatments, respectively), for two growing seasons (2006—2007 and 2007—2008) in a rice-winter wheat rotation system typical in China. A split-plot design was adopted. A root exclusion method was used to partition soil respiration (RS) into heterotrophic respiration (RH) and autotrophic respiration (RA).

Atmospheric CO_2 enrichment increased seasonal cumulative RS by 11.8% at low N and 5.6% at normal N when averaged over two growing seasons. Elevated CO_2 significantly enhanced ($P < 0.05$) RS (12.7%), mainly due to the increase in RH (caused by decomposition of larger amounts of rice residue under elevated CO_2) during a relative dry season in 2007—2008. Higher N supply also enhanced RS under ambient and elevated CO_2. In the 2007—2008 season, normal N treatment had a significant positive effect ($P < 0.01$) on seasonal cumulative RS relative to low N treatment when averaged across CO_2 levels (16.3%). A significant increase in RA was mainly responsible for the enhanced RS under higher N supply. The correlation (r^2) between RH and soil temperature was stronger ($P < 0.001$) than that between RS and soil temperature when averaged across all treatments in both seasons. Seasonal patterns of RA may be more closely related to the plant phenology than soil temperature. The Q_{10} (the multiplier to the respiration rate for a 10℃

---

① The paper published in *Pedosphere*, Volume 23, Issue 6, December 2013, Pages 752-766

② Supported by the National Natural Science Foundation of China (No. 41171191), the National Key Technologies Research and Development Program of China during the 11th Five-Year Plan Period (No. 2008BAD95B05), the Knowledge Innovation Program of the Chinese Academy of Sciences (Nos. KZCX2-YW-Q1-07, KZCX2-EW-409 and KZCX3-SW-440), and the International Science and Technology Cooperation Program of China (No. 2010DFA22770).

③ Corresponding author. E-mail: jgzhu@issas. ac. cn.

increase in soil temperature) values of RS and RH were not affected by elevated CO_2 or higher N supply. These results mainly suggested that the increase in RS at elevated CO_2 depended on the input of rice residue, and the increase in RS at higher N supply was due to stimulated root growth and concomitant increase in RA during the wheat growing portion of a rice-winter wheat rotation system.
Key words: autotrophic respiration, carbon dynamics, heterotrophic respiration, N fertilization, soil temperature

1 Introduction

The atmospheric carbon dioxide (CO_2) concentration is increasing and is predicted to double in the mid- to late 21st century (Houghton *et al.*, 1996; IPCC, 2007). The elevated CO_2 concentration in the atmosphere generally stimulates plant growth and biomass production (Amthor, 2001; Ainsworth and Long, 2005), especially in C_3 plants with high nutrient availability (Curtis and Wang, 1998; Daepp *et al.*, 2001). However, elevated CO_2 could decrease plant tissue N concentrations(Cotrufo, *et al.*, 1998; Coûteaux *et al.*, 1999) and increase its C/N ratio (de Graaff *et al.*, 2004). The changes in quantity and quality of plant residue by elevated CO_2 may modify decomposition, thereby altering soil respiration (RS).

RS is considered to be the second largest flux following the gross primary production in the global C cycle (Schlesinger and Andrews, 2000). Previous studies have shown that atmospheric CO_2 enrichment stimulates RS in many plant-soil systems (King *et al.*, 2004; Ross *et al.*, 2004; Kou *et al.*, 2007; Wan *et al.*, 2007). Cheng and Johnson (1998) pointed out that the increase in RS under elevated CO_2 mainly can be attributed to the increase in autotrophic respiration (RA) by enhancing root biomass and rhizodeposition. During wheat growth under a rice-winter wheat rotation system, Kou *et al.* (2007) suggested that enhanced heterotrophic respiration (RH, equal to RS minus root respiration) at elevated CO_2 caused most of the stimulation in RS under higher CO_2 conditions. They ascribed the enhanced RH by elevated CO_2 chiefly to the higher rhizosphere respiration, as 83% of the variation in RH could be explained by root biomass. The stimulation in root growth and turnover under elevated CO_2 played an important role in increasing RS. Moreover, the decomposition of soil native organic C was enhanced by higher C input (rhizodeposition) under elevated CO_2 (van Kessel *et al.*, 2000; Xie *et al.*, 2005), which may further increase RS. However, Cheng and Johnson (1998) demonstrated that the change in soil organic C at elevated CO_2 depended on the soil N status. Therefore, different N fertilizer application levels may potentially alter the response of RS to elevated CO_2.

It is well known that N is the most important nutrient limiting plant growth in most terrestrial ecosystems, especially in agroecosystems. It influences RS by altering plant growth and microbial acitivity. Higher N supply could increase root growth and microbial biomass (Yuan *et al.*, 2005), consequently stimulating RS (Liljeroth *et al.*, 1990; Peng *et al.*, 2011). However, neutral (Søe *et al.*, 2004; Kou *et al.*, 2007; Ding *et al.*, 2010) or negative (Bowden *et al.*, 2004; Al-Kaisi *et al.*, 2008) effects of higher N supply on RS have also been observed in many studies when compared to lower N supply. These authors attributed the reductions in RS at higher N supply to the decreases in fine root production, microbial biomass or soil organic C decomposition. The disagreement among the results makes it urgent for us to clearly understand the effect of different N fertilizer application levels on RS under elevated CO_2 as a potential positive reinforcement to the global warming.

Rice-winter wheat rotations are popular at the experiment site. Rice plants are typically transplanted on about June 15th and harvested on October 20th, followed by sowing wheat in early November and harvesting in early June of the next year. As rice or winter wheat is harvested, residue is plowed into the soil. Plant residue decomposition may influence RS in the following season. Xie *et al.* (2002) reported that elevated CO_2 enhanced total biomass and C/N ratio, but reduced N concentration of rice plants under China free-air CO_2 enrichment (FACE) system. However, a number of studies have demonstrated that atmospheric CO_2 enrichment has little effect on plant residue decomposition (Booker *et al.*, 2000; Norby *et al.*, 2001; Ross *et al.*, 2002), especially in the long term. Furthermore, Booker *et al.* (2000) suggested that C cycling might simply be affected *via* the increases in plant biomass production rather than the changes in plant residue quality under elevated CO_2. Therefore, the contribution of straw from a previous season to RS must be studied for a better understanding of the soil C cycling in such systems.

The objective of our study was to evaluate the effects of elevated CO_2 and different N fertilizer application levels on RS in wheat seasons within a typical rice-winter wheat rotation system. The hypotheses were: that i) residue from a prior season may greatly influence RS under elevated CO_2; and that ii) higher N fertilization may increase RS by increasing RA.

2 Materials and Methods

2.1 Study Site

The China FACE system was established in Jiangdu City, Jiangsu Province, China (32°35′5″ N, 119°42′0″ E) and has been in operation since the rice season of 2004. A split-plot design was used with two levels of target CO_2 concentration and two application levels of N fertilization. This system consists of octagonal rings (about 14 m in diameter) with target CO_2 concentration about 200 $\mu mol \cdot mol^{-1}$ higher than ambient condition (FACE) and rectangle rings (14 m length × 13 m width) for ambient CO_2, each replicated three times. To avoid cross contamination, the distance between the FACE rings and the ambient rings was >90 m. Pure CO_2 was vented through small holes in the pipes located about 50—60 cm above the canopy, angled towards the FACE ring center. Within the FACE rings, the CO_2 concentration was elevated during daylight hours only from about re-greening to maturity stage of winter wheat (about 3 months). The average CO_2 concentration of seasonal (from re-greening to maturity stage) daytime was 588 (2006—2007) and 577 (2007—2008) $\mu mol \cdot mol^{-1}$ in the FACE rings and 398 (2006—2007) and 391 (2007—2008) $\mu mol \cdot mol^{-1}$ in the ambient rings. More details of the design, rationale, operation, and performance of the CO_2 exposure system can be found in Okada *et al.* (2001) and Liu *et al.* (2002).

The plow layer of the soil at the experimental site is about 15 cm in depth. The soil chemical and physical properties are as follows: soil organic C (SOC) 18.5 $g \cdot kg^{-1}$, total N 1.96 $g \cdot kg^{-1}$, bulk density 1.16 $g \cdot cm^{-3}$ at 0—15 cm, pH (1 : 2.5, w/v) 6.8, clay (<0.002 mm) 21.0%, silt (0.002—0.02 mm) 27.7%, and sand (0.02—2 mm) 51.3%. The soil is classified as a Shajiang-Aquic Cambosol (Cooperative Research Group on Chinese Soil Taxonomy, 2001) and has a clay-loamy texture according to the international classification systems.

Our study was carried out in the third (2006—2007) and fourth (2007—2008) winter wheat growing seasons since the FACE platform was established in 2004. The seasonal changes in daily precipitation and daily mean air temperature during the two winter wheat growing seasons are shown in Fig. 1. Over the 2006—2007 season, daily mean air temperature ranged from -0.5 to 26.4℃ (seasonal mean: 10.2℃), while it ranged from −3.9 to 26.8℃ (seasonal mean: 9.7℃) over the 2007—2008 season. The total amount of precipitation was 329.7 and 264.2 mm in the 2006—2007 and 2007—2008 winter wheat growing seasons, respectively.

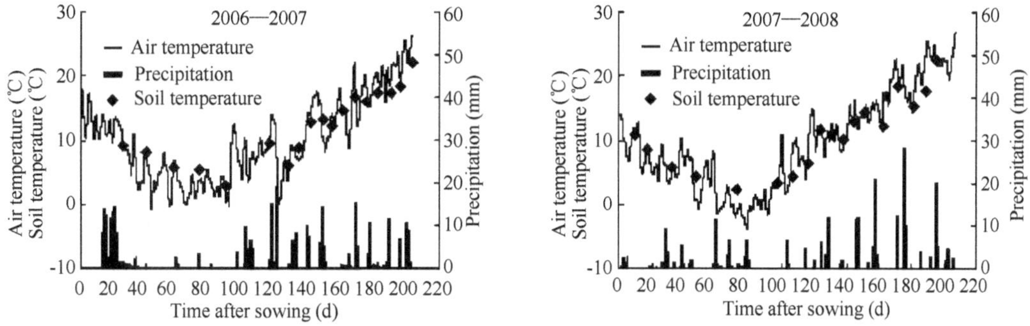

Fig. 1　Seasonal variations of daily precipitation, soil temperature and daily mean air temperature during 2006—2007 and 2007—2008 winter wheat growing seasons.

The 2006—2007 and 2007—2008 seasons began on November 4th, 2006 and November 14th, 2007, and ended on May 26th (duration: 203 d) and June 8th (duration: 207 d) of the following year, respectively.

2.2 N Fertilizer Application

Winter wheat (*Triticum aestivum* L. cv. Yangmai 14) was sown in early November at a density of 225 seedlings · m^{-2} with a row space of 20 cm, and harvested in early June of the following year. Nitrogen fertilizer was applied to wheat as compost($N : P_2O_5 : K_2O =$ 15 : 15 : 15, w/w/w) and as urea (46.3% of N) at rates of 225 and 112.5 kg N · hm^{-2} in the NN (normal N level) and LN (low N level) subtreatments, respectively. Each N subtreatment received phosphorous (P_2O_5) and potassium (K_2O) fertilizers applied in the compost at a same rate of 75 kg · hm^{-2}. Fifty percent of the NPK fertilizer was applied as a base, 10% at stem-elongation, and 40% at heading, regardless of N level. The weight ratios of compost to urea were about 6.1 and 1.5 for LN and NN treatments, respectively, at every fertilizer application time, respectively.

As rice was harvested (on October 21st and 24th in the 2006 and 2007 rice seasons, respectively), the rice residue (including root and stubble) left in the field was dug out, cut to small pieces, and then uniformly plowed into the soil. This was done 1 to 2 d before sowing wheat. The amount of rice residue incorporated into the soil accounted for about 22% and 15% of total residue production averaged over all treatments in the 2006—2007 and 2007—2008 seasons, respectively. Insecticide and herbicide application, weeding, and other agricultural practices were in accordance with local farming practices.

2.3 Measurements

Soil respiration was measured by a static chamber sampling method and gas chromatography. The samples for determining RS were taken using a chamber, which consists of a PVC (polyvinyl chloride) base frame (15 cm length × 14 cm width × 5.5 cm height) and a PVC lid (13.5 cm length×12.5 cm width×10 cm height). The chamber for measuring RH is composed of a stainless steel base frame (15 cm length ×14 cm width × 20 cm height) and a PVC lid identical to that used in measuring RS. Three PVC base frames for RS and three stainless steel ones for RH were inserted into soil approximately 3 cm and 15 cm deep, respectively, between rows in each N subtreatment. Two gas samples were collected from each chamber at 30 min intervals at each gas sampling timing. The gas samples were injected to 18.5 mL evacuated brown vials and then taken to the laboratory for measurement of CO_2 concentrations as soon as possible. All samples were taken between 07:00 and 13:00 Beijing time on the same day. The sampling frequency was once fortnightly during the first half of the wheat season and once a week over the second half of the wheat season.

Gas samples were analyzed using a Varian CP-3380 gas chromatograph (Varian Inc., USA) with a thermal conductivity detector (TCD) for CO_2. Carbon dioxide separations were carried out by Chromosorb 102 column (2.0 m length ×2.0 mm diameter). Operating conditions for the gas chromatograph were: 50℃ injector temperature, 40℃ oven temperature and 100℃ detector temperature. The carrier gas was ultra-high purity N_2 (99.999%). Data processing was performed using Varian Star Chromatography Workstation (version 6.3) software.

At maturity stage, the wheat aboveground portion was harvested with an area of 2 m^2. After the aboveground portions of rice and wheat were removed from the field, the belowground biomass of rice (root and stubble) and wheat (root) in each N subtreatment were dug out to a depth of 15 cm and an area of 0.5 m^2, washed. Above- and belowground portions were oven-dried at 80℃ for 72 h and weighed.

Soil temperature was measured by a soil thermometer (JM222L, Tianjin Jinming Instrument Ltd., China) with a silicon semiconductor sensing probe. The temperature determination range is -50 to 100℃. The sensing probes were buried in the soil at the depth of 10 cm before the beginning of the season. There was a temperature probe in each N subtreatment. The soil temperature was determined during the process of gas sampling.

2.4 Calculation and Statistical Analysis

When the soil respiration rate was calculated for the two CO_2 concentrations, the assumption was that the gas concentration in the headspace increased linearly with time within a certain period. Thus the equation for calculating soil respiration rate was as follows:

$$R = \frac{\Delta C \cdot M \cdot V}{22.4 \cdot \Delta t \cdot A} \cdot 10^{-3} \tag{1}$$

where R is the soil respiration rate ($mg \cdot m^{-2} \cdot h^{-1}$), ΔC is the differences between the concentrations at 30 min intervals ($\mu g \cdot g^{-1}$), M is the gas molecular weight ($g \cdot mol^{-1}$), V is the effective volume of chamber (L), 22.4 is the occupied volume per mole gas under standard state ($L \cdot mol^{-1}$), Δt is the time intervals (h), and A is the area the chamber covered (m^2).

The seasonal cumulative soil respiration was estimated by this equation:

$$RS_{cum} = \sum_{1}^{t} \frac{(R_{i+1} + R_i)}{2} \cdot (t_{i+1} - t_i) \cdot 24 \cdot 10^{-5} \tag{2}$$

where RS_{cum} is the cumulative soil respiration in an entire season ($t \cdot hm^{-2}$), R_i and R_{i+1} are the soil respiration rates at the ith and ($i+1$)th gas sampling, respectively ($mg \cdot m^{-2} \cdot h^{-1}$), and t is the sampling time (d after sowing).

The autotrophic respiration (RA) was assumed to be equal to the difference between soil respiration (RS) and heterotrophic respiration (RH). Thus the equation for calculating RA was as follows:

$$RA = RS - RH \tag{3}$$

The relationship between soil temperature (T) and soil respiration rate was modeled using an exponential function (Lloyd and Taylor, 1994):

$$R = a \cdot \exp(bT) \tag{4}$$

where a and b are the regression coefficients. To assess the sensitivity of RS to soil temperature, the Q_{10} function was used. Q_{10} represents the multiplier to the respiration rate for a 10℃ increase in soil temperature:

$$Q_{10} = \exp(10b) \tag{5}$$

A split-plot statistical design was used with CO_2 as the main plot factor ($n=2$) and N level ($n=2$) as the subplot factor. The effects of a single factor and interaction between factors were analyzed by one-way analysis of variation (ANOVA) and univariate analysis, respectively. The ANOVA method was applied in the effect of CO_2 on soil respiration rate and belowground biomass using least significant difference (LSD) tests. All statistical analyses were performed using SPSS 13.0 software, with a confidence level of 95%.

3 Results

3.1 Plant Below- and Aboveground Biomass

The increased atmospheric CO_2 concentration significantly enhanced the rice belowground (residue) biomass by 33.3% ($P< 0.05$) and 24.2% ($P< 0.01$) in the NN treatment in 2006 and 2007 rice growing seasons, respectively (Fig. 2). However, the impact of elevated CO_2 on rice belowground biomass was not significant in the LN treatment, although the corresponding stimulations were 21.3% and 3.6% for 2006 and 2007, respectively. No significant differences were detected in rice belowground biomass between LN and NN treatments under both CO_2 levels during two rice growing seasons (Fig. 2). However, when averaged across all treatments, a significantly larger ($P< 0.001$) rice belowground biomass was found in 2006 than 2007.

In the 2006—2007 wheat season, no significant differences in belowground (root) biomass between treatments were found (Fig. 2). However, a significant enhancement ($P< 0.05$) in aboveground biomass by elevated CO_2 was detected when averaged across N levels (Fig. 2). Atmospheric CO_2 enrichment stimulated aboveground biomass by 14.7% at LN and 11.2% at NN ($P> 0.05$). The aboveground biomass was similar between LN and NN treatments under either ambient or elevated CO_2 conditions (Fig. 2).

In the 2007—2008 wheat season, elevated CO_2 increased belowground biomass by 39.4% at LN and 17.0% at NN ($P>0.05$, Fig. 2). However, averaged across both N levels, a significant positive effect ($P< 0.05$) on belowground biomass by elevated CO_2 was observed in the 2007—2008 wheat season. Atmospheric CO_2 enrichment stimulated

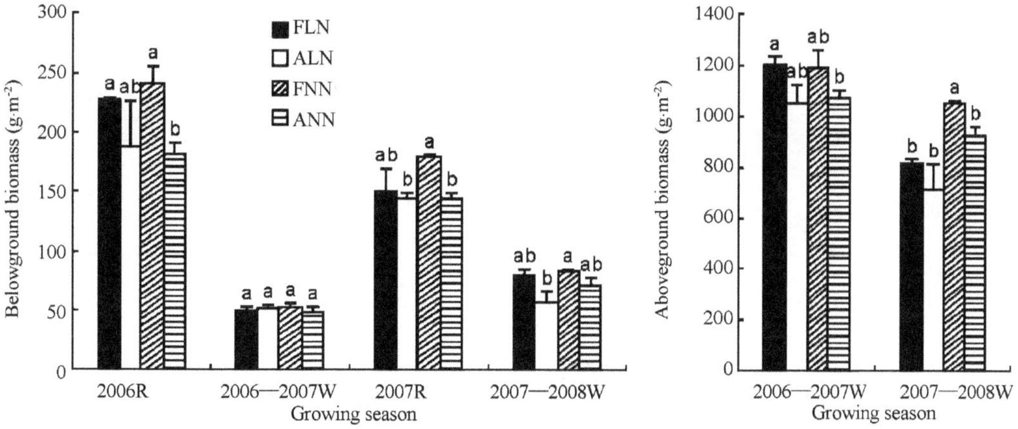

Fig. 2　Below- and aboveground biomass at maturity stage in rice (R) and/or winter wheat (W) growing seasons. FLN, ALN, FNN and ANN represent free-air CO_2 enrichment (FACE)-low nitrogen (LN), ambient-LN, FACE-normal N (NN), and ambient-NN treatments, respectively. Vertical bars represent the standard errors of the means ($n=3$). Bars with the same letter(s) are not significantly different among the treatments at $P=0.05$.

aboveground biomass by 14.0% ($P>0.05$) and 13.8% ($P<0.05$) at LN and NN, respectively (Fig. 2). Higher N supply stimulated above- and belowground biomass under ambient and elevated CO_2. Additionally, in comparison with LN treatment, NN treatment significantly enhanced ($P<0.001$) aboveground biomass by 29.3% under elevated CO_2 condition(Fig. 2).

When averaged over all treatments, a lower belowground (root) biomass and a higher aboveground (shoot) biomass, thereby a significantly lower ($P<0.001$) root/shoot ratio of wheat was observed in 2006—2007 than 2007—2008.

3.2　Seasonal Changes of Soil Respiration Rates in Winter Wheat

When averaged across all treatments, RS decreased ($P<0.05$) from initiation to 89 d after sowing in 2006—2007 (Fig. 3), but RH was constant during this period (Fig. 4). However, both RS and RH reduced ($P<0.001$) from initiation to 71 d after sowing in 2007—2008 (Figs. 3 and 4). RS and RH in both seasons increased afterwards until 161 d (heading stage, 2006—2007) and 137 d (stem elongation stage, 2007—2008) after sowing. Thereafter, RS declined intermittently while RH still exhibited an upward trend until the end (Figs. 3 and 4). RA patterns resembled those of RS, but with a faster decrease in RA when compared to RS late in the season (Fig. 5).

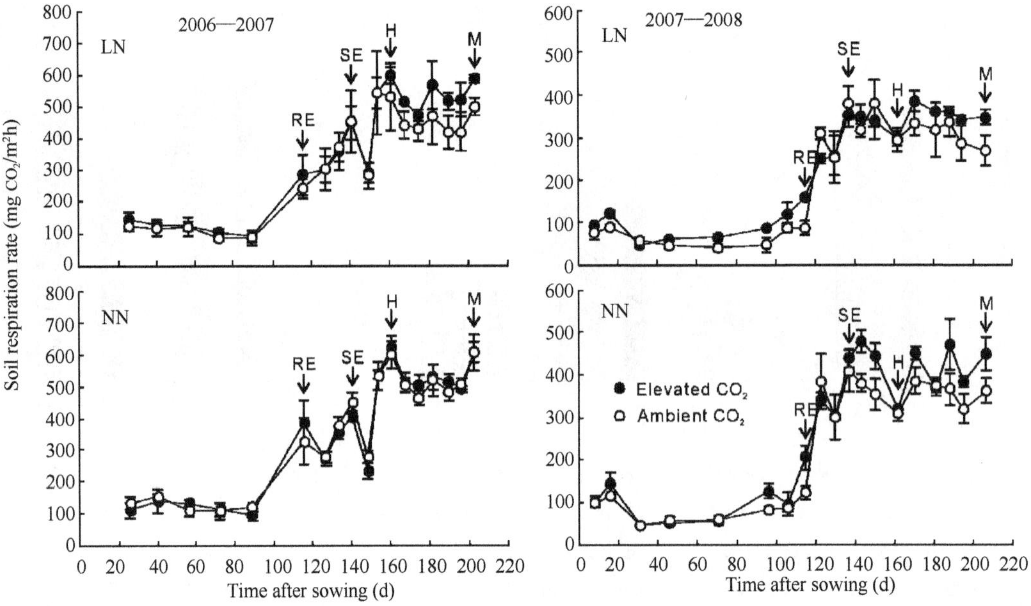

Fig. 3 Soil respiration rates over the 2006—2007 and 2007—2008 winter wheat growing seasons under elevated and ambient CO_2 conditions. LN and NN represent low N and normal N treatments, respectively. RE, SE, H and M represent re-greening, stem elongation, heading and maturity stages, respectively. Vertical bars represent the standard errors of the means ($n=3$).

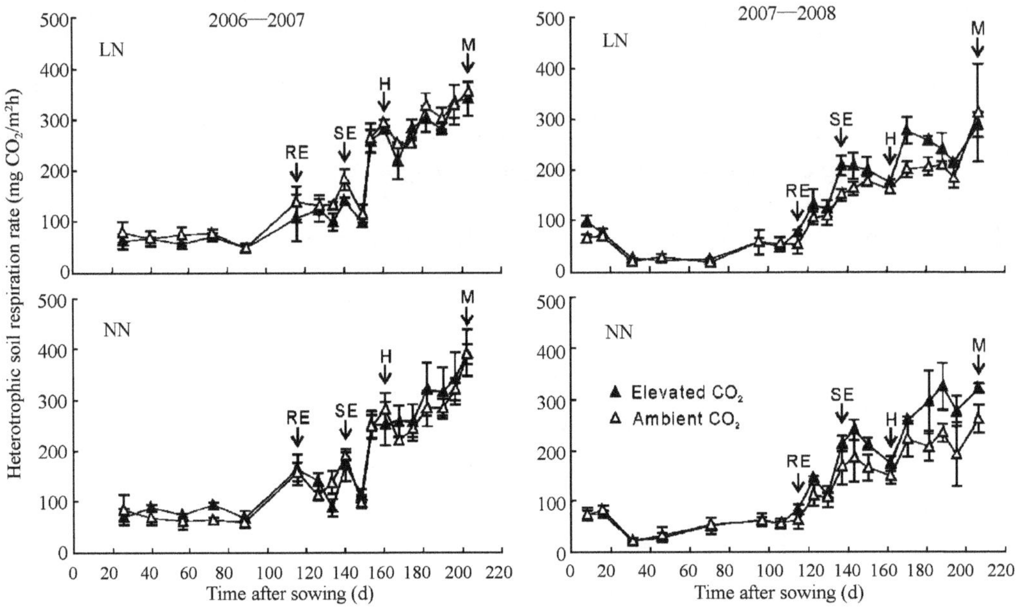

Fig. 4 Heterotrophic respiration rates over the 2006—2007 and 2007—2008 winter wheat growing seasons under elevated and ambient CO_2 conditions. LN and NN represent low N and normal N treatments, respectively. RE, SE, H and M represent re-greening, stem elongation, heading and maturity stages, respectively. Vertical bars represent the standard errors of the means ($n=3$).

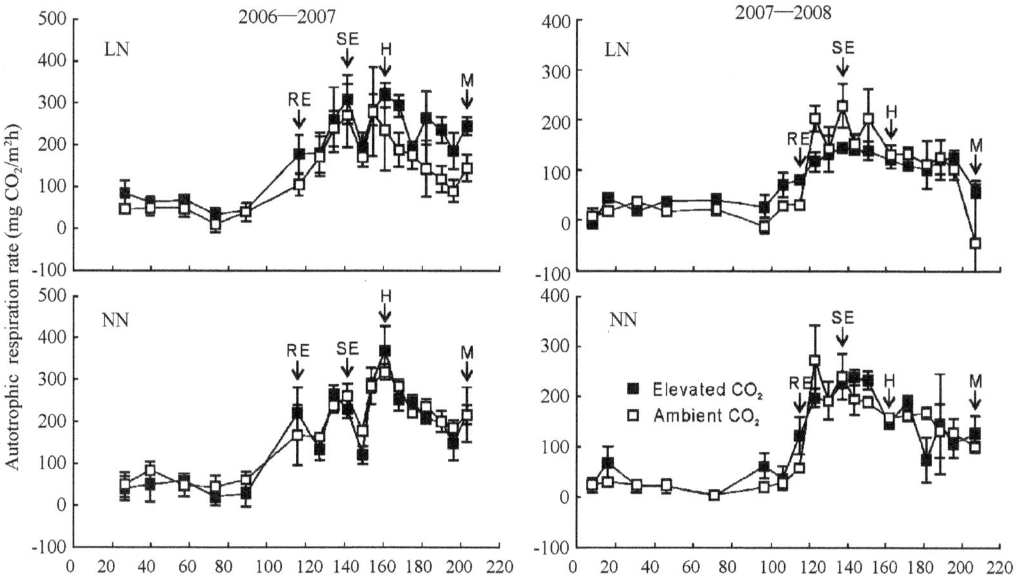

Fig. 5　Autotrophic respiration rates over the 2006—2007 and 2007—2008 winter wheat growing seasons under elevated and ambient CO_2 conditions. LN and NN respresent low N and normal N treatments, respectively. RE, SE, H and M represent re-greening, stem elongation, heading and maturity stages, respectively. Vertical bars represent the standard errors of the means ($n=3$).

3.3　Seasonal Cumulative Soil Respiration in Response to Elevated CO_2

Positive impacts of elevated CO_2 on RS or RH mainly occurred during the second half of each season (from re-greening to maturity stage), especially in 2007—2008, while there was little effect during the beginning of the winter wheat seasons (Figs. 3 and 4). Elevated CO_2 had little effect on RA under NN treatment in both seasons (Fig. 5). Nevertheless, under LN treatment, RA was enhanced by elevated CO_2 through the entire 2006—2007 season, although no significant effect was detected (Fig. 5). However, RA was apparently increased during the first half but decreased during the second half of the 2007—2008 season under elevated CO_2 condition (Fig. 5).

Over the entire growing seasons, significant differences ($P<0.05$) in RS between elevated CO_2 and ambient CO_2 were found on 203 d after sowing in 2006—2007, and on 16, 115, and 123 d after sowing in 2007—2008 at the LN level (Fig. 3). However, at the NN level, significant increase ($P<0.05$) in RS under elevated CO_2 was only found on 143 d after sowing in 2007—2008 (Fig. 3). Averaged over N levels, a significant positive effect ($P<0.05$) of elevated CO_2 on seasonal cumulative RS was only observed in 2007—2008 (12.7%, Table 1). Elevated CO_2 enhanced seasonal cumulative RS by 12.0% and 13.4%, respectively, at LN and NN in 2007—2008 ($P>0.05$). In 2007—2008, significant temporary changes in RS at early (seedling to re-greening stage) and late growing season (heading to maturity stage) were the main contributors for enhanced RS by elevated CO_2

(Table 2). A positive, but not statistically significant increase in RS by elevated CO_2 was found in 2006—2007. When averaged over the two growing seasons, atmospheric CO_2 enrichment increased seasonal cumulative RS by 11.8% at LN and 5.6% at NN ($P>0.05$).

Table 1 Seasonal cumulative soil respiration (RS_{cum}), heterotrophic respiration (RH_{cum}) and autotrophic respiration(RA_{cum}) during 2006—2007 and 2007—2008 winter wheat growing seasons

Treatment[a)]	2006—2007				2007—2008			
	RS_{cum} t CO_2 hm^{-2}	RH_{cum} t CO_2 hm^{-2}	RA_{cum} t CO_2 hm^{-2}	fRA[b)] %	RS_{cum} t CO_2 hm^{-2}	RH_{cum} t CO_2 hm^{-2}	RA_{cum} t CO_2 hm^{-2}	f_{RA} %
FLN	13.4±1.1[c)]	6.3±0.3	7.2±1.4	52.3±6.0	9.4±0.4	5.8±0.5	3.6±0.2	38.3±2.6
ALN	12.0±1.4	7.0±0.3	5.1±1.3	41.1±5.7	8.4±0.6	5.0±0.4	3.4±0.4	40.8±3.0
FNN	13.5±0.8	7.2±0.6	6.3±1.2	46.1±6.4	11.0±0.1	6.4±0.4	4.6±0.5	41.8±3.9
ANN	13.5±0.4	6.8±0.6	6.7±0.7	49.4±4.3	9.7±0.5	5.3±0.8	4.4±0.5	45.4±6.1
CO_2	NS[d)]	NS	NS	NS	*	NS	NS	NS
N	NS	NS	NS	NS	* *	NS	*	NS
CO_2 × N	NS	NS	NS	NS	NS	NS	NS	NS

*, * * Significant at $P=0.05$ and 0.01 levels, respectively.

a) FLN = free-air CO_2 enrichment (FACE)-low nitrogen (LN), ALN = ambient-LN, FNN = FACE-normal N (NN), ANN = ambient-NN.

b) Contribution of autotrophic respiration (RA) to soil respiration (RS).

c) Means ± standard errors ($n=3$).

d) Not significant at $P=0.05$.

Table 2 Mean soil respiration rates at wheat periodic developmental stages[a)] during the 2007—2008 growing season

Treatment[b)]	S to RE mg CO_2 m^{-2} · h^{-1}	RE to SE mg CO_2 m^{-2} · h^{-1}	SE to H mg CO_2 m^{-2} · h^{-1}	H to M mg CO_2 m^{-2} · h^{-1}	S to M mg CO_2 m^{-2} · h^{-1}
			Soil respiration		
FLN	84.3±5.4[c)]	221.1±19.6	346.8±31.8	348.5±7.0	230.8±11.2
ALN	62.8±4.5	215.8±22.0	359.0±31.5	305.9±28.5	210.5±14.8
FNN	87.8±9.1	283.0±6.2	454.1±25.5	407.1±15.5	277.3±4.7
ANN	76.9±3.9	268.6±26.3	381.0±32.1	353.1±13.4	242.4±12.4
CO_2	*	NS[d)]	NS	*	*
N	NS	*	NS	*	* *
CO_2 × N	NS	NS	NS	NS	NS
			Heterotrophic respiration		
FLN	50.9±0.2	134.4±17.1	204.6±23.7	241.9±11.3	144.8±11.1
ALN	45.2±3.9	89.3±15.4	163.6±11.1	212.2±24.5	123.6±11.0
FNN	53.5±5.4	112.5±2.5	221.2±14.5	274.6±20.5	159.1±7.7
ANN	54.6±4.8	94.1±19.3	172.0±35.4	210.8±30.5	128.7±19.6

(Continued)

Treatment[b)]	S to RE mg CO_2 m^{-2} · h^{-1}	RE to SE mg CO_2 m^{-2} · h^{-1}	SE to H mg CO_2 m^{-2} · h^{-1}	H to M mg CO_2 m^{-2} · h^{-1}	S to M mg CO_2 m^{-2} · h^{-1}
			Heterotrophic respiration		
CO_2	NS	NS	NS	NS	NS
N	NS	NS	NS	NS	NS
CO_2× N	NS	NS	NS	NS	NS
			Autotrophic respiration		
FLN	33.4±4.8	111.2±11.8	142.2±8.4	106.6±4.3	86.0±2.8
ALN	17.6±3.9	126.5±7.1	195.5±37.4	93.7±9.5	86.9±9.5
FNN	34.3±11.3	170.5±8.6	232.9±11.1	132.5±35.7	118.2±11.8
ANN	22.3±2.4	174.5±24.6	209.0±24.4	142.3±17.3	113.7±12.6
CO_2	NS	NS	NS	NS	NS
N	NS	* *	NS	NS	*
CO_2× N	NS	NS	NS	NS	NS

*, * * Significant at P=0.05 and 0.01 levels, respectively.

[a)] S=seedling stage, RE=re-greening stage, SE=stem elongation stage, H=heading stage, M=maturity stage.

[b)] FLN =free-air CO_2 enrichment (FACE)-low nitrogen (LN), ALN= ambient-LN, FNN =FACE-normal N (NN), ANN =ambient-NN.

[c)] Means± standard errors (n=3).

[d)] Not significant at P=0.05.

RH was significantly enhanced ($P<0.05$) by elevated CO_2 on 73 d after sowing at NN in 2006—2007, and on 181 d after sowing at LN in 2007—2008 (Fig. 4). Elevated CO_2 had no significant effects on seasonal cumulative RH in both seasons regardless of N level (Table 1). Overthe 2007—2008 season, elevated CO_2 stimulated RH by 16.0% and 20.8% at LN and NN levels, respectively ($P>0.05$). No consistent effects of elevated CO_2 on seasonal cumulative RH were detected at LN and NN in 2006—2007.

Significant stimulations ($P<0.01$) in RA by elevated CO_2 were observed on 46 and 115 d after sowing at LN in 2007—2008 (Fig. 5). No significant differences in RA between elevated CO_2 and ambient CO_2 at either N level through the entire 2006—2007 growing season (Fig. 5). Elevated CO_2 had no significant impacts on seasonal cumulative RA in either season as well (Table 1). Elevated CO_2 had a slight, but not significant stimulation in RA at both N levels over the 2007—2008 season (5.9% and 4.5% at LN and NN levels, respectively).

On a seasonal timescale, the fractions of RA in RS (f_{RA}) were not affected by elevated CO_2 regardless of N level, which ranged from 38.3% to 52.3% in two growing seasons (Table 1). On average, the f_{RA} values were 47.2% and 41.6% in 2006—2007 and 2007—2008, respectively. Additionally, the averaged seasonal cumulative RS, RH and RAacross all treatments were significantly higher ($P<0.01$ or $P<0.001$) in 2006—2007 than in

2007—2008 (Table 1).

3.4 Seasonal Cumulative Soil Respiration in Response to *N* Fertilization

Few significant enhancements in RS by higher N supply were noted in 2007—2008 on 16 d after sowing under ambient CO_2 condition ($P<0.05$) and on 123 and 143 d after sowing under elevated CO_2 ($P<0.01$) (Fig. 3). Averaged across CO_2 levels, a significant increase ($P<0.01$) in seasonal cumulative RS at NN was found in 2007—2008 when compared to LN (16.3%, Table 1). The NN treatment significantly increased ($P<0.05$) RS by 17.0% under elevated CO_2 relative to the LN treatment in 2007—2008. However, under ambient CO_2 condition, the positive effect of higher N supply on seasonal cumulative RS was not significant (15.5%). A transient positive effect ($P<0.05$) of higher N supply on RS was detected at re-greening to stem elongation stages and heading to maturity stages in 2007—2008 (Table 2). In 2006—2007, higher N supply increased ($P>0.05$) RS under ambient and elevated CO_2 conditions either (Table 1).

Among daily measurements, significant differences ($P<0.05$) in RH between LN and NN treatments were only found on 40 and 57 d after sowing under elevated CO_2 and on 168 d after sowing under ambient CO_2 in 2006—2007 (Fig. 4). Higher N supply had no significant effects on seasonal cumulative RH (Table 1). Consistent positive effects of higher N supply on seasonal cumulative RH were only observed under elevated CO_2 condition, but no significant effects were detected in both seasons (Table 1).

In 2006—2007, in comparison with LN, the NN significantly increased ($P<0.05$) RA on 175 and 196 d after sowing under elevated and ambient CO_2, respectively (Fig. 5). Significant differences ($P<0.05$ or $P<0.01$) in RA between LN and NN treatments were also recorded on 46, 137, 150, and 171 d after sowing during the 2007—2008 season (Fig. 5). Seasonal cumulative RA was significantly enhanced ($P<0.05$) by higher N supply when averaged over CO_2 levels in 2007—2008 (28.6%), but no significant effect was observed at individual CO_2 levels (Table 1). A temporary but significant increase ($P<0.01$) in RA under NN treatment was detected from re-greening to stem elongation stages relative to LN treatment in 2007—2008 (Table 2). No significant effects of higher N supply on RA were observed in 2006—2007 (Table 1).

The autotrophic contribution (*i.e.*, f_{RA}) was not affected by higher N supply either (Table 1). No interactions between atmospheric CO_2 concentrations and N fertilizer application levels were observed in seasonal cumulative RS, RH and RA in either growing season (Table 1).

3.5 Relationships between Soil Respiration and Soil Temperature

Soil respiration followed seasonal trends in soil temperature, especially for RS and RH (Figs. 1, 3 and 4). RS and RH in all treatments were strongly and exponentially correlated ($P<0.001$) with soil temperature during both seasons (Table 3). However, the correlation between RA and soil temperature was weak (Table 3). When averaged over all treatments in both seasons, the correlation between soil respiration and soil temperature were in the order of RH ($r^2=0.72$) > RS ($r^2=0.63$) > RA ($r^2=0.32$) ($P<0.001$). Furthermore, the Q_{10} value of RH (3.1) was significantly greater ($P<0.05$) than that of RS (2.8) or RA (2.5). Significant differences in r^2 and Q_{10} values of RS or RH were observed neither at elevated CO_2 nor at higher N supply.

Table 3　Relationships among soil respiration, heterotrophic respiration or autotrophic respiration and soil temperature at 10 cm depth during the 2006—2007 and 2007—2008 wheat growing seasons

Treatment [a]	2006—2007				2007—2008			
	a [b]	b [b]	r^2 [c]	Q_{10} [d]	a	b	r^2	Q_{10}
				Soil respiration				
FLN	86.574	0.1062	0.70±0.04 [e]	2.92±0.29	64.615	0.0940	0.66±0.06	2.57±0.11
ALN	83.910	0.0977	0.62±0.04	2.67±0.16	44.130	0.1093	0.57±0.03	3.00±0.20
FNN	86.831	0.1065	0.66±0.03	2.93±0.27	63.949	0.1076	0.61±0.04	2.94±0.14
ANN	91.139	0.1014	0.67±0.03	2.77±0.22	57.397	0.1031	0.59±0.03	2.80±0.06
				Heterotrophic respiration				
FLN	32.971	0.1191	0.74±0.04	3.30±0.17	28.768	0.1187	0.74±0.05	3.30±0.26
ALN	41.557	0.1101	0.79±0.06	3.02±0.19	24.396	0.1175	0.70±0.02	3.28±0.36
FNN	45.711	0.1038	0.74±0.05	2.84±0.24	32.592	0.1176	0.71±0.02	3.26±0.26
ANN	38.102	0.1120	0.73±0.04	3.04±0.12	32.797	0.1002	0.66±0.05	2.73±0.09
				Autotrophic respiration				
FLN	47.509	0.1017	0.48±0.05	2.81±0.36	41.074	0.0574	0.24±0.07	1.80±0.23
ALN	49.740	0.0641	0.20±0.06	1.93±0.23	20.309	0.1048	0.39±0.04	2.87±0.21
FNN	45.588	0.0920	0.29±0.10	2.51±0.12	43.640	0.0720	0.22±0.01	2.09±0.26
ANN	56.596	0.0811	0.34±0.09	2.25±0.06	17.724	0.1243	0.39±0.05	3.52±0.44

[a] FLN = free-air CO_2 enrichment (FACE)-low nitrogen (LN), ALN = ambient-LN, FNN = FACE-normal N (NN), ANN = ambient-NN.

[b] a and b are the regression coefficients in Eq. of $R=a\cdot\exp(bT)$, where R and T are the soil respiration rate and soil temperature.

[c] Coefficient of determination.

[d] Multiplier to the respiration rate for a 10℃ increase in soil temperature.

[e] Means ± standard errors ($n=3$).

4　Discussion

4.1　Cumulative Soil Respiration in Different Growing Seasons

Although the patterns of RS, RH and RA were similar in both wheat growing seasons (Figs. 3, 4 and 5), the seasonal cumulative RS, RH and RA were significantly higher

($P<0.01$ or $P<0.001$) in 2006—2007 than in 2007—2008 when averaged across all treatments (Table 1). Higher organic C inputs from incorporated rice residue and more favorable soil moisture conditions likely were the main reasons for the higher seasonal cumulative RS and RH in 2006—2007. There was also a larger cumulative RA, but smaller root biomass in all treatments in 2006—2007 compared with 2007—2008. The total precipitation amount in 2007—2008 (264.2 mm) was much lower than that in 2006—2007 (329.7 mm). The relative dry conditions of 2007—2008 may have promoted the photosynthates allocated belowground toward root growth (Schall *et al.*, 2012) and increased the wheat's root/shoot ratio (Fig. 2), which is in agreement with the results reported by Liu *et al.* (2004). Plants may have adjusted biomass allocation in response to lower availability of soil water. However, a reduction in ion uptake by roots due to lower soil moisture results in a decrease in root respiration (Poorter *et al.*, 1991). Moreover, relatively lower soil moisture could reduce microbial biomass (Kassem *et al.*, 2008), which may decrease microbial activity, thereby decrease RA. In contrast, in 2006—2007, relatively favorable soil moisture might have enhanced soil N mineralization rates (Hungate *et al.*, 1997) and increased N availability to wheat. Higher N availability can increase microbial biomass and stimulate microbial activity in the wheat rhizosphere (Yuan *et al.*, 2011), which in turn would have enhanced RA.

Averaged over all treatments, autotrophic contributions to soil respiration were 47.2% in 2006—2007 and 41.6% in 2007—2008. These values seem to be much lower than those obtained by Kuzyakov and Cheng (2001) in a previous study with spring wheat. They reported a f_{RA} value of 88% in a laboratory experiment with wheat seedlings. However, they noted that the 1 kg of soil they used for five wheat plants would create a rooting density much higher than in a field setting. They calculated f_{RA} values for field conditions of 35%—40%, which compare well with our results. The depth of plow layer at the experiment site is about 15 cm. The compacted plough pan under the plow layer in these soils would have prevented most wheat roots from growing deeper in soil according to our observation, thus the stainless steel base frames inserted into the soil to a 15 cm depth would have prevented roots from growing into the frames from outside. Therefore, the measured CO_2 flux from the soil covered by the frames can well represent RH just as reported by Zheng *et al.* (2006). Interestingly, Ding *et al.* (2010) investigated the effects of different N fertilizer application levels on soil respiration with maize and obtained comparable autotrophic contribution of about 54% at 0 kg N • hm^{-2} and 49% at 150 kg N • hm^{-2}. Similarly, in grassland ecosystems, Kuzyakov *et al.* (2001) and Heinemeyer *et al.* (2012) showed that the autotrophic contribution was about 40% and 46%, respectively. The averaged autotrophic contribution was higher in a relatively wet season (2006—2007) than in a relatively dry season (2007—2008) in the present experiment, although the difference was not significant. Under soil moisture deficiency, the reduction in root respiration (Poorter *et al.*, 1991) and microbial activity in the rhizosphere (Yuan *et al.*, 2011) may have been stronger than the decrease in RH, thus the autotrophic contribution was relatively smaller in 2007—2008.

4.2 Effect of Elevated CO_2 on Soil Respiration

Soil respiration stimulated by elevated CO_2 has been reported in many studies with various plants (Janssens *et al.*, 1998; Allen *et al.*, 2000; Pendall *et al.*, 2001; Wan *et al.*, 2007). Similarly, in this study, RS was enhanced by elevated CO_2 in both seasons, which was in line with previous investigations, but a significant effect was only found in 2007—2008 ($P < 0.05$). The degree of this increase (12.0%—13.4%) was within the reported range of the responses of soil respiration to elevated CO_2 (-10% to 162%, Zak *et al.*, 2000).

The increase in CO_2 concentration inside the chamber and subsequent reduction in diffusion gradient may account for the underestimation (Nay *et al.*, 1994; Healy *et al.*, 1996; Davidson *et al.*, 2002). The cumulative RS by closing 30 min in our experiment was comparable with the results reported by Kou *et al.* (2007), who collected three gas samples from each chamber at a 5 min interval.

The differences in RS or RH between elevated CO_2 and ambient CO_2 mainly occurred at the second half season (Figs. 3 and 4). The soil microbial organisms became active in these warmer days (Fig. 1). If more C sources were supplied to them under elevated CO_2, this would lead to higher CO_2 release compared with ambient CO_2.

It has also been well documented that elevated CO_2 stimulates wheat photosynthesis (Osborne *et al.*, 1998; Wall *et al.*, 2000; Pal *et al.*, 2005), leaf area index (Brooks *et al.*, 2000; Pendall *et al.*, 2001) and aboveground biomass (Ma *et al.*, 2005; Kou *et al.*, 2007). Accordingly, in the present experiment, elevated CO_2 increased aboveground biomass by 11.2%—14.7% across both wheat growing seasons, irrespective of N level (Fig. 2). The enhanced aboveground biomass may have a higher potential for the assimilated C to be allocated belowground. Generally, plants grown under elevated CO_2 conditions exhibit increased photosynthesis and growth, disproportionately large increase in C allocation to roots (Norby *et al.*, 1986; Pregitzer *et al.*, 1995), and stimulated rhizodeposition (Hungate *et al.*, 1996; Ineson *et al.*, 1996; Cheng and Johnson, 1998). The fate of this enhanced C input into the soils and its subsequent influence on soil C storage bears important implications for global C cycles. There are two possible outcomes for the increased soil C resulting from elevated CO_2: first, this C source can be utilized by soil microorganisms and a portion of it converted into soil organic matter, thereby enhancing soil C storage; second, this labile C source alters soil microbial processes, thereby stimulating or suppressing decomposition of native soil organic matter (Kuzyakov, 2002; Allard *et al.*, 2006). Cheng and Johnson (1998) reported that averaged across two N treatments, the impact of elevated CO_2 on the enhancement of wheat autotrophic respiration (60%) was much higher than on root biomass (26%). In our study, higher degrees of enhancement in belowground biomass (17.0%—39.4%) than in aboveground biomass (13.8%—14.0%) at elevated CO_2 was also observed in 2007—2008 regardless of N level (Fig. 2). However, elevated CO_2 had little effect on RA (4.5%—5.9%), although significantly enhanced RS in 2007—2008. This illustrates that elevated CO_2 greatly depresses RA on the basis of root biomass. The increased photosynthates allocated belowground under elevated CO_2 apparently are used for root growth or for soil organic C increase rather than RA. In the 2007—2008 wheat season, Zhong *et al.* (2009) found that elevated CO_2 significantly increased soil organic C at jointing stage, and had a stimulative effect on soil organic C during the fallow period after wheat harvest regardless of N level. Therefore, the increased C input into soils by elevated CO_2 in this season may be converted into soil organic matter rather than utilized by soil microorganisms as RA. Additionally, soil moisture within the stainless steel frames for measuring RH may have been higher compared to that within PVC frames for measuring RS, due to the lack of root uptake, as suggested by Syer and Tanner (2010). A potentially greater decomposition in rice residue and/or soil organic C in the stainless steel frames due to higher soil moisture may have resulted in a higher RH, thereby leading to a small response of RA to elevated CO_2 when calculating RA by substracting RH from RS.

Kou*et al.* (2007) investigated the responses of root respiration of winter wheat to elevated CO_2 using a special plant growth chamber at the same site, and concluded that increasing atmospheric CO_2 stimulated soil respiration by 15.1% and 14.8% at NN and LN treatment, respectively. Moreover, they suggested that heterotrophic microbial respiration contributed much more than root respiration to the rise in total soil respiration with atmospheric CO_2 enrichment. Our results from 2007—2008 partly confirm their conclusion. On a seasonal timescale, the increase in CO_2 release at elevated CO_2 was clearly larger in RH than that in RA, regardless of N level (Table 1). This indicates that RH was the main contributor to enhanced soil respiration under elevated CO_2 in winter wheat. Elevated CO_2 cannot directly affect soil microbial activity, because the CO_2 concentration in soil is several orders of magnitude higher than that in the atmosphere (van Veen *et al.*, 1991; Blagodatskaya *et al.*, 2010). The elevated CO_2 most likely affected RH *via* the quantity and quality of rice belowground residue. Low N concentrations and increased C/N ratios in rice residue under elevated CO_2 had been reported by Xie *et al.* (2002) and Yang *et al.* (2008) under China FACE system. In a meta-analysis by Norby *et al.* (2001) on the effects of elevated CO_2 on litter chemistry and decomposition, they concluded that no differences existed in plant litter respiration rates between elevated CO_2 and ambient CO_2. Incorporating the surplus plant residue produced under elevated CO_2 potentially leads to higher release of CO_2 to the atmosphere(Booker *et al.*, 2000). Elevated CO_2 significantly enhanced rice belowground biomass production under NN at the end of the 2006 and 2007 rice growing seasons (Fig. 2). Accordingly, higher seasonal cumulative RH in FNN treatment was observed in both seasons relative to ANN treatment (Table 1). Therefore, residue from a prior season most likely influenced soil respiration under elevated CO_2.

4.3 Effect of N Fertilizer Application Levels on Soil Respiration

Fertilization is an important agricultural practice that could affect many soil processes such as microbial activity, root growth and turnover, and soil respiration (Lee and Jose, 2003). N is an important limiting factor for plants in many agricultural soils (Kou *et al.*, 2007). Numerous studies had been carried out to investigate the effects of different N fertilizer application levels on soil respiration (Craine *et al.*, 2001; Olsson *et al.*, 2005; Lagomarsino *et al.*, 2009; Ding *et al.*, 2010). Higher N supply can positively (Liljeroth *et al.*, 1990; Peng *et al.*, 2011), neutrally (Søe *et al.*, 2004; Kou *et al.*, 2007; Ding *et al.*, 2010) or negatively (Olsson *et al.*, 2005; Al-Kaisi *et al.*, 2008) affect soil respiration. The positive effect of higher N supply on soil respiration is attributed to relatively more photosynthesized C release into the soil as rhizodeposition or root respiration rather than higher soil organic C decomposition (Liljeroth *et al.*, 1990; Liljeroth *et al.*, 1994). The reduction in soil respiration by higher N supply was due to decreases in microbial biomass and activity (Lee and Jose, 2003; Bowden *et al.*, 2004), or suppression of decomposition of soil organic C (Foereid *et al.*, 2004). In our study, NN treatment enhanced RS during both seasons relative to LN treatment (Table 1). In 2007—2008, a significant increase ($P < 0.01$) in seasonal cumulative RS (16.3%) by higher N supply to a great extent derived from the substantial increase ($P < 0.05$) in RA (28.6%). This agrees well with the results of Liljeroth *et al.* (1990) and Peng *et al.* (2011). In agroecosystems, soil microorganisms in the rhizosphere may shift from C limited to N limited due to N uptake and exudation of high C/N ratio organic compounds by plants (Liljeroth *et al.*, 1994). In the present experiment, higher N supply (NN treatment) could have stimulated plant growth, increased root biomass (though not significantly), enhanced soil microbial activity in the wheat rhizosphere, and consequently promoted the RA. Therefore, higher N supply likely stimulated soil respiration by enhancing autotrophic respiration.

4.4 Effect of Elevated CO_2 on Temperature Sensitivity of Soil Respiration

Significant exponential relationships ($P < 0.01$ or $P < 0.001$) between RS, RH or RA and soil temperature were detected in all treatments in both seasons (Table 3). Numerous studies have also suggested a high dependence of RS, RH or RA on soil temperature (Boone *et al.*, 1998; Fang and Moncrieff, 2001; Fang *et al.*, 2005; Jenkins and Adams, 2011). In general, seasonal variations in soil respiration rates follow the patterns of soil temperature, especially in upland fields. In our study, when averaged across all treatments of both seasons, about 63% and 72% of seasonal variations in RS and RH could be explained by soil temperature, which are in accordance with the results reported by Boone *et al.* (1998) and Fang and Moncrieff (2001). However, soil temperature accounted for only 32% of seasonal variations in RA. A significant correlation between RA and plant root biomass (Ding *et al.*, 2010) or green leaf area index (Pendall *et al.*, 2001) had been reported, thus seasonal patterns of RA may be more closely related to the plant phenology than soil temperature. This supports previous findings that the temperature sensitivity of autotrophic respiration cannot be accurately estimated from seasonal soil respiration data (Schindlbacher *et al.*, 2009).

The Q_{10} values averaged across all treatments of both seasons were higher ($P< 0.05$) in RH than RS. Soil organic C is always conceptually divided into labile organic C and recalcitrant organic C. The temperature sensitivity of decomposition in labile organic C may be greater than (Boone *et al.*, 1998), equivalent to (Bååth and Wallander, 2003; Fang *et al.*, 2005), or less than (Knorr *et al.*, 2005; Xu *et al.*, 2010) that in recalcitrant organic C. Thus, Q_{10} values for soil respiration greatly depend on soil organic C availability (Kirschbaum, 2006). In the present experiment, the decomposition of relatively recalcitrant organic C (RH) was more sensitive to soil temperature than decomposition of relatively labile organic C (RS). This is in accordance with kinetic theory based on chemical reactions (Fissore *et al.*, 2009) and is supported by previous studies (Melillo *et al.*, 2002; Xu *et al.*, 2010).

In a forest ecosystem, Butnor*et al.* (2003) found higher soil respiration Q_{10} values under elevated CO_2 conditions regardless of fertilizer treatment. However, they did not give any explanation. Elevated CO_2 may have enhanced tree litter production but reduced it availability in their experiment. The increased recalcitrant organic C on the soil surface at elevated CO_2 may have promoted the temperature sensitivity of soil respiration (Q_{10}). Nevertheless, in our study, elevated CO_2 had no significant effects on Q_{10} values of RS or RH in either season, irrespective of N level (Table 3). The soil organic C availability in this study may have not been affected by elevated CO_2, thus the temperature sensitivity of RS or RH was not altered. Similarly, Lagomarsino *et al.* (2009) suggested that the Q_{10} values of soil respiration were not modified by elevated CO_2. However, they showed that N fertilization significantly increased Q_{10} values of soil respiration, and they ascribed the effect to the changes in microbial communities. In contrast, Wang *et al.* (2004) and Craine *et al.* (2007) reported that soil labile organic C increased with N addition, thus causing a decrease in Q_{10} values of soil respiration. In the present experiment, no effects of higher N supply were observed on Q_{10} values of soil respiration. In 2007—2008, the stimulation in RS at higher N supply was due to the enhancement in RA *via* stimulating soil labile organic C (rhizodeposition). Because of a weak link between RA and soil temperature, Q_{10} values of RS were not affected by N fertilizer application levels.

5 Conclusions

Although atmospheric CO_2 enrichment and higher N application level had little effect on seasonal fluctuations in RS, seasonal cumulative RS was promoted in both 2006—2007 and 2007—2008 wheat growing seasons. The significant enhancement in seasonal cumulative RS by elevated CO_2 in 2007—2008 was mainly due to significant periodic increases in RS at seedling to re-greening and heading to maturity stages of wheat. Elevated CO_2 enhanced RS mainly by increasing RH. Decomposition of larger amounts of rice residue might account for the increased CO_2 emissions from soil under elevated CO_2 condition. However, higher N supply increased RS mainly by stimulating RA.

Acknowledgements

We wish to express our gratitude to Prof. Sean Bloszies from North Carolina State University, USA for his help in paper revision. We are also highly grateful to anonymous reviewers and editors for their valuable suggestions in improving this manuscript. The main instruments and apparatus of the FACE system were supplied by Japan National Institute for Agro-Enviromental Sciences (NIAES) and Japan Agricultural Research Center for Tohoku Region (NARCT).

References

Ainsworth E A, Long S P. 2005. What have we learned from 15 years of free-air CO_2 enrichment (FACE)? A meta-analytic review of the responses of photosynthesis, canopy properties and plant production to rising CO_2. *New Phytol*. **165**: 351-372.

Al-Kaisi M M, Kruse M L, Sawyer J E. 2008. Effect of nitrogen fertilizer application on growing season soil carbon dioxide emission in a corn-soybean rotation. *J. Environ. Qual.* **37**: 325-332.

Allard V, Robin C, Newton P C D, *et al*. 2006. Short and long-term effects of elevated CO_2 on *Lolium perenne* rhizodeposition and its consequences on soil organic matter turnover and plant N yield. *Soil Biol. Biochem.* **38**: 1178-1187.

Allen A S, Andrews J A, Finzi A C, *et al*. 2000. Effects of free-air CO_2 enrichment (FACE) on belowground processes in a *Pinus taeda* forest. *Ecol. Appl.* **10**: 437-448.

Amthor J S. 2001. Effects of atmospheric CO_2 concentration on wheat yield: review of results from experiments using various approaches to control CO_2 concentration. *Field Crop. Res.* **73**: 1-34.

Bååth E, Wallander H. 2003. Soil and rhizosphere microorganisms have the same Q_{10} for respiration in a model system. *Glob. Change Biol*. **9**: 1788-1791.

Blagodatskaya E, Blagodatsky S, Dorodnikov M, *et al*. 2010. Elevated atmospheric CO_2 increases microbial growth rates in soil: results of three CO_2 enrichment experiments. *Glob. Change Biol*. **16**: 836-848.

Booker F L, Shafer S R, Wei C M, *et al*. 2000. Carbon dioxide enrichment and nitrogen fertilization effects on cotton (*Gossypium hirsutum* L.) plant residue chemistry and decomposition. *Plant Soil*. **220**: 89-98.

Boone R D, Nadelhoffer K J, Canary J D, *et al*. 1998. Roots exert a strong influence on the temperature sensitivity of soil respiration. *Nature*. **396**: 570-572.

Bowden R D, Davidson E, Savage K, *et al*. 2004. Chronic nitrogen additions reduce total soil respiration and microbial respiration in temperate forest soils at the Harvard forest. *Forest Ecol. Manag*. **196**: 43-56.

Brooks T J, Wall G W, Printer P J Jr, *et al*. 2000. Acclimation response of spring wheat in a free-air CO_2 enrichment (FACE) atmosphere with variable soil nitrogen regimes. 3. Canopy architecture and gas exchange. *Photosynth. Res.* **66**: 97-108.

Butnor J R, Johnsen K H, Oren R, *et al*. 2003. Reduction of forest floor respiration by fertilization on both carbon dioxide-enriched and reference 17-year-old loblolly pine stands. *Glob. Change Biol*. **9**: 849-861.

Cheng W X, Johnson D W. 1998. Elevated CO_2, rhizosphere processes, and soil organic matter decomposition. *Plant Soil*. **202**: 167-174.

Cooperative Research Group of Chinese Soil Taxonomy. 2001. Chinese Soil Taxonomy. Science Press, Beijing, New York.

Cotrufo M F, Ineson P, Scott A. 1998. Elevated CO_2 reduces the nitrogen concentration of plant tissues. *Glob. Change Biol*. **4**: 43-54.

Coûteaux M M, Kurz C, Bottner P, *et al*. 1999. Influence of increased atmospheric CO_2 concentration on quality of plant material and litter decomposition. *Tree Physiol*. **19**: 301-311.

Craine J M, Morrow C, Fierer N. 2007. Microbial nitrogen limitation increases decomposition. *Ecology*. **88**: 2105-2113.

Craine J M, Wedin D A, Reich P B. 2001. The response of soil CO_2 flux to changes in atmospheric CO_2, nitrogen supply and plant diversity. *Glob. Change Biol*. **7**: 947-953.

Curtis P S, Wang X Z. 1998. A meta-analysis of elevated CO_2 effects on woody plant mass, form, and physiology. *Oecologia*. **113**: 299-313.

Daepp M, Nösberger J, Lüscher A. 2001. Nitrogen fertilization and developmental stage alter the response of *Lolium perenne* to elevated CO_2. *New Phytol*. **150**: 347-358.

Davidson E A, Savage K, Verchot L V, *et al*. 2002. Minimizing artifacts and biases in chamber-based measurements of soil respiration. *Agr. Forest Meteorol*. **113**: 21-37.

de Graaff M-A, Six J, Harris D, *et al*. 2004. Decomposition of soil and plant carbon from pasture systems after 9 years of exposure to elevated CO_2: impact on C cycling and modeling. *Glob. Change Biol*. **10**: 1922-1935.

Ding W X, Yu H Y, Cai Z C, *et al*. 2010. Responses of soil respiration to N fertilization in a loamy soil under maize cultivation. *Geoderma*. **155**: 381-389.

Fang C, Moncrieff J B. 2001. The dependence of soil CO_2 efflux on temperature. *Soil Biol. Biochem*. **33**: 155-165.

Fang C M, Smith P, Moncrieff J B, *et al*. 2005. Similar response of labile and resistant soil organic matter pools to changes in temperature. *Nature*. **433**: 57-59.

Fissore N, Giardina C P, Swanston C W, *et al*. 2009. Variable temperature sensitivity of soil organic carbon in North American forests. *Glob. Change Biol*. **15**: 2295-2310.

Foereid B, de Neergaard A, Høgh-Jensen H. 2004. Turnover of organic matter in a *Miscanthus* field: effect of time in *Miscanthus* cultivation and inorganic nitrogen supply. *Soil Biol. Biochem*. **36**: 1075-1085.

Healy R W, Striegl R G, Russell T F, *et al*. 1996. Numerical evaluation of static-chamber measurements of soil-atmosphere gas exchange: identification of physical processes. *Soil Sci. Soc. Am. J*. **60**: 740-747.

Heinemeyer A, Tortorella D, Petrovičová, *et al*. 2012. Partitioning of soil CO_2 flux components in a temperate grassland ecosystem. *Eur. J. Soil Sci*. **63**: 249-260.

Houghton J T, Meira Filho L G, Callander B A, *et al*. 1996. Climate Change 1995: the Science of Climate Change. Cambridge University Press, Cambridge.

Hungate B A, Canadell J, Chapin F S Ⅲ. 1996. Plant species mediate changes in soil microbial N in response to elevated CO_2. *Ecology*. **77**: 2505-2515.

Hungate B A, Chapin F S III, Zhong H, *et al*. 1997. Stimulation of grassland nitrogen cycling under carbon dioxide enrichment. *Oecologia*. **109**: 149-153.

Ineson P, Cotrufo M F, Bol R, *et al*. 1996. Quantification of soil carbon inputs under elevated CO_2: C_3 plants in a C_4 soil. *Plant Soil*. **187**: 345-350.

IPCC. 2007. Climate Change 2007: Synthesis Reports. Contribution of Working Groups I, II and III to the Fourth Assessment Report of the Intergovernmental Panel on Climate Change. IPCC, Geneva, Switzerland.

Janssens I A, Crookshanks M, Taylor G, *et al*. 1998. Elevated atmospheric CO_2 increases fine root production, respiration, rhizosphere respiration and soil CO_2 efflux in Scots pine seedlings. *Glob. Change Biol*. **4**: 871-878.

Jenkins M E, Adams M A. 2011. Respiratory quotients and Q_{10} of soil respiration in sub-alpine Australia reflect influences of vegetation types. *Soil Biol. Biochem*. **43**: 1266-1274.

Kassem I I, Joshi P, Sigler V, *et al*. 2008. Effect of elevated CO_2 and drought on soil microbial communities associated with *Andropogon gerardii*. *J. Integr. Plant Biol*. **50**: 1406-1415.

King J S, Hanson P J, Bernhardt E, *et al*. 2004. A multiyear synthesis of soil respiration responses to elevated atmospheric CO_2 from four forest FACE experiments. *Glob. Change Biol*. **10**: 1027-1042.

Kirschbaum M U F. 2006. The temperature dependence of organic-matter decomposition: still a topic of debate. *Soil Biol. Biochem*. **38**: 2510-2518.

Knorr W, Prentice I C, House J I, *et al*. 2005. Long-term sensitivity of soil carbon turnover to warming. *Nature*. **433**: 298-301.

Kou T J, Zhu J G, Xie Z B, *et al*. 2007. Effect of elevated atmospheric CO_2 concentration on soil and root respiration in winter wheat by using a respiration partitioning chamber. *Plant Soil*. **299**: 237-249.

Kuzyakov Y. 2002. Review: Factors affecting rhizosphere priming effects. *J. Plant Nutr. Soil Sci*. **165**: 382-396.

Kuzyakov Y, Cheng W. 2001. Photosynthesis controls of rhizosphere respiration and organic matter decomposition. *Soil Biol Biochem*. **33**: 1915-1925.

Kuzyakov Y, Ehrensberger H and Stahr K. 2001. Carbon partitioning and below-ground translocation by *Lolium perenne*. *Soil Biol. Biochem*. **33**: 61-74.

Lagomarsino A, De Angelis P, Moscatelli M C, *et al*. 2009. The influence of temperature and labile C substrates on heterotrophic respiration in response to elevated CO_2 and nitrogen fertilization. *Plant Soil*. **317**: 223-234.

Lee K H, Jose S. 2003. Soil respiration, fine root production, and microbial biomass in cottonwood and loblolly pine plantations along a nitrogen fertilization gradient. *Forest Ecol. Manag*. **185**: 263-273.

Liljeroth E, Kuikman P, van Veen J A. 1994. Carbon translocation to the rhizosphere of maize and wheat and influence on the turnover of native soil organic matter at different soil nitrogen levels. *Plant Soil*. **161**: 233-240.

Liljeroth E, van Veen J A, Miller H J. 1990. Assimilate translocation to the rhizosphere of two wheat lines and subsequent utilization by rhizosphere microorganisms at two soil nitrogen concentrations. *Soil Biol. Biochem*. **22**: 1015-1021.

Liu G, Han Y, Zhu J G, *et al*. 2002. Rice-wheat rotational FACE platform. I. System structure and control. *Chin. J. Appl. Ecol*. (in Chinese). **13**: 1253-1258.

Liu H S, Li F M, Xu H. 2004. Deficiency of water can enhance root respiration rate of drought-sensitive but not drought-tolerant spring wheat. *Agr. Water Manage*. **64**: 41-48.

Lloyd J, Taylor J A. 1994. On the temperature dependence of soil respiration. *Funct. Ecol*. **8**: 315-323.

Ma H L, Zhu J G, Xie Z B, *et al*. 2005. Effects of free-air carbon dioxide enrichment on growth and uptake of nitrogen in winter wheat. *Acta Agron. Sin*. (in Chinese). **31**: 1634-1639.

Melillo J M, Steudler P A, Aber J D, *et al*. 2002. Soil warming and carbon-cycle feedbacks to the climate system. *Science*. **298**: 2173-2176.

Nay S M, Mattson K G, Bormann B T. 1994. Biases of chamber methods for measuring soil CO_2 efflux demonstrated with a laboratory apparatus. *Ecology*. **75**: 2460-2463.

Norby R J, Cotrufo M F, Ineson P, *et al*. 2001. Elevated CO_2, litter chemistry, and decomposition: a synthesis. *Oecologia*. **127**: 153-165.

Norby R J, O'Neill E G, Luxmoore R J. 1986. Effects of atmospheric CO_2 enrichment on the growth and mineral nutrition of *Quercus alba* seedlings in nutrient-poor soil. *Plant Physiol*. **82**: 83-89.

Okada M, Lieffering M, Nakamura H, *et al*. 2001. Free-air CO_2 enrichment (FACE) using pure CO_2 injection: system description. *New Phytol*. **150**: 251-260.

Olsson P, Linder S, Giesler R, Högberg P. 2005. Fertilization of boreal forest reduces both autotrophic and heterotrophic soil respiration. *Glob. Change Biol*. **11**: 1745-1753.

Osborne C P, LaRoche J, Garcia R L, *et al*. 1998. Does leaf position within a canopy affect acclimation of photosynthesis to elevated CO_2? *Plant Physiol*. **117**: 1037-1045.

Pal M, Rao L S, Jain V, *et al*. 2005. Effects of elevated CO_2 and nitrogen on wheat growth and photosynthesis. *Biol. Plantarum*. **49**: 467-470.

Pendall E, Leavitt S W, Brooks T, *et al*. 2001. Elevated CO_2 stimulates soil respiration in a FACE wheat field. *Basic Appl. Ecol*. **2**: 193-201.

Peng Q, Dong Y S, Qi Y C, *et al*. 2011. Effects of nitrogen fertilization on soil respiration in temperate grassland inInner Mongolia, China. *Environ. Earth Sci*. **62**: 1163-1171.

Poorter H, van der Werf A, Atkin O K, *et al*. 1991. Respiratory energy requirements of roots vary with the potential growth rate of a plant species. *Physiol. Plantarum*. **83**: 469-475.

Pregitzer K S, Zak D R, Curtis P S, *et al*. 1995. Atmospheric CO_2, soil nitrogen and turnover of fine roots. *New Phytol*. **129**: 579-585.

Ross D J, Newton P C D, Tate K R. 2004. Elevated [CO_2] effects on herbage production and soil carbon and nitrogen pools and mineralization in a species-rich, grazed pasture on a seasonally dry sand. *Plant Soil*. **260**: 183-196.

Ross D J, Tate K R, Newton P C D, *et al*. 2002. Decomposability of C3 and C4 grass litter sampled under different concentrations of atmospheric carbon dioxide at a natural CO_2 spring. *Plant Soil*. **240**: 275-286.

Sayer E J, Tanner E V J. 2010. A new approach to trenching experiments for measuring root-rhizosphere respiration in a lowland tropical forest. *Soil Biol. Biochem*. **42**: 347-352.

Schall P, Lödige C, Beck M, *et al*. 2012. Biomass allocation to roots and shoots is more sensitive to shade and drought in European beech than in Norway spruce seedlings. *Forest Ecol. Manag*. **266**: 246-253.

Schindlbacher A, Zechmeister-Boltenstern S, Jandl R. 2009. Carbon losses due to soil warming: Do autotrophic and heterotrophic soil respiration respond equally? *Glob. Change Biol*. **15**: 901-913.

Schlesinger W H, Andrews J A. 2000. Soil respiration and the global carbon cycle. *Biogeochemistry*. **48**: 7-20.

Søe A R B, Giesemann A, Anderson T H, *et al*. 2004. Soil respiration under elevated CO_2 and its partitioning into recently assimilated and older carbon sources. *Plant Soil*. **262**: 85-94.

van Kessel C, Horwatt W R, Hartwig U, *et al*. 2000. Net soil carbon input under ambient and elevated CO_2 concentrations: isotopic evidence after 4 years. *Glob. Change Biol*. **6**: 435-444.

van Veen J A, Liljeroth E, Lekkerkerk L J A. 1991. Carbon fluxes in plant-soil systems at elevated atmospheric CO_2 levels. *Ecol. Appl*. **1**: 175-181.

Wall G W, Adam N R, Brooks T J, *et al*. 2000. Acclimation response of spring wheat in a free-air CO_2 enrichment (FACE) atmosphere with variable soil nitrogen regimes. 2. Net assimilation and stomatal conductance of leaves. *Photosynth Res*. **66**: 79-95.

Wan S Q, Norby R J, Ledford J, *et al*. 2007. Responses of soil respiration to elevated CO_2, air warming, and changing soil water availability in a model old-field grassland. *Glob. Change Biol*. **13**: 2411-2424.

Wang W J, Baldock J A, Dalal R C, *et al*. 2004. Decomposition dynamics of plant materials in relation to nitrogen availability and biochemistry determined by NMR and wet-chemical analysis. *Soil Biol. Biochem*. **36**: 2045-2058.

Xie Z B, Cadisch G, Edwards G, *et al*. 2005. Carbon dynamics in a temperate grassland soil after 9 years exposure to elevated CO_2 (Swiss FACE). *Soil Biol. Biochem*. **37**: 1387-1395.

Xie Z B, Zhu J G, Zhang Y L, *et al*. 2002. Responses of rice (*Oryza sativa*) growth and its C, N and P composition to FACE (free-air carbon dioxide enrichment) and N, P fertilization. *Chin. J. Appl. Ecol*. (in Chinese). **13**: 1223-1230.

Xu X, Zhou Y, Ruan H H, *et al*. 2010. Temperature sensitivity increases with soil organic carbon recalcitrance along an elevational gradient in the Wuyi Mountains, China. *Soil Biol. Biochem*. **42**: 1811-1815.

Yang L X, Wang Y L, Kobayashi K, *et al*. 2008. Seasonal changes in the effects of free-air CO_2 enrichment (FACE) on growth, morphology and physiology of rice root at three levels of nitrogen fertilization. *Glob. Change Biol*. **14**: 1844-1853.

Yuan L, Bao D J, Jin Y, *et al*. 2011. Influence of fertilizers on nitrogen mineralization and utilization in the rhizosphere of wheat. *Plant Soil*. **343**: 187-193.

Yuan X X, Chu H Y, Lin X G, *et al*. 2005. Responses of soil microbial biomass and respiration to elevated atmospheric CO_2. *Plant Nutr. Fert. Sci*. (in Chinese). **11**: 564-567.

Zak D R, Pregitzer K S, King J S, *et al*. 2000. Elevated atmospheric CO_2, fine roots and the response of soil microorganisms: a review and hypothesis. *New Phytol*. **147**: 201-222.

Zheng X H, Zhou Z X, Wang Y S, *et al*. 2006. Nitrogen-regulated effects of free-air CO_2 enrichment on methane emissions from paddy rice fields. *Glob. Change Biol*. **12**: 1717-1732.

Zhong S, Liang W J, Lou Y L, *et al*. 2009. Four years of free-air CO_2 enrichment enhance soil C concentrations in a Chinese wheat field. *J. Environ. Sci*. **21**: 1221-1224.

环境因子对土壤水分空间异质性的影响研究——以北京市怀柔区为例①②

蔡庆空[1)③] 蒋金豹[1)④] 崔希民[1),2)] 郭 徽[3)] 马开锋[1),4)] 洪雪倩[1)]

[1)] 中国矿业大学(北京)地球科学与测绘工程学院,北京 100083;
[2)] 中国矿业大学(北京)煤炭资源与安全开采国家重点实验室,北京 100083;
[3)] 北京师范大学 资源学院,北京 100875;
[4)] 华北水利水电学院 资源与环境学院,郑州 450011

摘要

为了探讨山区表层土壤水分的空间分布格局及其影响因素,本文以北京市怀柔区为研究区域,联合使用 ALOS/PALSAR 微波数据和 Landsat-5 遥感影像反演得到研究区的土壤水分数据,运用旋转主成分分析法分析了高程、坡度、坡向和植被盖度 4 个环境因子对土壤水分的影响情况及分布规律,并确定相应的主控因子。结果表明:高程和坡度是影响山区表层土壤水分空间变异的主控因子,植被盖度次之,坡向的影响最弱。对主控因子(高程和坡度)的单因素分析表明,土壤含水量随着高程的增加而逐渐减少,随着坡度的增加,土壤含水量总体上呈现先增加(坡度小于 3°)后减小(坡度大于 3°)的趋势,对深入研究山区土壤水分分布特性和水土保持具有指导意义。

关键字:ALOS/PALSAR;土壤水分;环境因子;旋转主成分分析

1 引言

土壤水分是连接大气圈与生物圈的重要纽带,是气候系统中不可或缺的一个关键参数,其在空间上的分布受到土地利用(植被)、剖面曲率、高程、气象因素、地形、土壤和人为活动等多因子综合作用(邱扬等,2001;邱扬等,2007)。目前国内外对土壤水分的研究主要集中在以下 3 个方面(王军德等,2001;张秀英等,2005):1)采用统计学方法进行水平方向和垂直方向上土壤水分的时空变异性研究(李海滨等,2001;Western A W *et .al.*,1998);2)土壤水分分布的影响因素研究(邱扬等,2000);3)从分布式水文模型的角度出发定量研究土壤水分时空分布的情况(Gomez-Plaza A *et al.*,2001;李保国等,2000;Rodrguez-Iturbe *et al.*,1999;Strasser u I *et al.*,2001)。相比较而言,以往研究中土壤水分数据的获取方法多是通过在研究区布设少量监测点,用 TDR 测定监测点的土壤水分,利用克吕格插值法计算得到土壤水分空间分布图,由于克吕格插值法单纯依据样点数据进行空间插值,仅能反映空间平

---

① 本文发表于山地学报,**33**(5):294-299.2013.

② 基金项目(Foundation item):国际科技合作项目(2010DFA32920)资助。

③ 作者简介(Biography):蔡庆空(1986-),男,河南南阳人,博士生,研究方向:遥感与 GIS 应用研究。

④ 通讯作者(Communication author):蒋金豹(1978-),男,博士,副教授。E-mail:ahdsjjb@126.com

面位置与土壤水分的线性关系，不能反映环境因子对土壤水分的空间异质性（赫晓慧等，2008）。由于微波遥感具有穿云透雾的能力和光学遥感具有信息丰富的特性，在植被覆盖区域，文中联合使用 ALOS/PALSAR 微波数据和同时相的 Landsat-5 遥感数据，充分发挥两者自身的优势，首先从光学遥感影像中提取归一化植被水分指数，利用“水-云模型”去除植被层在土壤水分后向散射中的贡献，之后构建后向散射系数与土壤水分反演模型得到研究区的土壤水分数据，周鹏等人的研究结果表明采用 HH 极化方式的土壤后向散射系数与土壤含水量的相关系数可以达到 0.5227（周鹏等，2010），张友静等人的研究结果表明反演的土壤含水量和准同步实测数据的均方根误差为 3.83%（张友静等，2010）。在反演得到研究区土壤水分数据的基础上，以旋转主成分分析法和地统计学方法为分析工具，对环境因子与土壤水分的时空变异特征之间的关系进行研究，并确定相应的主控因子，对于深入研究山区土壤水分的变化特性和水土保持具有重要意义。

2　研究区概况与研究方法

2.1　研究区概况

怀柔区地处北京市东北部（图 1），地理坐标介于东经 116°17′—116°53′E，40°14′—40°04′N 之间，地形南北狭长，呈哑铃状，南北长 128 km，东西最窄处 11 km。境内地势北高南低，其中山区面积 88.7%。气候属典型的暖温带半湿润大陆性季风气候，四季分明，年平均气温 9～13℃，年平均降水在 600～700 mm，主要集中在 6—8 月份。植被类型多样，以常绿阔叶林和落叶阔叶林为主。全区地处华北褐土带，主要土壤类型有棕壤、褐土、潮土、水稻土四大土类，土壤质地平原区为轻壤和沙壤质，山区多为壤质和沙壤质，pH 值在 5.9～8 之间。研究区海拔在 14～1660 m，坡度范围在 0.34°～76.28°，主要的土地利用类型是林地、草地、建设用地、农用地，人类活动对研究区的影响相对较小。

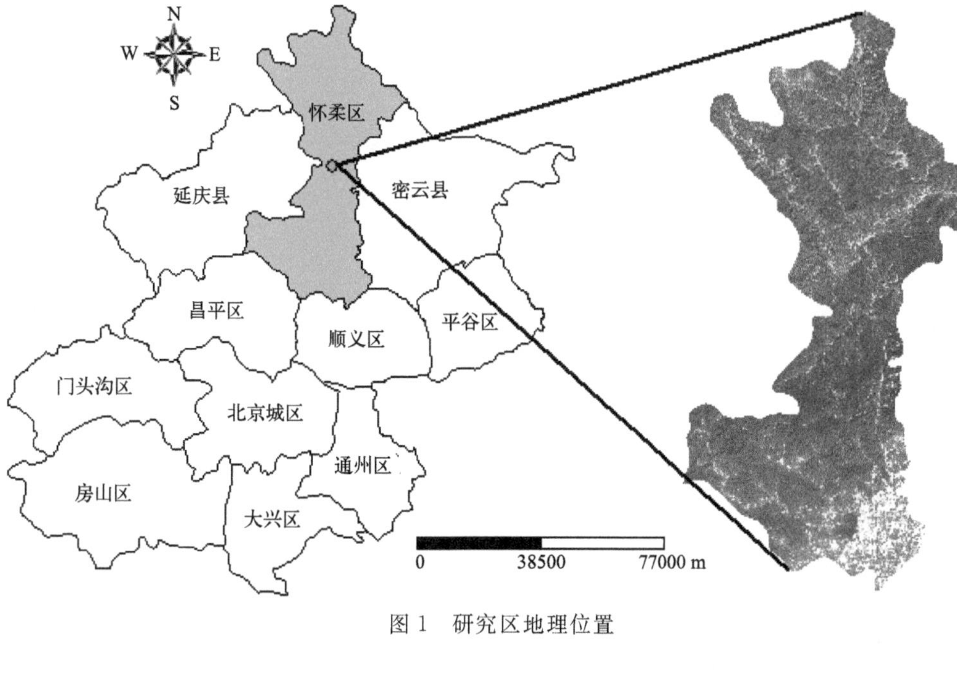

图 1　研究区地理位置

2.2 研究方法

2.2.1 数据源

数据源包括 ALOS/PALSAR 微波数据、Landsat-5 卫星遥感数据、研究区的 DEM、坡度和坡向数据。ALOS/PALSAR 微波数据获取时间是 2010 年 6 月 4 日，数据级别是 Level 1.5 级，有 HH 和 HV 两种极化方式，影像分辨率为 12.5 m，入射角为 34.3°；Landsat-5 卫星遥感数据的获取时间是 2010 年 6 月 5 日，影像分辨率是 30 m，通过查阅相关气象资料，在此期间没有出现下雨和自然灾害天气，因此可以认为两者准同步，通过联合使用 ALOS/PALSAR 微波数据与 Landsat-5 遥感数据反演最终得到研究区的土壤水分数据；研究区 DEM 和坡度数据是利用 ASTER GDEM 第一版本(V1)数据进行加工得来；坡向数据是利用研究区数字高程数据基于 ARCGIS9.3 中的函数模型加工得来；植被盖度数据是利用 Landsat-5 遥感数据，经大气校正、辐射定标后，采用 20 世纪末张仁华等提出的植被盖度与植被指数的模型计算得来(张仁华等，1996)。数据源如图 2 所示：

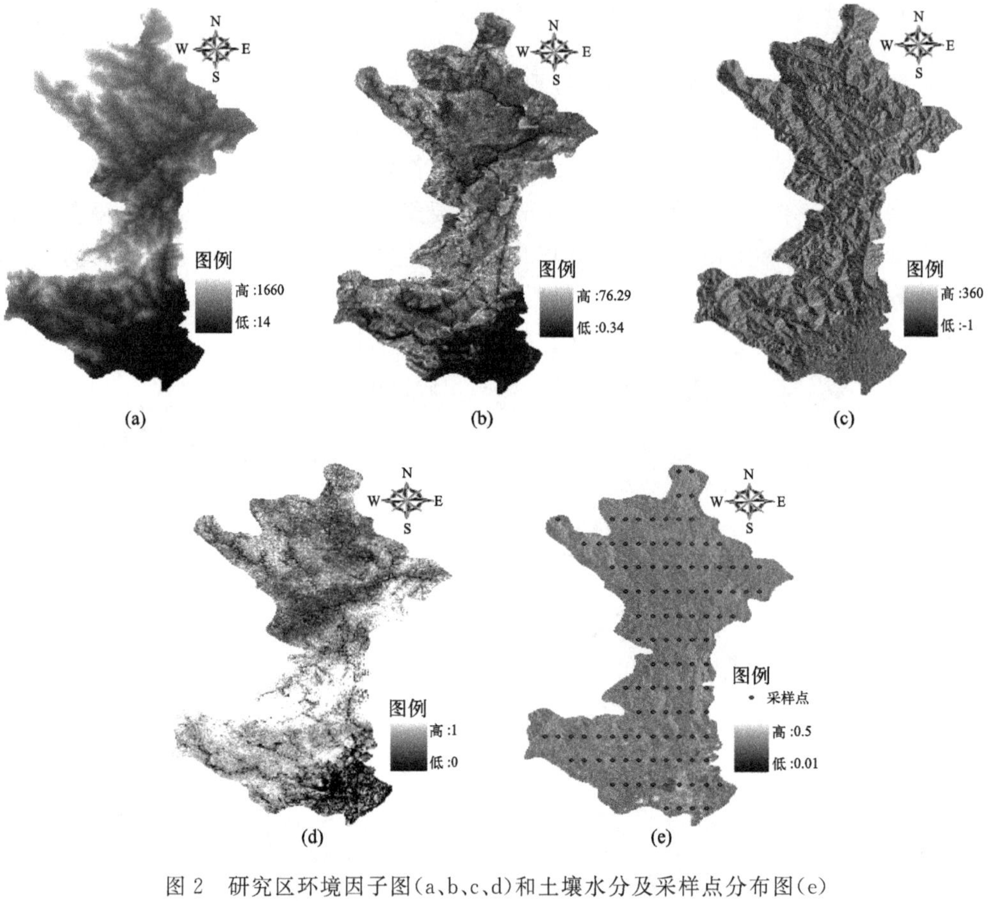

图 2 研究区环境因子图(a、b、c、d)和土壤水分及采样点分布图(e)

(a)研究区高程图；(b)研究区坡度图；(c)研究区坡向图；(d)研究区植被盖度图；(e)研究区土壤水分和采样点分布图

2.2.2 分析方法

主成分分析法是一种将原始数据中包含的多个变量压缩为几个不相关的分量，由于特征向量的空间正交性是一种过强而非希望的约束条件，造成其结果依赖于数据覆盖的空间范围。旋转主成分分析法可以有效的改进这些缺陷，并且能够突出主成分，得到一些简明而易于理解的空间模态（王军德等，2001；John D H *et al.*，薛薇，2001）。本次研究采用旋转主成分分析法对土壤水分与环境因子之间的关系进行研究，通过将研究区的高程、坡向、坡度分布图和植被盖度图与土壤水分分布图进行叠加，得到环境因子（高程、坡向、坡度和植被覆盖度）与土壤水分叠加后的数据库，基于此数据库，使用 SPSS 统计学分析软件进行旋转主成分分析，得出主成分因子和各因子的贡献率，在此基础上基于主成分因子与土壤水分叠加的数据库，进行单因子与土壤水分作用规律的分析。

3 结果与分析

3.1 环境因子对土壤水分影响的旋转主成分分析

土壤水分的空间分布是由多种因素综合作用的结果，本文结合研究区的地形和土地利用信息，在研究区采用均匀网格法选取测定样点（如图 2e 所示），共选取 115 个测点进行旋转主成分分析，在进行旋转主成分分析之前对四个环境因子进行 KMO（Kaiser-Meyer-Olkin）检验，以判断是否适于用旋转主成分分析。结果表明，四个环境因子的 KMO 均值为 0.636，大于 0.5，可以进行旋转主成分分析。采用 SPSS 软件进行旋转主成分分析提取特征根大于 1 主成分因子，最终得到两个主成分因子。各主成分因子的贡献率和主成分因子与环境因子的负荷矩阵如表 1 和表 2 所示。

表 1　变量共同度分析表

主成分	原始变量的特征值			提取的主成分		
	特征根	贡献率%	累计贡献率%	特征根	贡献率%	累计贡献率%
F1	1.997	49.914	49.914	1.997	49.914	49.914
F2	1.061	26.527	76.442	1.061	26.527	76.442
F3	0.514	12.843	89.285			
F4	0.429	10.715	100			

表 2　旋转主成分负荷矩阵

主成分	高程	坡度	植被盖度	坡向
F1	0.858	0.811	0.660	0.215
F2	0.023	0.029	-0.390	0.869

由土壤水分与环境因子的旋转主成分分析结果（表 1 和表 2）可以看出，对于使用微波数据反演的土壤水分来说，影响土壤水分的环境因子被归类为两个主成分，其中第一主成分主要有高程、坡度、植被盖度因子组成，占第一主成分的负荷分别是 0.858、0.811、0.660，与土

壤水分都表现为正相关，并且第一主成分对土壤水分的贡献率为 49.914%；第二主成分主要由坡向因子组成，对土壤水分影响的贡献率为 26.527%，坡向因子在第二主成分中所占负荷最大，达到 0.889，除植被盖度与第二主成分呈负相关以外，其余环境因子与土壤水分都表现为正相关，植被盖度、坡度和高程在第二主成分中所占比例依次减小，两个主成分的累计贡献率达 76.442%，以 4 个环境因子在主成分中所占的比例大小来确定环境因子对土壤水分的影响大小依次为：高程>坡度>植被盖度>坡向，高程和坡度是影响山区土壤水分空间变异的主控因子，植被盖度次之，坡向的影响最弱。这与 Hawley(1983) 和 Henninger(1976) 等人研究的结论一致，但与刘鑫(2007)等人在晋西黄土区的研究结论不同，究其原因作者认为黄土区由于植被较少，表层土壤水分受太阳辐射的影响较大，而坡向可以最大程度的反映太阳辐射的差异，最终使得坡向成为影响黄土区土壤水分变异的主控因子，而本文中的研究区域是有植被覆盖的山区，由于植被的影响，导致太阳辐射对表层土壤水分影响能力明显削弱，导致坡向对表层土壤水分的影响程度降低，而高程和坡度可以最大程度的影响山区土壤水分的流动和保蓄，使得其成为影响表层土壤水分变异的主控因子，其次植被盖度和坡向也是影响土壤水分分布的因素，因此对土壤水分差异的形成有一定的作用(葛翠萍等，2008)。

3.2 主控因子(高程和坡度)与土壤水分状况的关系

根据旋转主成分分析结果，基于主控因子(高程和坡度)与土壤水分的叠加数据库，用 Excel 进行高程和坡度对土壤含水量作用规律分析，结果如图 3 所示

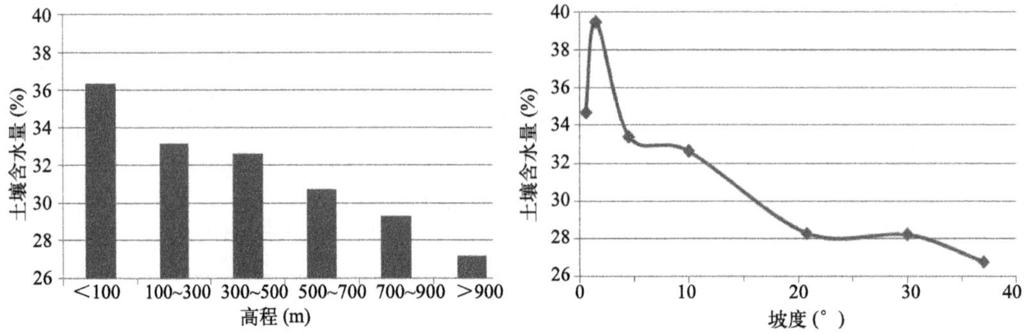

图 3 高程、坡度与土壤水分的关系曲线

从图 3 可以看出，土壤含水量随着高程的增加而逐渐减少，总体上呈现出高的地方土壤含水量低，低的地方土壤含水量高的态势；随着坡度的增加，土壤含水量总体上呈现先增加后减小的趋势，坡度在(0°～3°)时，土壤含水量随着坡度的增加而不断增大，坡度在 3°附近时土壤含水量最大，当坡度大于 3°以后，随着坡度的增加，土壤含水量逐渐减少。究其原因，作者认为研究区的低海拔区域由于地形平坦，有利于土壤水分的入渗和保蓄，使得土壤含水量较高，随着海拔的升高，山区地形由于不利于土壤水分的入渗，使得高海拔地区土壤含水量减少；再者研究区低坡度区域大多位于城区，由于城区混凝土地面不利于土壤水分的下渗，而且混凝土地面蒸发较大，致使城区土壤含水量较低，随着坡度的增加，城区减少，低坡度区域由于有利于土壤水分的入渗，土壤含水量逐渐增加，坡度在 3°左右土壤含水量达到最大，此后随着坡度的增加由于地形陡峭，地表径流增大，不利于山区土壤水分的入渗和保蓄，导致土壤含水量较低，这也是最终导致了土壤含水量随着高程和坡度的增加而不断减少的原因所在。

4　结论

通过对北京市怀柔区土壤水分与高程、坡度、植被盖度和坡向等环境因子的分析可知：

(1)经旋转主成分分析得出，影响北京市山区土壤水分分布的因素可以分为两大主要因子：依次是高程、坡度和植被盖度因子、坡向因子，其中高程和坡度对山区土壤水分的影响最为突出，植被盖度次之，坡向的影响最弱，因此，在山区进行植树造林工作应充分考虑这些因子对土壤水分分布特征的影响。

(2)高程和坡度作为影响研究区土壤水分分布的主控因子，对土壤水分的影响可表述为：土壤含水量随着高程的增加而逐渐减少，总体上呈现出高的地方土壤含水量低，低的地方土壤含水量高的态势；随着坡度的增加，土壤含水量总体上呈现先增加(坡度小于3°)后减小(坡度大于3°)的趋势，坡度在(0°～3°)时，土壤含水量随着坡度的增加而不断增大，坡度在3°附近时土壤含水量最大，当坡度大于3°以后，随着坡度的增加，土壤含水量逐渐减少。

(3)对于各环境因子对山区土壤水分的具体影响机制亟待进一步研究，这将有利于掌握山区土壤水分的变化特性，对于解决山区水源涵养和生态环境保护有重要意义。

参考文献

葛翠萍，赵军，王秀峰，等. 2008. 东北黑土区坡耕地地形因子对土壤水分和容重的影响. 水土保持通报，**28**(6)：16-19.

赫晓慧，温仲明. 2008. 小流域地形因子影响下的土壤水分空间变异性研究[J]. 水土保持研究，**15**(2)：80-83.

李保国，龚元石. 2000. 农田土壤水的动态模型及应用[M]. 北京：科学出版社.

李海滨，林忠辉，刘苏峡. 2001. Kriging方法在区域土壤水分插值中的应用. 地理研究，**20**(4)：446-452.

刘鑫，毕华兴，李笑吟，等. 2007. 晋西黄土区基于地形因子的土壤水分分异规律研究. 土壤学报，**44**(3)：411-417.

邱扬，傅伯杰，等. 2001. 黄土丘陵小流域土壤水分的空间异质性及其影响因子[J]. 应用生态学报，**12**(5)：715-720.

邱扬，傅伯杰，王军，等. 2000. 黄土丘陵小流域土壤水分时空分异与环境关系的数量分析[J]. 生态学报，**20**(5)：741-747.

邱扬，傅伯杰，王军，等. 2007. 土壤水分时空变异及其与环境因子的关系[J]. 生态学杂志，**26** (1)：100-107.

王军德，王根绪，等. 2001. 高寒草甸土壤水分的影响因子及其空间变异研究[J]. 冰川冻土，**28**(3)：428-433.

薛薇. 2001. 统计分析与SPSS的应用[M]. 北京：中国人民大学出版社.

张仁华. 1996. 实验遥感模型及地面基础[M]. 北京：科学出版社.

张秀英，冯学智，赵传燕. 2005. 基于GIS的黄土高原小流域水分时空分布模拟——以定西安家沟为例. 自然资源学报，**20**(1)：132-138.

张友静，王军战，鲍艳松. 2010. 多源遥感数据反演土壤水分方法[J]. 水科学进展. **21**(2)：222-228.

周鹏，丁建丽，王飞，等. 2010. 植被覆盖地表土壤水分遥感反演[J]. 遥感学报，**14**(5)：966-973.

Gomez-Plaza A, Martinea-Meca M, Albaladejo J, *et al*. 2001. Factors regulatingspatial distribution of soil water content in small semiarid catchments[J]. *Journal of Hydrology*, **253**: 211-226.

Hawley M E, Jackson T J, McCuen R H. 1983. Surface soil moisture on a small agricultural watershed[J]. *Journal of Hydrology*, **62**: 179-200.

Henninger D L, Petersen G W, Engman E T. 1976. Surface soil moisture within a watershed: Variations, factors influencing, and relationship to surface runoff[J]. *Soil Science Society of America Journal*, **40**: 773-776.

John D H. 1981. A rotated principal component analysis of the interannual variability of the northern hemisphere 500 hPa high field[J]. *Mon. Wea. Rev.*, **109**: 2080-2092.

Rodrguez-Iturbe I, Odorico P D, Porporato A, *et al*. 1999. On the spatial and temporal links between vegetation, climate and soil moisture[J]. *Water Resources Research*, **35**(12): 3709-3722.

Strasser U I, Mauser W. 2001. Modelling the spatial and temporal variations of the water balance for the Water catchment 1965—1994[J]. *Journal of Hydrology*, **254**(1): 199-214.

Western A W, Bloschl G, Grayson R B. 1998. Geostatistical characterization of soil moisture patterns in the Tarawarra catchment[J]. *Journal of Hydrology*, **205**(1): 20- 37.

INFLUENCES OF ENVIRONMENTAL FACTORS ON SPATIAL HETEROGENEITY OF SOIL MOISTURE—A CASE STUDY IN HUAIROU DISTRICT OF BEIJING CATCHMENT

CAI Qingkong[1), JIANG Jinbao[1), CUI Ximin[1),2), GUO Zheng[3), MA Kaifeng[1,4), HONG Xueqian[1)

[1) College of Geoscience and Surveying Engineering, China University of Mining and Technology, Beijing 100083, China

[2) State Key Laboratory of Coal Resources and Safe Mining, China University of Mining and Technology, Beijing 100083, China

[3) College of Resources Science Technology, Beijing Normal University, Beijing 100875, China

[4) College of Resources and Environment, North China University of Water Resources and Electric Power, Zhengzhou 450011, China

Abstract

In order to discuss spatial distribution pattern and influence factors of surface soil moisture in mountainous area, this paper took Huairou district of Beijing as the research area, the soil moisture of study area was inversed by the combined use of ALOS/PALSAR microwave data and Landsat-5 remote sensing data, we analyse the impact which bring about by four environment factors on soil moisture and the law of distribution, such as elevation, slope, aspect and vegetation coverage, and also determine the corresponding main control factor. Results indicate that the main control factors which influence mountain surface soil moisture spatial variability are elevation and slope, followed by vegetation coverage, the last is aspect. Single factor analysis on main control factor (elevation and slope) shows that soil moisure decreases gradually as height increases, with increase of slope, soil water content appears to increase first (slope less than 3°) and then decreases (slope greater than 3°) on the whole, it has guide significance for further study of the distribution characteristics of soil moisture and water and soil conservation.

Key words: ALOS/PALSAR; soil moisture; environmental factor; rotated principal component analysis

第三部分

气候变化对策的相关研究与进展

Part Three

The Related Study and Development on the Countermeasures to the Climate Change

EFFECTIVE PHOTOELECTROCATALYSIS DEGRADATION OF MICROCYSTIN-LR ON $Ag/AgCl/TiO_2$ NANOTUBE ARRAYS ELECTRODE UNDER VISIBLE LIGHT IRRADIATION①

LIAO Wenjuan and ZHANG Yanrong

Environmental Science Research Institute, Huazhong University of Science and Technology, Wuhan 430074, China

Abstract

$Ag/AgCl/TiO_2$ nanotube arrays electrode with enhanced visible-light photoactivity is synthesized by a two-step approach including an electrochemical anodization process and an electrodeposition process. $Ag/AgCl/TiO_2$ is used as an effective visible-light-driven (VLD) photoelectrocatalytic (PEC) electrode for the degradation of microcystin-LR (MC-LR). The reactive oxidative species (ROS) were derived from h^+ and e^- in the PEC process. The mechanism of the PEC degradation of MC-LR is investigated using different scavengers and found that the photogenerated h^+, $\cdot OH$ and $O_2^{\cdot -}$, are the major reactive species for the degradation of MC-LR. The effect of the operational parameters, such as the initial pH of MC-LR and the type of anions, are studied during the degradation process. The results show that the degradation of MC-LR was highly promoted in the NaCl electrolyte, and the low pH is contributed to the degradation of MC-LR. The experimental results reveal the feasibility of the PEC process to remove MC-LR.

Key words: $Ag/AgCl/TiO_2$ nanotube arrays; photoelectrocatalytic; microcystin-LR; reactive oxidative species

1 Introduction

Cyanobacterial toxins produced and released by cyanobacteria in freshwater source are well documented(Carmichael *et al.*, 2006; Sivoven, 1996). Microcystins (MC) are the most common toxigenic cyanobacteria found in lakes, rivers, ponds, tap water and mineral water(Liu I *et al.*, 2009), as well as being the most responsible for poisoning animals and humans who come into contact with toxic blooms and contaminated water (Codd *et al.*, 1989).

About 80 MC variants have been identified so far, with six variants, such as MC-LR, MC-LA, MC-YR, MC-RR, MC-LF and MC-LW being the most common. MC-LR is the

① The paper published in *Chem. Eng. J.*, **231**: 455-463. 2013 most abundant, around 46.0% − 99.8% of the total concentration of MCs in natural blooms (Monks *et al.*, 2007). As shown in Fig. 1 MC-LR contains three D-amino acids (alanine (Ala), methylaspartic acid (MeAsp), and glutamic acid (Glu)), two unusual amino acids (N-methyldehydroalanine (Mdha) and 3-amino-9-methoxy-2,6,8-trimethyl-10-phenyldeca-4,6-dienoic acid (Adda)), and two L-amino acids (leucine (Leu) and arginine (Arg)). The common Adda side chain shared in MC variants is mainly responsible for their toxicity, which interacts with and irreversibly inhibits the eukaryotic serine / threonine protein phosphatases 1 (PP1) and 2A (PP2A) (Ito *et al.*, 2001; Westrick *et al.*, 2010). Acute or chronic exposure has been shown to invoke hepatotoxicity and neurotoxicity, kidney impairment, gastrointestinal disorders and decreased testosterone levels (Monks *et al.*, 2007; Ito *et al.*, 2010). Nevertheless, MCs are very stable under natural sunlight and resist to high temperatures and UV radiation due to their cyclic structures (Apeldoorn *et al.*, 2007). Conventional treatment techniques are found to be inefficient in removing the toxins from the potable water system (Harada *et al.*, 1996; Keijola *et al.*, 1988; Bandala *et al.*, 2004).

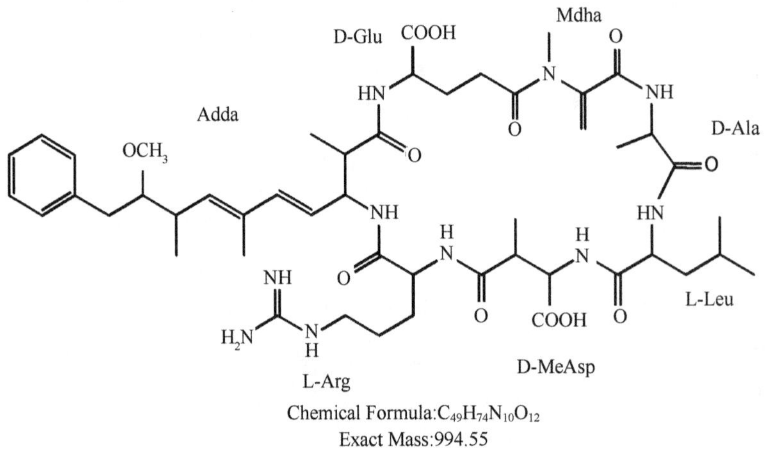

Fig. 1 Structure of microcystin-LR, where L-Leu is L-leucine, D-Me-Asp is D-erythro-β-methlaspartic acid, L-Arg is L-arginine, Adda is (2s,3s,8s,9s)-3-amino-9-methoxy-2,6,8-trimethyl-10-phenyldeca-4(E),6(E)-dienoic acid, D-Glu is D-glutamic acid, Mdha is N-methyl-dehydroalanine, and D -Ala is D-alanine

Advanced oxidation processes (AOPs) are considered as promising techniques in the destruction of microcystins. Previous investigations have shown that TiO_2 photocatalysis (Lawton *et al.*, 2004), Fenton and Fenton-like techniques (Bandala *et al.*, 2004), and UV/H_2O_2 (Qiao *et al.*, 2005) can effectively destroy microcystin-LR in aqueous solution, among which the photocatalytic technique based on TiO_2 photocatalyst has been broadly studied. For TiO_2 photocatalytic system, the major reactive species responsible for degradation of MC-LR are often proposed to be hydroxyl radical (•OH), which preferentially oxidizes the aromatic ring, the C_4-C_5/C_6-C_7 diene bond of the Adda chain (Liu *et al.*, 2003; Antoniou *et al.*, 2010), the double bond of the Mdha chain, and the two free carboxylic groups in D-Glu and D-MeAsp(Fang *et al.*, 2001). However, TiO_2 could only be activated under UV radiations due to its large optical band gap (3.0—3.2 eV). To overcome this drawback, numerous studies have been performed to enhance the photocatalytic efficiency and to extend the absorption of TiO_2 into the visible region (impurity doping, metallization and sensitization) (Sunada *et al.*, 2003; Elahifard *et al.*, 2007). $Ag/AgX/TiO_2$ (X=Cl, Br and I) has been reported as an efficient photocatalyst driven by visible light (Elahifard *et al.*, 2007; Yu *et al.*, 2009). In 2008, Wang *et al.* reported the fabrication of a highly efficient and stable plasmonic photocatalyst Ag/AgCl under visible-light illumination, suggesting that the electron-hole separation might occur smoothly in the presence of Ag^0 species on the surface of AgCl (Wang *et al.*, 2008). According to previous research, the surface plasmon resonance (SPR) effect of silver and silver halide makes it feasible to synthesize a new type of active and stable photocatalyst by combining the advantages of silver and silver halide nanoparticles with TiO_2 (Guo *et al.*, 2012). However, the drawback of the powder-form catalysts that is difficult to separate and recycle from the reaction media hinders their practical application.

In this work, new visible-light-driven plasmonic photocatalyst $Ag/AgCl/TiO_2$-NTs were prepared. The present study is to evaluate the photoelectrocatalytic activity of $Ag/AgCl/TiO_2$ nanotube towards the degradation of MC-LR, and the prepared samples show high visible-light photoelectrocatalytic (PEC) activity toward the degradation of MC-LR in aqueous solution. In order to clarify the mechanism of MC-LR degradation by PEC, the reactive species (h^+, e^-, $\cdot OH$, $O_2^{\cdot -}$ and H_2O_2) for destruction of MC-LR were discussed. In addition, to find the optimal conditions of PEC, the main operating variables, such as the initial pH, and the type of anions were studied in detail.

2 Experimental

2.1 Preparation of $Ag/AgCl/TiO_2$ Nanotube Arrays

The highly ordered TiO_2 Nanotube Arrays (TiO_2-NTs) was synthesized by anodic oxidation in a NH_4F electrolyte, following the method described previously (Gong *et al.*, 2012). AgCl nanoparticles were deposited into the ordered TiO_2-NTs by using an electrodeposition method. A conventional three-electrode system was employed with TiO_2-NTs as the cathode, a Pt sheet as the anode and a saturated calomel electrode (SCE) as the reference electrode. First, the foil of TiO_2-NTs was subjected to +1 V (*vs* SCE) in a 0.5 M HCl aqueous solution for 10 min at room temperature, which made Cl^- transfer onto the surface of the TiO_2-NTs. The resulted Cl^-/TiO_2-NTs samples were dried at 70℃. The foil of Cl^-/TiO_2-NTs with a constant voltage of -0.6 V (*vs* SCE) in a 0.01 M $AgNO_3$ aqueous solution for 10 min at room temperature, in which process AgCl nanoparticles were deposited into the interior tubes of TiO_2-NTs (denoted as $AgCl/TiO_2$-NTs), and Ag nanoparticles were simultaneously formed. Such an electrodeposition/electroreduction cycle was repeated several times, typically between 1 and 3 cycles. The prepared samples were irradiated by an 18 W UV lamp for 10 min to reduce Ag^+ ions to Ag species by photochemical decomposition of AgCl via TiO_2 photocatalytic reduction (Yu *et al.*, 2009) and finally formed the Ag/AgCl/TiO_2 nanotube arrays.

2.2 Characterization

The morphology observation was performed on a field emission scanning electron microscope (FE-SEM, NANOSEM 450, FEI) operating an accelerating voltage of 30 kV. The phase purity and crystal structure of the obtained samples were examined by X-ray diffraction (XRD) using Rigaku Ultima IV X-ray diffraction (Rigaku Corporation, Japan) equipped with Cu-Kα radiation (40 kV, λ=1.5406 Å), the 2θ scanning angle range was 20°—80° with a step of 0.05° s^{-1}. UV-visible diffuse reflectance spectra (DRS) was obtained on a UV-visible spectrophotometer (UV-3600, Japan). The analyzed range was 250—800 nm, and $BaSO_4$ was used as a reflectance standard.

2.3 Photoelectrocatalytic Degradation of MC-LR under Visible Light

Experiments were carried out with a standard three-electrode configuration in one glass cell with a working solution volume of 100 mL. All the experiments were performed with magnetic stirring, using 0.01 M $NaNO_3$ as the electrolyte, and the volume was 40 mL, which was placed on a magnetic stirrer fixed a constant stirring rate of 350 rpm. The initial concentration of the MC-LR aqueous solution was 1 mg/L by using an electrochemical workstation (EC-510, China) Ag/AgCl/TiO_2-NTs electrode with an exposed area of 3×5 cm^2 to irradiation served as a photo-anode, Pt foil as the counter electrode and a standard calomel electrode (SCE) as the reference electrode, respectively. The cylindrical cell was cooled by flowing tap water in the process of measurements to keep the temperature of the interior cell around 298 K. Light source was from a 60 W incandescent lamps(PHILPS, Netherlands) and projected on the surface of the Ag/AgCl/TiO_2-NTs electrode with a light intensity of 36610 Lux measured by a visible-light radiometer (testo 540, Germany). Prior to irradiation, the solution was magnetically stirred in the dark for ca. 30 min to establish adsorption/desorption equilibrium between the MC-LR and the surface of Ag/AgCl/TiO_2-NTs under room air equilibrated conditions. At given irradiation time intervals, 300 μL aliquots were collected. The 1 M HNO_3 and 1 M NaOH solutions were used to adjust pH to the desired values and the HNO_3 solution did not cause any removal of MC-LR. The concentration of MC-LR was measured byHPLC (Shimadzu, Japan) equipped with a UV detector and a C_{18} reverse-phase column (4.6 mm i.d. × 150 mm Agilent, USA). The mobile phase was water containing 0.05% trifluoroacetic acid (v/v) and methanol, with a ratio of 32∶68 (v/v). The injection volume of the sample was 20 μL and the flow rate was 1 mL/min. The wavelength of the UV absorbance detector was fixed at 238 nm. To further determine the mineralization of MC-LR, changes in total organic carbon (TOC) were determined using TOC analyzer (Multi N/C, Jena, Germany). To check the effectiveness and the reproducibility of the photoelectrocatalytic process, the measurements of all parameters were repeated in triplicate.

2.4 Hydroxyl Radical Measurements

The formation of hydroxyl radicals (•OH) atphoto-irradiated Ag/AgCl/TiO_2-NTs electrode/water interface was detected by the photoluminescence (PL) technique using terephthalic acid as a probe molecule. Terephthalic acid reacts with •OH to produce highly fluorescent product 2-hydroxyterephthalic rapidly and specifically, which showed the PL signal at 425 nm excited by 315 nm light(Ishibashi *et al.*, 2000). Experimental procedures were similar to the measurement of photoelectrochemical degradation except that MC-LR aqueous solution was replaced by the1 $\times 10^{-3}$ M terephthalic acid aqueous solution with a concentration of 4×10^{-3} M NaOH. PL spectra of generated 2-hydroxyterephthalic acid were measured on a fluorescence spectrophotometer (Jasco FP-6200, Japan).

3 Results and Discussion

3.1 Morphology and Structure of NTs Film

Fig. 2(a, b) shows the top-views of the TiO_2-NTs and the Ag/AgCl deposited TiO_2-NTs, respectively. After the electrodeposition/electroreduction and photoassisted reduction process, Ag/AgCl nanoparticles distributed on the TiO_2 nanotube arrays.

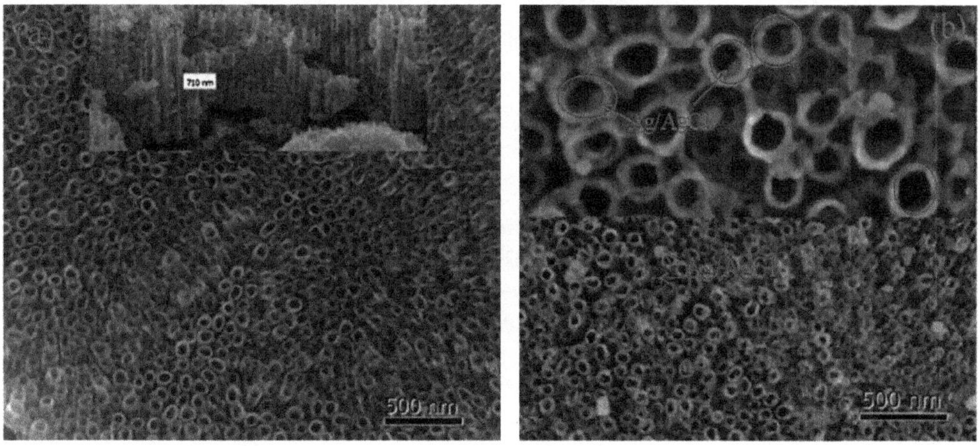

Fig. 2 SEM top-view images of (a) TiO_2- NTs and (b) Ag/AgCl/TiO_2-NTs, the inset in (a) shows the cross-sectional SEM image of TiO_2- NPs

The highly ordered nanotubular structure of TiO_2 still retainedits integrity after Ag/AgCl NPs deposition, suggesting that the deposition process did not damage the ordered TiO_2 matrix. The inset gives a cross-sectional image of the TiO_2 nanotube arrays, from which the nanotubes with length and wall thickness of about 710 nm and 20 nm were observed vertically oriented on the Ti substrate. The XRD results show that both anatase TiO_2 and AgCl are present in the fresh Ag/AgCl /TiO_2- NTs (Fig. 3).

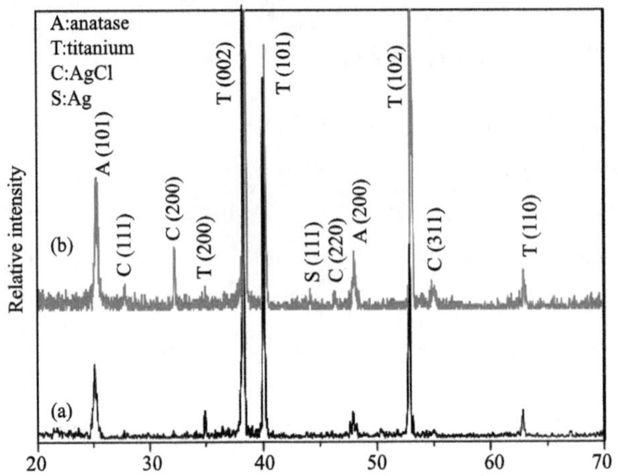

Fig. 3 XRD patterns of (a) TiO_2- NTs and (b) Ag/AgCl/TiO_2- NTs

The intensive peak of the A (101) crystal face at 2θ of 25. 2° indicates a fine preferential growth of the TiO_2-NTs in the 101 orientation. Compared to the XRD pattern of the TiO_2-NTs, additional peaks at 2θ of 27. 8°, 32. 2°, 54. 8° appeared on the XRD pattern of Ag/AgCl/TiO_2-NTs (curve bin Fig. 3), which could be attributed to the cubic phase of AgCl with lattice constant of 5. 5491 Å (JCPDS file no. : 31-1238) marked with C in curve b. A diffraction peak assigned to metallic Ag is observed at 2θ of 44. 2°, probably due to its low content and high dispersity. Additionally, the DRS (Fig. 4) investigations verify that the loading of Ag/AgCl nanoparticles extends the absorption of the TiO_2 nanotube from the UV light of 380 nm to the visible light of 400—550 nm, andthe band gap

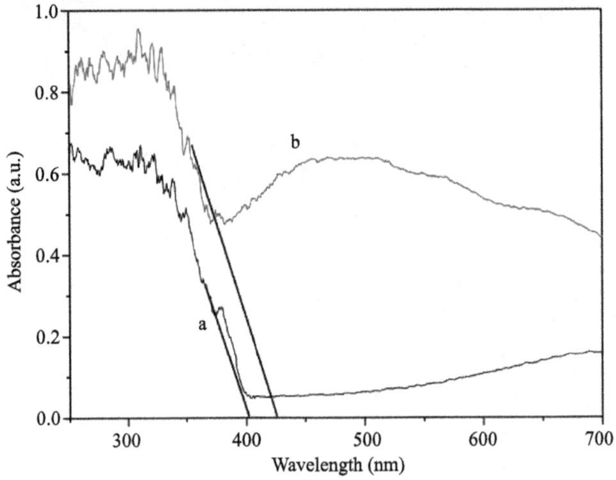

Fig. 4 UV-visible diffuse reflectance spectra of the TiO_2 (a) and Ag/AgCl/TiO_2 (b)

value of $Ag/AgCl/TiO_2$ was 2.3—3.0 eV, which was ascribed to the existence of Ag NPs produced by electroreduction and photochemical decomposition of AgCl.

3.2 PEC Degradation of MC-LR

Fig. 5 compared the degradation of MC-LR on the $Ag/AgCl/TiO_2$-NTs electrode under different processes, including direct photolysis (DP), electrochemical oxidation (EO), photocatalysis (PC) and photoelectrocatalysis (PEC) at the initial pH = 7.0 solution. The direct photolysis with incandescent light (the wavelength of light larger than 450 nm) alone shows no activity for the MC-LR degradation in 5 h. The stability of the molecule at this wavelength is because MC-LR does not absorb light of this wavelength range. The similar inactivity to MC-LR is found in the electrochemical process, affirming that the bias potential of + 0.6 V (*vs* SCE) was too low to induce MC-LR removal. However, compared to direct photolysis, the $Ag/AgCl/TiO_2$-NTs foil (without a bias potential) destructs 17% of MC-LR during the same time, implying that the photocatalyst was effective for the removal of MC-LR under incandescent light irradiation. Clearly, when the foil was subjected to a bias potential of +0.6 V, the MC-LR with the initial concentration of ca. 1.00 mg L^{-1} is progressively reduced in the proceeding of photoelectrocatalysis and completely degraded within 5 h under irradiation. The PEC degradation of the MC-LR follows a pseudo-first-order reaction with a rate constant of 0.4919 h^{-1}. Moreover, the removal efficiency (~ 92%) of MC-LR in the photoelectrocatalysis process is considerably larger than the sum of electrochemical process (0%) and photocatalysis (~ 17%), which was consistent with the results previously reported(Hou *et al.*, 2012). The given bias potential supplied to the $Ag/AgCl/TiO_2$-NTs contributes to the separation of the photo-generated electronsand the holes. The charge

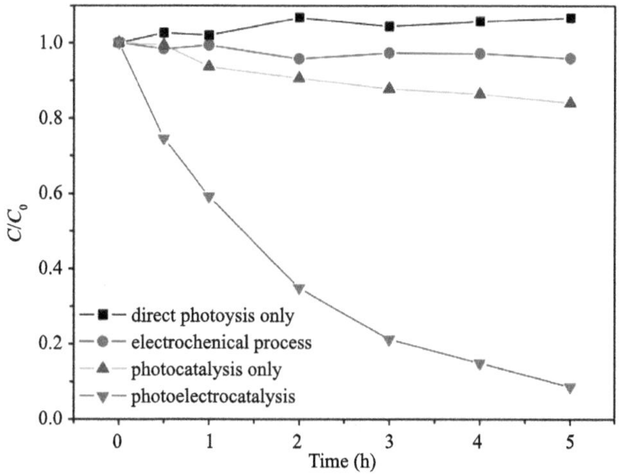

Fig. 5　Degradation of MC-LR by direct photolysis, electrochemical process, photocatalysis and photoelectrocatalysis, respectively, over the $Ag/AgCl/TiO_2$—NTs electrode under visible light irradiation (λ > 450 nm, 36610 Lux, 0.6 V vs SCE).

separation prevents the recombination of photo-generated hole-electron pairs. Thus, the electronsand holes have more opportunities to participate in degradation reaction, which causes to increase the quantity of active species. Consequently, the degradation efficiency increases in the photoelectrocatalytic process.

Moreover, Fig. 5b also shows that the total organic carbon of the solution continuously decreased. However, the TOC results indicated that the rate of MC-LR reduction was slower than that of PEC degradation of MC-LR. After 5 h of irradiation for $Ag/AgCl/TiO_2$-NTs, the decrease in the TOC of solution is about 65.8%. This is not difficult to understand because the complete mineralization of MC-LR requires a long time.

We find that other organic pollutants and microorganism such as rhodamine B, phenol and E. *coli* could also be decomposed by the prepared composite films under visible light irradiation, indicating that the prepared $Ag/AgCl/TiO_2$-NTs film was an efficient visible-light photocatalyst in the process of PEC. As a catalyst, the stability is very important for its practical application, recycling experiments were therefore performed. As shown in Fig. 6, after four cycles of PEC degradation of rhodamine B, the sample didn't show any significant loss of PEC activity, which indicated that the catalyst could keep stable during the PEC reaction.

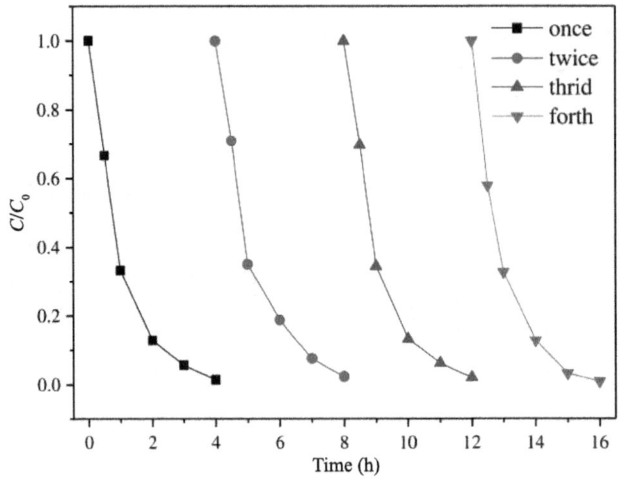

Fig. 6 Irradiation-time dependence of the relative concentration of RhB (1×10^{-5} mol L^{-1}, 50 mL), over $Ag/AgCl/TiO_2$ during repeated PEC composition experiments under visible light irradiation(19610 Lux).

3.3 Effect of pH

To explore the effect of pH, the PEC degradation of MC-LR with an initial concentration of about ~ 1.00 mg L^{-1} on the $Ag/AgCl/TiO_2$-NTs electrode was investigated at five pH conditions (3.5, 5.4, 7.0, 9.0 and 10.6). The initial degradation rates have been found to be significantly influenced by pH as shown in Fig. 7a. The degradation rate of MC-LR increases with the decrease of initial pH, which was in agreement with previous research(Pelaez *et al.*, 2011). The highest initial reaction rate is observed for the most acidic media of pH 3.5 (rate constant 0.696 h^{-1} as simulated by the pseudo-first-order reaction) and decreases at pH 5.4 (0.616 h^{-1}). With the solution pH adjusted to neutral and alkaline, the degradation rate further decreases and the lowest removal is observed in alkaline media of pH 10.6 (0.335 h^{-1}).

The degradation of organic chemicals is highly dependent on the heterogeneous interaction between the catalyst and the reactants in the PEC process. The effect of pH can be estimated from the change of the interaction, by affecting either the species distribution ratio of reactants in the solution or the surface charge of catalyst(Wang *et al.*, 2009; Malato *et al.*, 2009; Ojani *et al.*, 2012). It has been determined that the pKa values of MC-LR are 2.09, 2.19 and 12.48 with the ionization of two carboxylic groups and one free amino group, hence in the range of pH 3.5－10.6, the dominant species of MC-LR-$(COO^-)_2(NH_2^+)$ with an overall charge of －1 can be formed by the deprotonation of two carboxylic groups and protonation of one free amino group(Maagd *et al.*, 1999). On the other hand, the surface charge of Ag/AgCl is independent on the pH, whereas at more acidic pH, the exposed TiO_2 catalyst surface carries a positive charge due to protonation ($TiOH_2^+$), while it carries a negative charge at basicity (pH＝9.0, 10.6) due to hydroxylation ($TiOH^-$) (Yuan *et al.*, 2012). Accordingly, in the low pH (3.5, 5.0), $TiOH_2^+$ and $MCLRH^-$ show a more favorable adsorptive property, so an increase of the removal rate is observed. When pH is over 7.0 (9.0, 10.6), $TiOH^-$ and $MCLRH^-$ (some $MCLR^{2-}$ could be formed) should repulse each other, make it difficult for the $MCLRH^-$ to approach the catalyst surface. Although the yield of hydroxyl radical in the PEC process increases as the pH of the aqueous solution increases(Liang *et al.*, 2008; Doong *et al.*, 2001), which should enhance the degradation rate of MC-LR, the former factor is evidently overwhelming and finally causes the decrease in the degradation of MC-LR with the increase of solution pH in the PEC process.

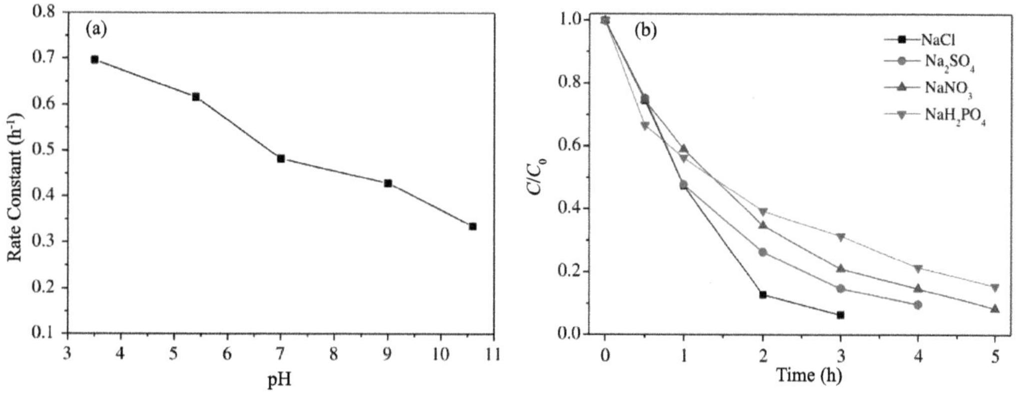

Fig. 7 (a) The PEC degradation efficiency of MC-LR at different pH, (b) Effects of different anions on the PEC degradation of MC-LR (the concentration of the electrolyte 0.01 M, initial pH 7.0)

3.4 Effects of Anions

Inorganic anions such as chloride, sulfate, nitrate and phosphate are considerably common in waste-waters and also in natural water. The importance of anions effect on the degradation of pollutants has been remarkably recognized (Liang *et al.*, 2008). Fig. 7b shows the effects of different anions (i. e. Cl^-, NO_3^-, $H_2PO_4^-$ and SO_4^{2-}) at the same concentration of 0.01 M and pH of 7.0. It is easily observed that the existence of $H_2PO_4^-$ reduced the MC-LR degradation to a certain degree, whereas the presence of Cl^- and SO_4^{2-} promoted the MC-LR degradation rate. Compared to the control test in the nitrate aqueous solutions, in which the complete degradation of the MC-LR was achieved after 5 h, in NaCl and Na_2SO_4 aqueous solution, the degradation rate of above 90% is achieved after 3 h and 4 h, respectively. However, in NaH_2PO_4 aqueous solution, the removal efficiency is only 85% after 5 h. In the case of Cl^-, the formed chloride radical, with a higher potential (+2.47 V) than valence band holes h^+ (+2.39 V), is capable of oxidizing organic compounds effectively (Hirakawa *et al.*, 2002). On the other hand, as shown in Fig. 8a, the more ·OH is produced in sodium nitrate aqueous solution (twice) than in sodium chloride aqueous solution in the PEC process. It will be explainable by the fact that chloride ions could act as scavengers of positive holes (h^+) and hydroxyl radical (·OH) (Liang *et al.*, 2008), and the by-products including chlorine atom, dichloride radical and free chlorine were formed (Eq. (1)-(3)) (Ghodbane *et al.*, 2010), which could add some unsaturated bonds of MC-LR (Eq. (4)) (Yuan *et al.*, 2012; Yuan *et al.*, 2011), such as the double bond of the Mdha chain and the C_4-C_5/C_6-C_7 diene bond of the Adda chain to generate chlorinated hydrocarbons (RCl). The halogenated organic compounds are easy to adsorb on the photoelectrocatalyst surface (Yuan *et al.*, 2011). Because the PEC process mainly occurs on the heterogeneous interface of the photoelectrocatalyst surface and the aqueous solution, when the halogenated organic compounds are in appropriate amount, it promotes the pollutants degradation (Yuan *et al.*, 2011).

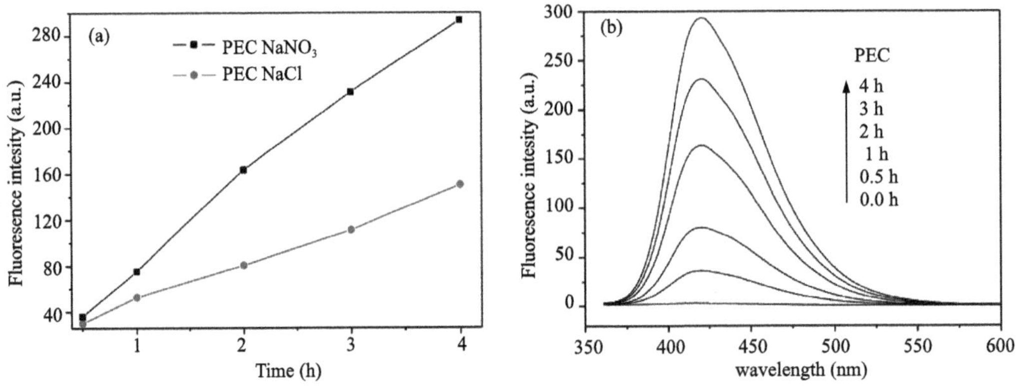

Fig. 8 (a) PL spectral changes with visible-light irradiation time on Ag/AgCl/TiO_2-NTs in 0.01 M $NaNO_3$ aqueous solutions and 0.01 M NaCl aqueous solutions; (b) Temporal PL spectra in the PEC process over the Ag/AgCl/TiO_2-NTs in a 1×10^{-3} M terephthalic acid with pH of under visible light irradiation

$$Cl^- + \cdot OH \rightarrow Cl\cdot + OH^- \quad (1)$$

$$Cl^- + h^+ \rightarrow Cl\cdot \quad (2)$$

$$Cl- + Cl^- \rightarrow Cl_2{}^{\cdot -} \quad (3)$$

$$Cl_2{}^{\cdot -} \text{ or } Cl\cdot + R_1C{=}CR_2 \rightarrow R_1ClC^{\cdot}{=}CR_2 \rightarrow\rightarrow AOX \quad (4)$$

In Na_2SO_4 aqueous solution (pH=7.0), the $SO_4{}^{2-}$ can be oxidized on Ag/AgCl/TiO_2 electrodes forming $S_2O_8{}^{2-}$ (+ 2.01V) and $SO_4^{\cdot -}$ finally(Canizares *et al.*, 2004; Chawla *et al.*, 1975). The $SO_4{}^{\cdot -}$ radical (+ 2.5 — + 3.1 V) is a stronger oxidant than the nitrate radicals and even stronger than $\cdot$OH (+ 1.89 — + 2.72 V)(Antoniou *et al.*, 2010). Many organic compounds react with $SO_4^{\cdot -}$ more efficiently than with $\cdot$OH, which not only react via hydrogen abstraction or addition similar to $\cdot$OH, but also take more specific reaction than by $\cdot$OH attacking(Neta *et al.*, 1988). The results from Antoniou *et al* (2010) showed that $SO_4{}^{\cdot -}$ could selectively remove the carboxylic group on the Glu or MeAsp in MC-LR, by which the toxin could be removed from the Glu or MeAsp sites.

As to the effect of $H_2PO_4{}^-$, it probably reacts with h^+ to form $H_2PO_4{}^-$, $HPO_4{}^{\cdot -}$ and $PO_4{}^{2\cdot -}$, all less reactive for the effective destruction of MC-LR(Liang *et al.*, 2008; Neta *et al.*,).

3.5 PEC Mechanism

Under visible light irradiation, TiO_2 cannot be excited due to its large band gap (3.2 eV), which was excited only with the wavelength less than 385 nm. On the contrary, Ag NPs can absorb visible light due to its surface plasmonic resonance (SPR) adsorption. Under visible light irradiation, SPR excited Ag nanoparticles ignite the photogenerated electron-hole pairs, in which the photoexcited electrons at the silver NPs are easily injected into the TiO_2 conduction band due to a similar Fermi level(Zhou *et al.*, 2011). Although the detailed mechanism of such transfer is still unclear at present, the electrons injection from Ag to TiO_2 on SPR excitation has been proven to occur in Ag/TiO_2 and Au/TiO_2 systems(Yu *et al.*, 2009; Tian *et al.*, 2005). Simultaneously, the leftover holes diffuse into the surface of the AgCl particles and cause the oxidation of Cl^- ions to Cl^0 atoms because the surface of AgCl particles was negatively charged and most likely terminated by Cl^- ions (Yu *et al.*, 2009). As Cl^0 atoms are reactive radical species, they are able to remove MC-LR and reduce Cl^0 atoms to Cl^- again. The holes can also directly react with MC-LR or interact with surface-bound H_2O or OH^- to produce the OH^- radical species(Chen *et al.*, 2011). The photoexcited electrons injected into the TiO_2 conduction band can be transferred to the ubiquitously present molecular oxygen to form first the super-oxide radical anions $O_2{}^{\cdot -}$, then upon protonation yields the HOO$\cdot$ radicals, which combines with the trapped electrons to produce H_2O_2 and finally form $\cdot$OH radicals(Yu *et al.*, 2009; Hou *et al.*, 2012). All these active species will be probably involved in the degradation and mineralization of MC-LR. Among the active species, it is evident that electrons and holes are the sources for other reactive species. In PEC process, visible light radiation activates $Ag/AgCl/TiO_2$-NTs photoanode by transferring electrons from the valance band (VB) to the conduction band (CB), while at the same time the applied external potential prevents the recombination of photo-generated electron-hole pairs. Thus, the efficiency of MC-LR degradation was enhanced. The major reaction steps in this plasmonic PEC degradation of MC-LR mechanism under visible light irradiation are summarized in Eqs. 5—13.

$$\text{Ag-NPs} + h\nu \rightarrow \text{Ag-NPs}^* \quad (5)$$

$$\text{Ag-NPs}^* + TiO_2 \rightarrow \text{Ag-NPs}^{\cdot +}(h^+) + TiO_2(e) \quad (6)$$

$$TiO_2(e) + O_2 \rightarrow TiO_2 + O_2^{\cdot -} \quad (7)$$

$$O_2^{\cdot -} + H^+ \rightarrow \cdot OOH \quad (8)$$

$$O_2^{\cdot -} + \text{MC-LR} \rightarrow\rightarrow CO_2 + H_2O \quad (9)$$

$$\cdot OOH + TiO_2(e) + H^+ \rightarrow TiO_2 + H_2O_2 \quad (10)$$

$$h^+ + OH^- \rightarrow \cdot OH \quad (11)$$

$$h^+ + (\text{MC-LR})\text{-COOH} \rightarrow [(\text{MC-LE})_{ad}\text{-COOH}]^{\cdot +} \rightarrow \text{fragment } (\text{MC-LR})^{\cdot +} + CO_2 \quad (12)$$

$$\cdot OH + \text{MC-LR} \rightarrow\rightarrow CO_2 + H_2O \quad (13)$$

3.6 Analysis of Reactive Species

To determine which reactive species were involved in the degradation of MC-LRover the $Ag/AgCl/TiO_2$ film in the PEC process, different scavengers were used individually to remove the specific reactive species and had no effect on the $Ag/AgCl/TiO_2$ film. The scavengers used in this study are sodium oxalate ($Na_2C_2O_2$) (0.5 mmol · L^{-1}) for h^+ (Chen *et al.*, 2011; Zhang *et al.*, 2010), isopropanol (1 mmol · L^{-1}) and sodium fluoride (NaF) (5 mmol · L^{-1}) for · OH in the bulk solution (· OH_{bulk}) and the surface of electrode (· OH_{ads})(Hou *et al.*, 2012), Cr (VI) (0.05 mmol · L^{-1}) for e^- (Chen *et al.*, 2005) and p-benzoquinone (0.5 mmol · L^{-1}) for $O_2^{\cdot -}$ (Hu *et al.*, 2009). It needs to mention that the applied concentration of each scavenger did not cause any removal of MC-LR by the corresponding control experiment.

As shown in Fig. 9a, b, in the absence of any scavenger, the degradation of MC-LR is 92% after 5 h by the PEC treatment, and the PEC degradation of MC-LR follows pseudo-first-order kinetics with the rate constant of 0.4819 h^{-1} as mentioned above. In the presence of $Na_2C_2O_2$ (a hole scavenger), the degradation of MC-LR is only 20.1% after 5 h, with the rate constant of 0.043 h^{-1}, indicating the involvement of · OH and Cl · generated by the oxidative pathway in the PEC degradation of MC-LR (Eqs. 2, 11). In addition, the photoinduced holes may directly react with MC-LR. By contrast, the photocurrent in the presence of the h^+ scavenger of $Na_2C_2O_2$ is rather higher than that without any scavenger, as shown in Fig. 10, which was probably due to an inhibition effect on the recombination of the photoinduced electrons and holes aroused by $Na_2C_2O_2$. With the addition of Cr (VI) (a electron scavenger), 86.5% of the MC-LR is removed in the same time with the rate constant of 0.40825 h^{-1}, which suggested that e^- also exerted some influence on the degradation of MC-LR to a certain degree. The results from the addition of Cr (VI) may be attributed to the facts: (1) the electron can reduce the absorbed oxygen to generate $O_2{}^{\cdot -}$, which undergoes protonation to yield HOO· radicals, and finally produces ROS including H_2O_2 and ·OH radicals(Yu *et al.*, 2009) by combining with the trapped electrons. The addition of Cr (VI) interrupts the pathway of oxygen reduction and therefore decreases the production of ROS, which restrains the removal of MC-LR; (2) Cr (VI) is also contributed to inhibit the recombination of the photoinduced electron and the hole, which promotes the reaction rate. The two factors play an inversely role in the removal of MC-LR, and evidently, the former is a dominate one. In addition, the introduction of the $O_2{}^{\cdot -}$ scavenger of p-benzoquinone markedly depresses the degradation of MC-LR as well, the degradation of MC-LR is only 53.8% in 5 h, slightly larger than that of the h^+, suggesting the involvement of $O_2{}^{\cdot -}$ in the reaction. Although the produced $O_2{}^{\cdot -}$ in water is very unstable, it will subsequently undergo facile disproportion to produce other reactive species including H_2O_2 and ·OH. The above results show that ROS generated by both the oxidative pathway and reductive pathway are responsible for the degradation of MC-LR in the PEC process mediated by the Ag/AgCl/ TiO_2-NTs.

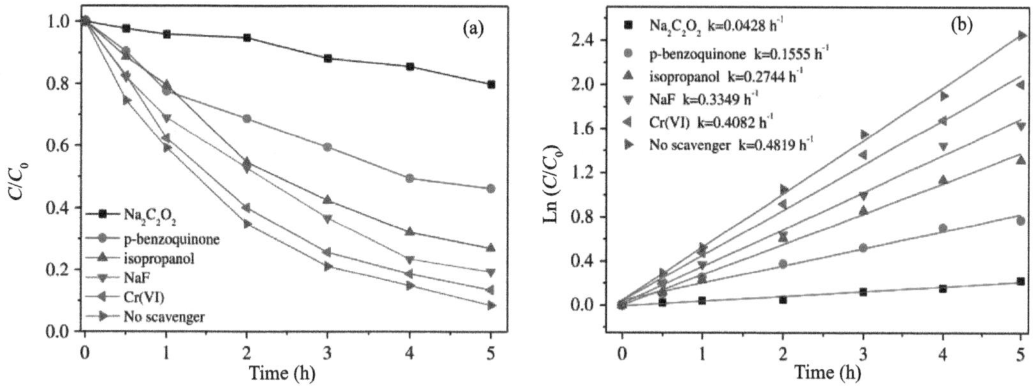

Fig. 9 Effects of different scavengers on the PEC degradation of MC-LR(0.05 mmol·L^{-1} Cr(VI): e-scavenger, 1 mmol·L^{-1} is opropanol: ·OH_{bulk} scavenger, 0.5 mmol·L^{-1} p-benzoquinone, 5 mmol·L^{-1} NaF: ·OH_{ads}, 0.5 mmol·L^{-1} sodium oxalate: h^+ scavenger)

The generation of ·OH was investigated through the method of photoluminescence with terephthalic acid. The PL intensity strengthens with increasing irradiation time in the PEC process (Fig. 8b), indicating that the amount of produced ·OH radicals increases with the proceeding of the photoelectrocatalysis. Previous studies have proved that hydroxyl radical attack is one major destruction pathway for MC-LR in the AOPs, by destroying some functional groups in MC molecular to form the dihyroxylated products. To clarify whether the destruction of MC-LR is induced by ·OH remaining bound to the surface or those diffused into the solution, isopropanol is employed at first as a diagnostic tool for the diffused • OH because it is easily oxidized by • OH (rate constant of $1.9 \times 10^9 M^{-1} \cdot s^{-1}$) and has a low affinity to semiconductor surfaces in aqueous media. The addition of isopropanol obviously inhibits the degradation of MC-LR at 73.1% after 5 h of irradiation and the rate constant is half as that without scavengers ($0.2744\ h^{-1}$), which suggests that diffused • OH is involved in the degradation of MC-LR. Subsequently, NaF is introduced to study the role of $\cdot OH_{abs}$, because the fluoride shows strong and stable adsorption on photocatalyst and minimizes the concentration of surface OH^- on the photocatalyst by fluoride-exchange. Upon the substitution of surface OH^- with the fluoride, the production of $\cdot OH_{abs}$ can be significantly reduced. With the addition of 5 mmol • F^-, the degradation efficiency is depressed to 80.5% with the rate - constant of $0.3369\ h^{-1}$. The surface of the Ag/AgCl/TiO_2-NPs is negatively charged in neutral solution, which is the same with surface charge of MC-LR. Thus, as the electrostatic repulsion and the motility of MC-LR, the $\cdot OH_{ads}$ has little chance to destruct the major portion of MC-LR than that by $\cdot OH_{bulk}$ in PEC.

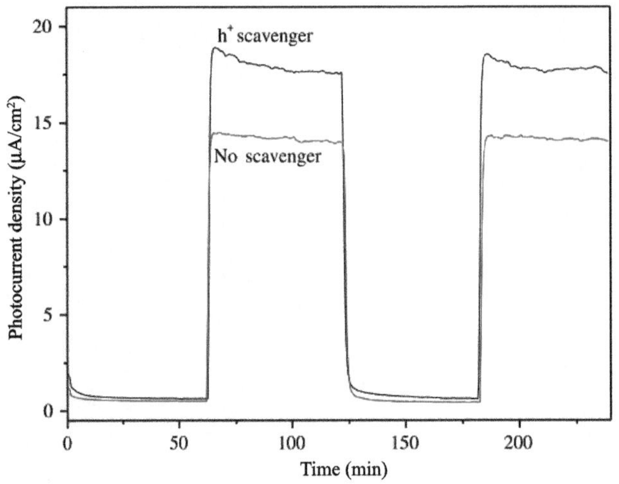

Fig. 10 Comparison of transient photocurrent response in the process of PEC degradation of MC-LR with and without h^+ scavenger(0.5 mmol • L^{-1} sodium oxalate) contained in the solution.

Fe^{2+} is generally used as a scavenger for H_2O_2 in various AOPs(Chen *et al.*, 2011, Zhang *et al.*, 2010), however, our experiment discovered that Fe^{2+} had a disaster effect on the Ag/AgCl/ TiO_2-NTs electrode. The introduction of Fe^{2+} in the reaction system will unrecoverably decrease the current density, and the recovery needs immersing the electrode into HCl aqueous solution for several hours. Therefore, the effect of H_2O_2 on the degradation of MC-LR in the PEC is evaluated by measuring the amount of produced H_2O_2 in the study. The measurements of H_2O_2 were performed by a photometric method using peroxidase (POD)(Bader *et al.*, 1988). Fig. 11 shows the absorbance (at 551 nm) of H_2O_2 against irradiation with MC-LR and without MC-LR systems, which clearly demonstrated that H_2O_2 was formed in both systems. In the MC-LR system, more H_2O_2 is generated in the first one hour, which could be due to an efficiently inhibition on the recombination of the photoexcited electron and hole by the reaction of the MC-LR with h^+. This also demonstrates that the H_2O_2 is mainly produced by the photoexcited electron. H_2O_2 is decreasing after 3 h, even less than without MC-LR system, which proved that H_2O_2 is the reactive species involved in the PEC degradation of MC-LR.

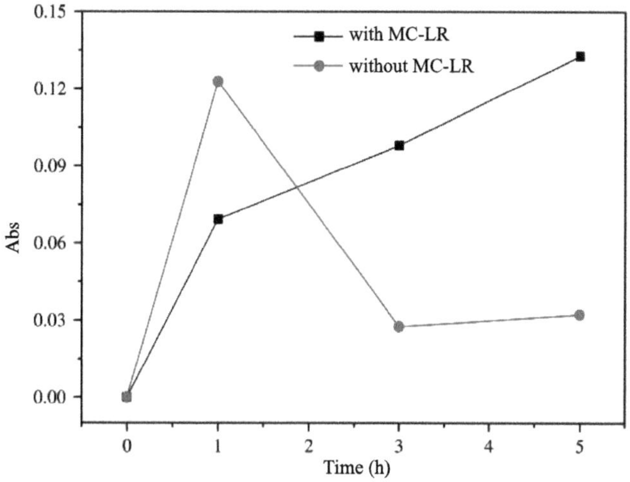

Fig. 11　Absorption intensity (at 551 nm) of DPD/POD after reaction with H_2O_2 against illumination time during the PEC process with and without MC-LR contained in the solution.

4　Conclusions

In this work, The $Ag/AgCl/TiO_2$－NTs electrode exhibits great potential for PEC degradation of MC-LR. The degradation of MC-LR is about 92%, and the decrease in the TOC of solution is about 65.8% after 5 h. Remarkable improvement of oxidizability for the $Ag/AgCl/TiO_2$－NTs electrode probably benefits from enhanced visible-light harvesting and reduced recombination of photogenerated electron-hole pairs due to the synergistic effect of Ag/AgCl nanoparticles and TiO_2－NTs. The mechanism studies show that the h^+, •OH and $O_2^{\cdot -}$ are mainly contributed to the PEC degradation of MC-LR, and H_2O_2 is also one of the reactive oxygen species (ROS) that contribute to the degradation of MC-LR during the photoelectrocatalytic process. For the four chosen anionic electrolytes, the influences on the degradation rates decrease in the order of $Cl^- > SO_4^{2-} > NO_3^-$ SO_4^{2-}, CO_3^{2-}, $> H_2PO_4^-$. Also, it is found that the pH value of solution have an obvious influence on the degradation of MC-LR. The toxic nature of MC-LR can effectively alleviatein PEC process, which has been confirmed by toxicity tests.

Acknowledgements

This work was supported by Ministry of Science and Technology of China (No. 2010DFA22770). The authors thank the Analytical and Testing Center of HUST for the use of SEM, XRD and DRS equipments.

References

AlHamedi F H, Rauf M A, Ashraf S S. 2009. Degradation studies of Rhodamine B in the presence of UV/H_2O_2, *Desalination*. **239**:159-166.

Antoniou M G, A A de la Cruz, Dionysiou D D. 2010a. Degradation of microcystin-LR using sulfate radicals generated through photolysis, thermolysis and e^- transfer mechanisms, *Appl. Catal. B: Environ.* **96**: 290-298.

Antoniou M G, A A de la Cruz, Dionysiou D D. 2010b. Intermediates and Reaction Pathways from the Degradation of Microcystin-LR with Sulfate Radicals, *Environ. Sci. Technol.* **44**:7238-7244.

Bader H, Sturzenegger V, Hoigné J. 1988. Photometric method for the determination of low concentrations of hydrogen peroxide by the peroxidase catalyzed oxidation of N, N-diethyl-p-phenylenediamine (DPD), *Water Res.* **22**:1109-1115.

Bandala E R, Martínez D, Martínez d, *et al*. 2004. Degradation of microcystin-LR toxin by Fenton and Photo-Fenton processes, *Toxicon*. **43**:829-832.

Cañizares P, Sáez C, Lobato J, *et al*. 2004. Electrochemical Treatment of 4-Nitrophenol-Containing Aqueous Wastes Using Boron-Doped Diamond Anodes, *Ind. Eng. Chem. Res.* **43**:1944-1951.

Carmichael W W, Li R. 2006. Cyanobacteria toxins in the Salton Sea, Saline systems **2**:5-18.

Chawla O P, Fessenden R W. 1975. Electron spin resonance and pulse radiolysis studies of some reactions of peroxysulfate (SO_4. 1,2), *J. Phys. Chem.*, *Am. Chem. Soc.* **79**:2693-2700.

Chen Y, Lu A, Li Y, *et al*. 2011. Naturally Occurring Sphalerite As a Novel Cost-Effective Photocatalyst for Bacterial Disinfection under Visible Light, *Environ. Sci. Technol.* **45**:5689-5695.

Chen Y, Yang S, Wang K, *et al*. 2005. Role of primary active species and TiO_2 surface characteristic in UV-illuminated photodegradation of Acid Orange 7, *J. Photochem. Photobiol. A: Chem.* **172**:47-54.

Codd G A, Bell S G, Brooks W P. 1989. Cyanobacterial toxins in water, *Water Sci. Technol.* **21**:1-13.

Doong R A, Chen C H, Maithreepala R A, *et al*. 2001. The influence of pH and cadmium sulfide on the photocatalytic degradation of 2-chlorophenol in titanium dioxide suspensions, *Water Res.* **35**: 2873-2880.

Elahifard M R, Rahimnejad S, Haghighi S, *et al*. 2007. Apatite-Coated Ag/AgBr/TiO_2 Visible-Light Photocatalyst for Destruction of Bacteria, *J. Am. Chem. Soc.* **129**:9552-9553.

Fang Y, Huang Y, Jing Y, *et al*. 2011. Unique Ability of BiOBr To Decarboxylate d-Glu and d-MeAsp in the Photocatalytic Degradation of Microcystin-LR in Water, *Environ. Sci. Technol.* **45**:1593-1600.

Ghodbane H, Hamdaoui O. 2010. Decolorization of antraquinonic dye, C. I. Acid Blue 25, in aqueous solution by direct UV irradiation, UV/H_2O_2 and UV/Fe(II) processes, *Chem. Eng. J.* **160**:226-231.

Gong J, Pu W, Yang C, *et al*. 2012. A simple electrochemical oxidation method to prepare highly ordered Cr-doped titania nanotube arrays with promoted photoelectrochemical property, *Electrochimica. Acta.* **68**:178-183.

Guo J F, Ma B, Yin A, *et al*. 2012. Highly stable and efficient Ag/AgCl@TiO_2 photocatalyst: Preparation, characterization, and application in the treatment of aqueous hazardous pollutants, *J. Hazard. Mater.*, 211-212:77-82.

Harada K I, Tsuji D, Watanabe M F, *et al*. 1996. Stability of microcystins from cyanobacteria-III. Effect of pH and temperature, *Phycologia*, **35**:83-88.

Hirakawa T, Nosaka Y. 2002. Properties of $O_2^{\cdot -}$ and OH· Formed in TiO_2 Aqueous Suspensions by

Photocatalytic Reaction and the Influence of H_2O_2 and Some Ions, *Langmuir*. **18**:3247-3254.

Hou Y, Li X, Zhao Q, *et al*. 2012. Role of Hydroxyl Radicals and Mechanism of Escherichia coli Inactivation on Ag/AgBr/TiO_2 Nanotube Array Electrode under Visible Light Irradiation, *Environ. Sci. Technol*. **46**:4042-4050.

Hu C, Peng T, Hu X, *et al*. 2009. Plasmon-Induced Photodegradation of Toxic Pollutants with Ag-AgI/Al_2O_3 under Visible-Light Irradiation, *J. Am. Chem. Soc*. **132**:857-862.

Ishibashi K-i, Fujishima A, Watanabe T, *et al*. 2000. Detection of active oxidative species in TiO_2 photocatalysis using the fluorescence technique, *Electrochem. Commun*. 2:207-210.

Ito E, Kondo F, Harada K I. 2001. Intratracheal administration of microcystin-LR, and its distribution, *Toxicon*. **39**. 265-271.

Keijola A M, Himberg K, Esala A L, *et al*. 1988. Removal of cyanobacterial toxins in water treatment processes: Laboratory and pilot-scale experiments, *Toxicity Assessment*, **3**:643-656.

Lawton L A, Robertson P K J, Cornish B J P A, *et al*. 1999. Detoxification of Microcystins (Cyanobacterial Hepatotoxins) Using TiO_2 Photocatalytic Oxidation, *Environ. Sci. Technol*. **33**:771-775.

Liang H C, Li X Z, Yang Y H, *et al*. 2008. Effects of dissolved oxygen, pH, and anions on the 2,3-dichlorophenol degradation by photocatalytic reaction with anodic TiO_2 nanotube films, *Chemosphere*. **73**:805-812.

Liu I, Lawton L A, Bahnemann D W, *et al*. 2009. The photocatalytic decomposition of microcystin-LR using selected titanium dioxide materials, *Chemosphere*. **76**:549-553.

Liu I, Lawton L A, Robertson P K J. 2003. Mechanistic Studies of the Photocatalytic Oxidation of Microcystin-LR: An Investigation of Byproducts of the Decomposition Process, *Environ. Sci. Technol*. **37**:3214-3219.

Malato S, Fernández-Ibáñez P, Maldonado M I, *et al*. 2009. Decontamination and disinfection of water by solar photocatalysis: Recent overview and trends, *Catal. Today*. **147**:1-59.

M E van Apeldoorn, H P van Egmond, Speijers G J A, *et al*. 2007. Toxins of cyanobacteria, *Mol. Nutr. Food Res*. **51**:7-60.

Monks N R, Liu S, Xu Y, *et al*. 2007. Potent cytotoxicity of the phosphatase inhibitor microcystin LR and microcystin analogues in OATP1B1- and OATP1B3-expressing HeLa cells, *Mole. Cancer Ther*. **6**: 587-598.

Neta P, Huie R E, Ross A B. 1988. Rate Constants for Reactions of Inorganic Radicals in Aqueous Solution, *J. Phy. Chem. Refer. Date*. **17**:1027-1284.

Ojani R, Raoof J B, Zarei E. 2012. Electrochemical monitoring of photoelectrocatalytic degradation of rhodamine B using TiO_2 thin film modified graphite electrode, *J. Solid State Electrochem*. **16**: 2143-2149.

Pelaez M, A A de la Cruz, K O'Shea, *et al*. 2011. Effects of water parameters on the degradation of microcystin-LR under visible light-activated TiO_2 photocatalyst. *Water Res*. **45**:3787-3796.

P G-J de Maagd, Hendriks A J, Seinen W, *et al*. 1999. pH-Dependent hydrophobicity of the cyanobacteria toxin microcystin-LR, *Water Res*. **33**:677-680.

Qiao R P, Li N, Qi X H, *et al*. 2005. Degradation of microcystin-RR by UV radiation in the presence of hydrogen peroxide, *Toxicon*. **45**:745-752.

Sivonen K. 1996. Cyanobacterial toxins and toxin production, *Phycologia*. **35**:12-24.

Sunada K, Watanabe T, Hashimoto K. 2003. Bactericidal Activity of Copper-Deposited TiO_2 Thin Film under Weak UV Light Illumination. *Environ. Sci. Technol*. **37**:4785-4789.

Tian Y, Tatsuma T. 2005. Mechanisms and Applications of Plasmon-Induced Charge Separation at TiO_2 Films Loaded with Gold Nanoparticles, *J. Am. Chem. Soc.* **127**:7632-7637.

Wang N, Li X, Wang Y, *et al.* 2009. Evaluation of bias potential enhanced photocatalytic degradation of 4-chlorophenol with TiO_2 nanotube fabricated by anodic oxidation method, *Chem. Eng. J.* **146**:30-35.

Wang P, Huang B, Qin X, *et al.* 2008. Ag@AgCl: A Highly Efficient and Stable Photocatalyst Active under Visible Light, *Angew. Chem. Int. Edi.* **47**:7931-7933.

Westrick J, Szlag D, Southwell B, *et al.* 2010. A review of cyanobacteria and cyanotoxins removal/inactivation in drinking water treatment, *Anal. Bioanal. Chem.* **397**:1705-1714.

Yuan R, Ramjaun S N, Wang Z, *et al.* 2011. Effects of chloride ion on degradation of Acid Orange 7 by sulfate radical-based advanced oxidation process: Implications for formation of chlorinated aromatic compounds, *J. Hazard. Mater.* **196**:173-179.

Yuan R, Ramjaun S N, Wang Z, *et al.* 2012. Photocatalytic degradation and chlorination of azo dye in saline wastewater: Kinetics and AOX formation, *Chem. Eng. J.* **192**:171-178.

Yu J, Dai G, Huang B. 2009. Fabrication and Characterization of Visible-Light-Driven Plasmonic Photocatalyst Ag/AgCl/TiO_2 Nanotube Arrays, *J. Phys. Chem. C* **113**:16394-16401.

Zhang L S, Wong K H, Yip H Y, *et al.* 2010. Effective Photocatalytic Disinfection of E. coli K-12 Using AgBr-Ag-Bi_2WO_6 Nanojunction System Irradiated by Visible Light: The Role of Diffusing Hydroxyl Radicals, *Environ. Sci. Technol.* **44**:1392-1398.

Zhou J, Cheng Y, Yu J. 2011. Preparation and characterization of visible-light-driven plasmonic photocatalyst Ag/AgCl/TiO_2 nanocomposite thin films, *J. Photochem. Photobiol. A: Chem.* **223**:82-87.

OPERATOR BASED ROBUST NONLINEAR CONTROL WITH SVM COMPENSATOR FOR A THERMAL PROCESS①

WEN Shengjun[1), WANG Dongyun[1), ZHANG Lei[1), DENG Mingcong[1),2)]

[1)] Department of Electronic Information, Zhongyuan University of Technology, Zhengzhou, China
[2)] Department of Electrical and Electronic Engineering, Tokyo University of Agriculture and Technology, Tokyo, Japan

Abstract

In this paper, robust nonlinear control with SVM compensator (Support Vector Machine) is presented for a Peltier actuated thermal process considering heat radiation. For the Peltier actuated thermal process, the heat radiation will affect evidently the temperature of the thermal process when temperature difference is large between the process and environment. That is the heat radiation is the fourth power of the temperature. Considering the heat radiation, a new nonlinear model is setup according to some thermal conduction laws, where the heat radiation is estimated by using a SVM predictive model. Radial basis kernel function based SVM method is proposed to setup the heat radiation model. To ensure the robust stability of the nonlinear system considering the heat radiation, operator based robust right coprime factorization design with SVM compensator is put forward. Finally, simulation results are given to show the effectiveness of the proposed method.

Key words: operator theory, robust control, nonlinear control, thermal process, SVM compensator

1 Introduction

Semiconductor refrigeration is also called electronic refrigeration, or thermoelectric refrigeration, which uses special semiconductor material to composite of P-N junction, forms of thermocouple and produces Peltier effect. That is a new type of refrigeration method through the direct current refrigerate, and the semiconductor refrigeration is becoming one of the three typical refrigeration ways in the world. Semiconductor refrigeration technology appeared in the early 1830s, but its performance is not satisfactory. With the rapid development of semiconductor materials in the 1950s, cooler is

---

① The paper published in *Mathematical Problems in Engineering*, 2014. in press.

gradually going from lab to engineering practice, such as national defense industry, agriculture, medical treatment and daily life and so on. And because of some advantages, such as no refrigerant, no pollution, no vibration, no noise, semiconductor refrigeration has been paid more attention in refrigeration industry(Chavez *et al*,2000). And the Peltier devices are usually used as the component of semiconductor refrigeration and aluminum plate is considered as the thermal sink.

Therefore, an aluminum plate thermal process with a Peltier actuator is often utilized to analyze the performance of the Peltier device. An aluminum plate thermal process with a Peltier device has been researched in some previous papers(Deng *et al.*, 2007a, 2007b, 2010, 2006; Bu and Deng, 2011). The Peltier actuated thermal process is a nonlinear control affine system and the modeling is based on the basis of the heat transfer balance. Heat transfer includes three methods: heat convection, heat conduction, heat radiation (Seifert *et al.*,2007). But in the previous papers (Deng *et al.*,2007a, 2007b, 2010, 2006; Bu and Deng, 2011), only the first two kind of heat transfer is considered because the thermal radiation is small comparing to the first two kind of transfer way under low temperature. Considering the heat radiation, a new modeling of the thermal process is presented in this paper. As the thermal radiation is drawn by Stefan - Boltzmann law, the new modeling will contain the fourth power of the temperature, which is not easy to be solved by using general differential equation solution. In the light of this problem, SVM (support vector machine) is used to establish the model of the thermal radiation part in this paper, and the model is used to estimate the thermal radiation heat and it is applied to design the compensator considering the heat radiation.

SVM is a new type of machine learning algorithm which is based on statistical learning theory and arose in recent years (Vapnik,1998). The learning process of SVMis based on structural risk minimization principle. Thus, the modeling algorithm of SVM is a convex quadratic optimization problem, which means that the local optimal solution is namely the global optimal one (Vapnik,1995; Drucker *et al.*,1997; Suykens *et al.*,2002; Wen *et al.*, 2012; Deng *et al.*,2009). That is, based on structural risk minimization criterion, SVM can effectively improve the abilities of the algorithm generalization, while having not a local minimum solution but an unique optimal solution (Drucker *et al.*,1997; Suykens *et al.*, 2002; Wen *et al.*,2012; Deng *et al.*,2009). Further, comparing with neural network and other machine learning methods, SVM has still some characteristics, such as small sample learning ability, automatically identifying the model parameters and the structure (Drucker *et al.*,1997; Suykens *et al.*,2002). In other words, SVM can solve the small sample, nonlinear, high dimension problem and has the merit of global optimal, simple structure, strongly generalized ability. These characteristics make SVM very suitable for the field of system classification and modeling. SVM based modeling can construct system model according to the input and output data of the system with unknown structure and parameters. Based on the above advantages of SVM, SVM is considered to establish the model of the thermal radiation part and the effectiveness of the model is checked in this paper.

The model of the Peltier actuated thermal process considering the heat radiation shows that the thermal process is a typical nonlinear control affine system, where the process temperature depends not only on input current, but also on the square of the input current. Also there exist some uncertainties. For the nonlinear control process with uncertainties, operator theory based nonlinear control method is confirmed to be effective to ensure robust stability of the nonlinear control system (Deng *et al.*, 2006; Chen and Han, 1998). So, operator theory is proposed to realize the temperature control for an aluminum plate thermal process considering heat radiation. Operator theory is a control theory based on an idea that a signal in the input space is mapped to the output space, (e. g. the operator $P = ND^{-1}$, the operation between N and D^{-1} means that the output of D^{-1} is the input of N). So far, a lot of researchers have paid attention to the theory, and some interesting results in regards to the theory have been concluded (Deng *et al.*, 2007a, 2007b, 2010, 2006; Bu and Deng, 2011; Chen and Han, 1998). An outstanding example is robust right coprime factorization design for a nonlinear feedback control system (Chen and Han, 1998), where all of the process and each controller are regarded as an operator, and signals are transferred from one space into another. Based on the design scheme, a robust tracking control system is proposed by Deng *et al.* (2006). Control and fault detection by using operator based design scheme are realized in (Deng *et al.*, 2007a, 2007b). Further, a control method for unstable plants with input constraints is given in (Deng *et al.*, 2007b), where the controlled plant with the input constraints satisfies the robust right coprime factorization conditions. However, operator based robust control has not been concerned for the Peltier actuated thermal process considering the heat radiation, where the heat radiation is estimated by using a SVM predictive model. It is difficult to process the uncertainties caused by the heat radiation. To ensure the robust stability of the nonlinear affine system considering the heat radiation, operator based robust nonlinear control with a SVM compensator is put forward in this paper.

The outline of this paper is as follows. Section 1 begins with the introduction to the principle of Peltier device and the models for thermal process and SVM are also shown in this part. Operator based on the robust right coprime factorization of the thermal process model and the controller design is presented for the thermal process in Section 2. Simulation results on robust control are illustrated to show the influence of heat radiation to the process and the effectiveness of the SVM modeling in Section 3. The conclusion is drawn in Section 4.

2 MODELING

2.1 Principle of Peltier Device

In general, the Peltier effect is used to realize refrigeration for Peltier devices. When a power is applied to the Peltier device, the current flows through a P type semiconductor and N type semiconductor. Energy moves opposite to the moving direction of the electron (Chavez *et al.*, 2000; Seifert *et al.*, 2007). For a metal conductor having the free movement electrons inside, its conductive reason is the movement of free electrons. Because the electrons are unable to transfer between the metal and semiconductor, it can only motion in the same object inside. So the mental conductor is only a conductive role. But, Peltier device mainly includes N type semiconductor and P type semiconductor. The basic working principle of the Peltier effect is that N type semiconductor is an impurities semiconductor and the concentration of free electrons is far greater than the holes. So the N type semiconductor is mainly conducted by free electrons. P type semiconductor is also an impurities semiconductor and the concentration of the holes is far greater than free electrons. So the P type semiconductor is mainly conducted by holes.

The principle diagram of semiconductor refrigeration is shown in Fig. 1. When the power is given, electrons in N type semiconductor and holes in P type semiconductor move from up to down. In the upper part, the internal energy convert into the potential energy which makes them moving down, and at the bottom, the potential energy converts into the internal energy to emit the heat. The holes of the P type semiconductor move from up to down, and its upper part absorbs heat due to the electrons breaking away from the holes, and the holes and electrons bond and produce heat at lower part. Similarly, the N type semiconductor's electron moves from up to down, which need to absorb energy to convert into potential energy in the upper part, and it emits the heat together with the hole at lower part.

$$\sigma = 5.6697 \times 10^{-8} \mathrm{W/(m^2 \cdot K^4)}$$

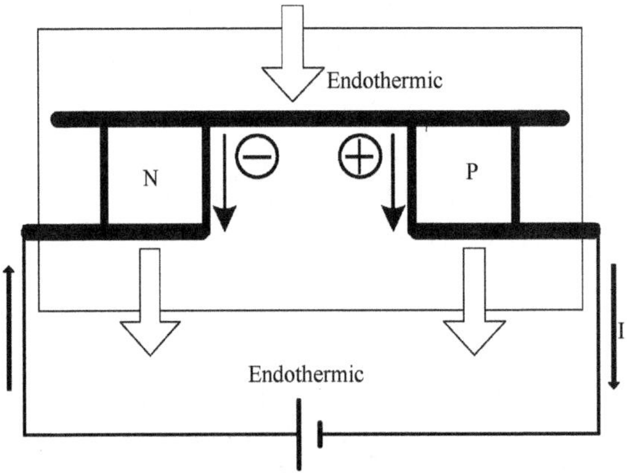

Fig. 1 The principle diagram of semiconductor refrigeration

2.2 Modeling on the Thermal Process

Some laws related to thermal exchange are shown as follows(Deng *et al.*, 2007b).

(1) Fourier's law concerning thermal conduction

$$q[\mathrm{W/m^2}] = -\lambda[\mathrm{W/(m \cdot K)}](\mathrm{d}\theta/\mathrm{d}n)[\mathrm{K/m}]$$

q: Heat flux, λ: Thermal conductivity, $(\mathrm{d}\theta/\mathrm{d}n)$: Temperature gradient of heat current

(2) Thermal conduction and Newton's cooling law

$$q[\mathrm{W/m^2}] = \alpha[\mathrm{W/(m \cdot K)}](\Delta T)[\mathrm{K}]$$

α: Thermal conductivity (Flowing air: 10～600[W/(m · K)]), ΔT: Temperature difference

(3) Stefan-Boltzmann law

$$\Phi[\mathrm{W}] = \varepsilon\sigma[\mathrm{W/(m^2 \cdot K^4)}]A[\mathrm{m^2}]T^4[\mathrm{K^4}]$$

ε: Emissivity, $\varepsilon < 1$, σ: Radiation coefficient

(4) Laws of thermal conduction

$$\mathrm{d}Q[\mathrm{J}] = c[\mathrm{J/(kg \cdot K)}]m[\mathrm{kg}]\mathrm{d}\theta[\mathrm{K}]$$

Q: Calorie, c: Specific heat, m: Mass

(5) Electrothermal amount by Peltier effect

$$Q[\mathrm{J}] = S[\mathrm{J/(K \cdot A)}]T[\mathrm{K}]I[\mathrm{A}]$$

S: Seebeck coefficient, I: Current

(6) Joule exothermic heat by current

$$Q[\mathrm{J}] = R[\mathrm{J/K}]I^2[\mathrm{K}]$$

R: Peltier's resistance

The aluminum plate thermal process with a Peltier device is shown in Fig. 2.

$$cm\frac{\mathrm{d}(T_0 - T_x)}{\mathrm{d}t} = u_d - \alpha(T_0 - T_x)(2S_4 + 2S_5 - S_6) - 2\lambda(T_0 - T_x)\frac{S_2}{d_1}$$

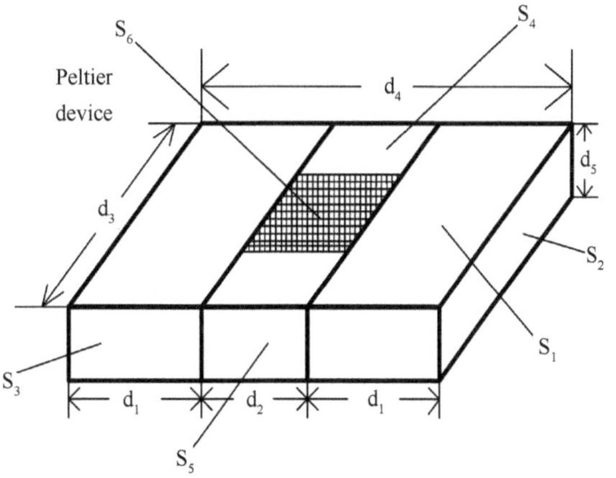

Fig. 2　The aluminum plate thermal process model

A differential equation based on the above laws is obtained for the thermal process

$$cm\frac{d(T_0 - T_x)}{dt} = u_d - \alpha(T_0 - T_x)(2S_4 + 2S_5 - S_6) - 2\lambda(T_0 - T_x)\frac{S_2}{d_1} - \varepsilon\sigma(4S_1 + 2S_2 + 4S_3 + 2S_4 + 2S_5 - S_6)(T_0{}^4 - T_x{}^4) \quad (1)$$

where T_0 is the environmental temperature, T_x is the temperature of the sensor, and u_d denotes the absorbed heat of the Peltier device. Some parameters of the thermal process and the aluminum plate are given in Table 1.

In the model, $\alpha(T_0 - T_x)(2S_4 + S_5 - S_6)$ denotes the energy produced by convection of the middle part of aluminum and air. $2\lambda(T_x - T_0)S_2/d_1$ is the thermal energy produced by the aluminum itself. And $\varepsilon\sigma(4S_1 + 2S_2 + 4S_3 + 2S_4 + 2S_5 - S_6)(T_0^4 - T_x^4)$ is thermal radiation produced by the all aluminum. The endothermic amount of the Peltier device u_d is shown as follows.

$$u_d = ST_1 I - K(T_h - T_1) - RI^2/2 \quad (2)$$

Here, T_1 is the temperature on endothermic side, T_h is the temperature on radiation side, I is the current flow in the circuit. ST_1I denotes the amount of heat that moves from the endothermic side to the radiation side by the Peltier effect, $K(T_h - T_1)$ shows the movement of heat caused by difference of temperature in the two sides of the device, and $RI^2/2$ is Joule heat which is generated by the input current through the device.

Table 1 Parameters of the aluminum plate thermal process

I	Current[A] (Input) 2.2[A] or 0[A]	d_1	0.086[m]	S_1	$d_1 \times d_3$[m^2]
u_d	Endothermic amount [W]	d_2	0.068[m]	S_2	$d_3 \times d_5$[m^2]
c	900[J·kg^{-1}·K^{-1}]	d_3	0.1[m]	S_3	$d_1 \times d_5$[m^2]
α	15[ω·m^{-2}·K^{-1}]	d_4	0.24[m]	S_4	$d_2 \times d_3$[m^2]
λ	238[ω·m^{-1}·K^{-1}]	d_5	0.005[m]	S_5	$d_2 \times d_5$m^2]
σ	5.6697×10^{-8}[W·m^{-2}·K^{-4}]	d	2700[kg/m^3]	S_6	9×10^{-4}[m^2]
K	0.63[W·K^{-1}]	S	0.053[V/K]	R	5.5[Ω]
m	$d_3 \times d_4 \times d_5 \times d$[kg]				

2.3 SVM Modeling on the Heat Radiation Part

In the above model, the thermal radiation part contains the fourth power of the temperature, which is not easy to solve by solving the differential equation directly. So, SVM technique is considered to estimate this part in this paper.

SVM is firstly put forward by Vapnik, like multilayer perception network and radial basis function network, which can be used for pattern classification and nonlinear regression (Vapnik, 1998). The main idea of SVM is to create a classification hyperplane as decision surface which make the isolation edge between positive examples and counter-example are maximized. SVM regression can construct the system identification model according to system input and output data under the condition of the system unknown parameters. SVM theory is based on statistical learning theory of VC (Vapnik-Chervonenkis), where a structural risk minimization principle is utilized to get a global optimal solution. It overcomes the neural network regression identification in the nonlinear optimization problems appeared in local minimum, process study and so on. It has a good promotion characteristic and is suitable for solving nonlinear, large time delay problem.

The detailed modeling on the heat radiation is shown as follows. Firstly, we define

$$A = [\alpha(2S_4 + 2S_5 - S_6) + 2\lambda S_2 / d_1]/(cm)$$

$$M_{svm} = \varepsilon\sigma(4S_1 + 2S_2 + 4S_3 + 2S_4 + 2S_5 - S_6)(T_0^4 - T_x^4)/(cm)$$

then the model equation is transformed as follows.

$$\frac{\mathrm{d}(T_0 - T_x)}{\mathrm{d}t} = \frac{u_d}{cm} - A(T_0 - T_x) - M_{svm} \tag{3}$$

It shows that M_{svm} denotes the heat radiation part, which contains the fourth power of the temperature. It cannot be solved by using the differential equation solution. Therefore, SVM is considered to set up the modelof M_{svm} in this paper, and it is used to estimate the heat radiation real time. SVM is concerned with estimating a real-valued function shown as follows.

$$f(\tilde{x}) = w^T \cdot \varphi(\tilde{x}) + b, w \in R^d, b \in R \tag{4}$$

The model is based on a finite number set of independent and distributed data $(\tilde{x}_i, \tilde{y}_i)$. x_i and y_i denote input and output respectively, w is weight vector and b is offset. In this paper, we make u_d and T_x as the input and the M_{svm} as the output when we build the SVM modeling. And w and b are obtained by the training. In Vapnik's ε-insensitive support vector regression, the aim is to find a function $(\tilde{x})$ which allows error of y_i to be no more than ε, and makes y_i flat for all the training data. Considering more interferential error, non-negative slack variables ξ and ξ^* are introduced. Then, the optimization problem on the modeling can be described as the following form

$$\min_{w,b} \frac{1}{2} w^T \cdot w + C \sum_{i=1}^{l} (\xi_i + \xi_i^*) \tag{5}$$

$$\text{s. t.} \begin{cases} \tilde{y}_i - (w^T \cdot \varphi(\tilde{x}_i) + b) \leqslant \varepsilon + \xi_i \\ (w^T \cdot \varphi(\tilde{x}_i) + b) - \tilde{y}_i \leqslant \varepsilon + \xi_i^* \\ \xi_i, \xi_i^* \geqslant 0, i = 1, \cdots, l \end{cases}$$

where C is a positive constant and to control the punishment to the samples beyond the error ε. Generalizing to kernel function based regression estimation by introducing Lagrange multipliers, we can arrive at the following optimization problem (Deng *et al.*, 2009).

$$\max_{\alpha,\alpha^*} W(\alpha, \alpha^*) = -\frac{1}{2} \sum_{i=1}^{l} \sum_{j=1}^{l} Q_{ij} (\alpha_i - \alpha_i^*)(\alpha_i - \alpha_j^*) + \sum_{i=1}^{l} \tilde{y}_i (\alpha_i - \alpha_i^*) - \sum_{i=1}^{l} \varepsilon(\alpha_i + \alpha_i^*)$$

$$s.\,t. \begin{cases} \alpha_i, \alpha_i^* \in [0, C], i = 1, \cdots, l \\ \sum_{i=1}^{l} (\alpha_i - \alpha_i^*) = 0 \\ K(\tilde{x}_i, \tilde{x}_j) = \varphi(\tilde{x}_i) \cdot \varphi(\tilde{x}_j) = Q_{ij} \end{cases}$$

Ultimately, the regression estimation takes the form

$$f(\tilde{x})=\sum_{i=1}^{l}(\alpha_i-\alpha_i^{*})(\tilde{x}_i\cdot\tilde{x})+b=\sum_{i=1}^{l}(\alpha_i-\alpha_i^{*})K(\tilde{x}_i\cdot\tilde{x})+b$$

where $K(\tilde{x}_i,\tilde{x}_j)=\varphi(\tilde{x}_i)\cdot\varphi(\tilde{x}_j)$ is kernel function that gives two vectors in input space and returns the dot product in feature space by using a nonlinear mapping φ. That is, linear regression in a high dimensional space corresponding to nonlinear regression in the low dimensional input space can be realized by mapping the input vectors into a feature space. That is, nonlinear regression problem is resolved by linear regression form.

There are many kernel functions used in SVM, such as linear kernel function, polynomial function, radial basis function and sigmoid function. Linear kernel can be looked as a special kind of form to RBF kernel function. The characteristic of sigmoid function is similar with RBF to a certain extent. In the radial basis function case, the support vector algorithm automatically determines centers, weights, and threshold such that we use radial basis kernel function in this paper, which is shown as follows.

$$K(\tilde{x}_i,\tilde{x})=\exp\left(-\frac{\|\tilde{x}_i-\tilde{x}\|^2}{2\sigma^2}\right) \tag{6}$$

Therefore, under the above case, some parameters of support vector algorithm also can be automatically determined, such as weight vector w, offset b, slack variables ξ_i, ξ_i^* and Lagrange multipliers α_i, α_i^*. Such that, SVM estimation accuracy depends on setting of parameters C, ε and σ. C is Penalizing constant, which determines the trade of between the model complexity and fatness. ε is error-accuracy parameter, and it controls the width of the ε-insensitive zone for fitting the training data. σ is kernel width parameter, which is appropriately selected to reflect the input range of the training data. By selecting the appropriate parameters C, ε and σ, the SVM model of the heat radiation M_{svm} can be obtained and the heat radiation M_{svm} can be estimated real time by the built SVM model. Then, operator based robust nonlinear control design using the model M_{svm} is presented for the thermal process by the following section.

3 Operator Based Robust Nonlinear Control Dedisgn

3.1 Operator Based Robust Right Coprime Factorization

The robustness of right coprime factorizations of a nonlinear system is to ensure that the system can be right coprime factorized and the system remains its right coprime factorizations while affecting by external interference. More specifically, we suppose the nonlinear system P has right factorization $P=ND^{-1}$, then if there are two operators S and R that satisfy Bezout identity $SN+RD=I$. That we can say the right coprime factorization factors of the system exist and the decomposition is right coprime factorization. Assume that the system P is disturbed by a perturbation ΔP. If $P+\Delta P$ has a right coprime factorization and satisfies some conditions, it shows that the system P has robustness right coprime factorization shown as Fig. 3.

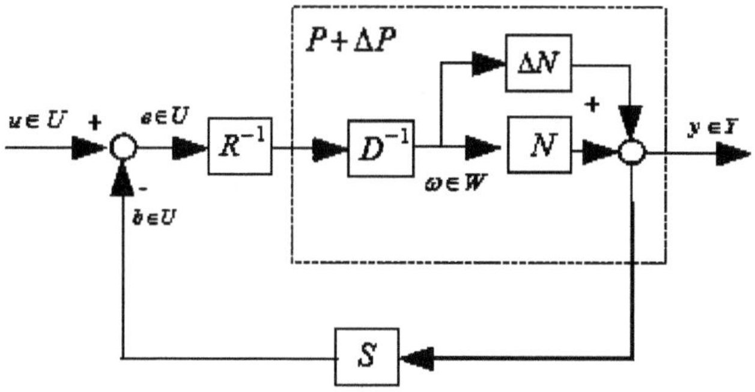

Fig. 3 Robust right coprime factorization

The system P without disturbance has the decomposition $P = ND^{-1}$. We define a disturbance operator as follows while P is disturbed by ΔP.

$$\Delta N = \Delta PD$$

So, for the stable operator $N+\Delta N$, $P+\Delta P$ can be factorized into

$$P + \Delta P = (N + \Delta N)D^{-1}$$

Therefore, the system P has the robustness right coprime factorizations under the following conditions of robust stability(Deng *et al.*, 2006).

$$SN + RD = L, \quad \text{where } L \text{ is a unit operator}$$

$$\| (S(N + \Delta N) - SN)L^{-1} \| < 1$$

Based on the above robust right coprime factorization technology, a robust nonlinear control method is proposed for the thermal process with a Peltier actuator.

3.2 Robust Nonlinear Control Scheme for the Thermal Process

For the thermal process shown in Equation (3), we definey $(t) = T_0 - T_x$, the model can be re-expressed as

$$\frac{\mathrm{d}y}{\mathrm{d}t} = \frac{u_d}{cm} - Ay - M_{svm} \tag{7}$$

where the term M_{svm} denotes the heat radiation part which is estimated real time by using SVM model. Considering the heat radiation as the perturbation ΔP, the thermal process is given as follows.

$$y = \frac{1}{cm}(e^{-At} + \Delta)\int u_d(t)e^{At}\,\mathrm{d}t \tag{8}$$

In order to cancel the affection of the heat radiation, a compensation operator P_{svm} is designed according to the SVM compensation operator M_{svm} and the transformer operator$T$. Robust right coprime factorization with compensation operator is designed shown in Fig. 4 for the perturbed thermal process.

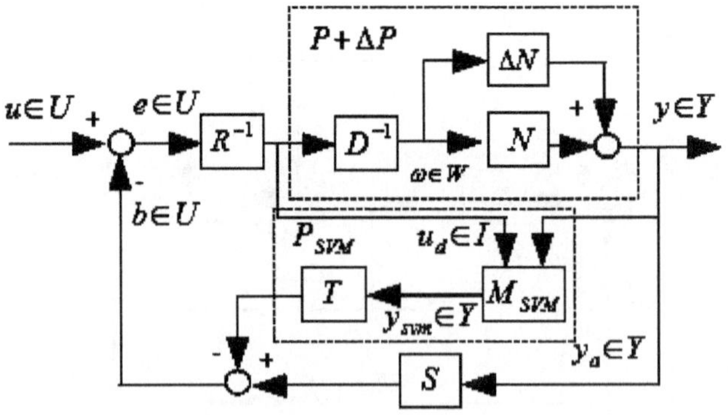

Fig. 4 Robust nonlinear control scheme with compensation operator

In Fig. 4, the operator M_{svm} is designed by using the SVM technology and T is designed to transform the signal y_{svm} from output space Y to input space U.

If we define $\Delta\widetilde{N}=S^{-1}TM(D,N)$, then the equivalent diagram of Fig. 4 is obtained in Fig. 5. The robust stability of the thermal process considering the heat radiation is ensured if the following equations and inequality are satisfied. The proof is omitted.

$$SN+RD=I \tag{9}$$

$$S(N+\Delta\widetilde{N})-TM(D,N)+RD=L \tag{10}$$

$$\|(S(N+\Delta\widetilde{N})-TM(D,N)-SN)\|<1 \tag{11}$$

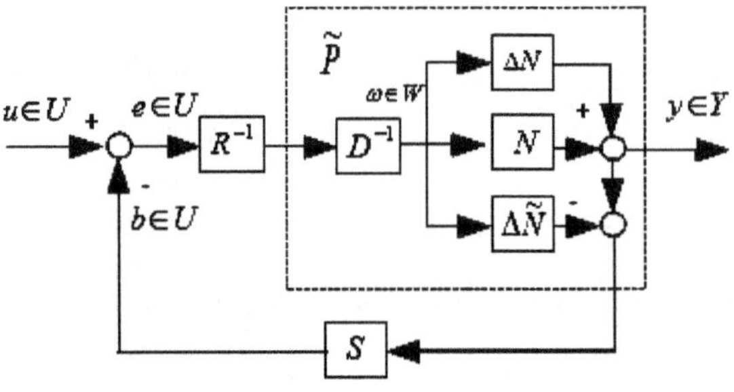

Fig. 5 The equivalent diagram of Fig. 4

The nominal thermal process is right divided into two parts with $P=ND^{-1}$ as follows.

$$D^{-1}(u_d)=\frac{u_d(t)}{cm} \tag{12}$$

$$N(\omega)(t)=e^{-At}\int e^{At}\omega(\tau)\mathrm{d}\tau \tag{13}$$

Considering the perturbation ΔP caused by the heat radiation, the corresponding perturbed operator $N+\Delta N$ can be expressed as

$$(N+\Delta N)(\omega)(t)=(e^{-At}+\Delta)\int e^{At}\omega(\tau)\mathrm{d}\tau$$

which is bounded and stable. The uncertain term Δ is related to the heat radiation M_{svm}.

For the nonlinear process, nonlinear operator controllers S and R are designed to satisfy the Bezout identity (9) and inequality (11). It is worth to mention that the initial state should also be considered, that is, $SN(\omega_0)(t_0)+RD(\omega_0)(t_0)=I(\omega_0)(t_0)$ should be satisfied. In this paper, we select the initial time $t_0=0$ and $\omega_0=\omega_0(t_0)$. By considering the structure of D, the controller R is designed as

$$R(u_d) = Bu_d(t)/cm \tag{14}$$

where B is a constant. Then we can get

$$RD(\omega) = R(D) = R(cm\omega(t)) = B\omega(t)$$

which shows the expression of the RD is simple. After that, it will be easy to get the expression of controller S. From Fig. 4, we know that

$$SN = S(N) = (1\text{-}B)\omega(t)$$

Based on $N(\omega)(t)=e^{-At}\int e^{At}\omega(\tau)\mathrm{d}\tau$, we can get the following operator

$$S(y) = (1-B)(Ay(t)+\mathrm{d}y(t)/\mathrm{d}t) \tag{15}$$

The SVM compensator operator M_{svm} is obtained by using the SVM technology introduced in Section 1. The transformer operator T is

$$T(y_{svm}) = (1-B)cmy_{svm}(t) \tag{16}$$

Furthermore, the tracking filter M is designed to make the process output y track the reference input r shown in Fig. 6. According to the Bezout identity (9), Fig. 6 is equivalent to Fig. 7. Using the result in (Bu and Deng, 2011), the tracking filter M is designed as follows.

$$M(r)(t) = (N+\Delta N)^{-1}(r)(t) \tag{17}$$

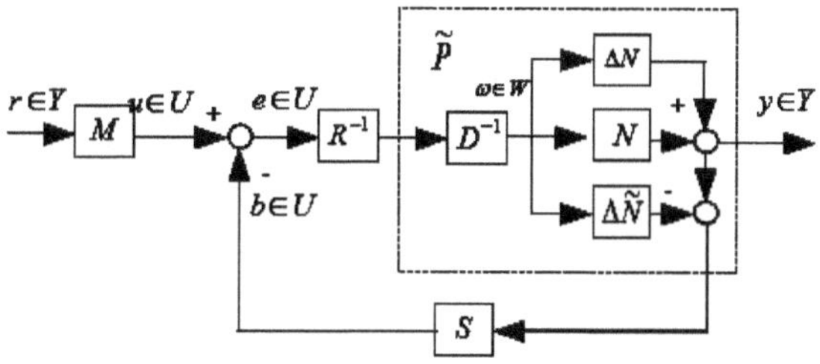

Fig. 6 Robust tracking control scheme

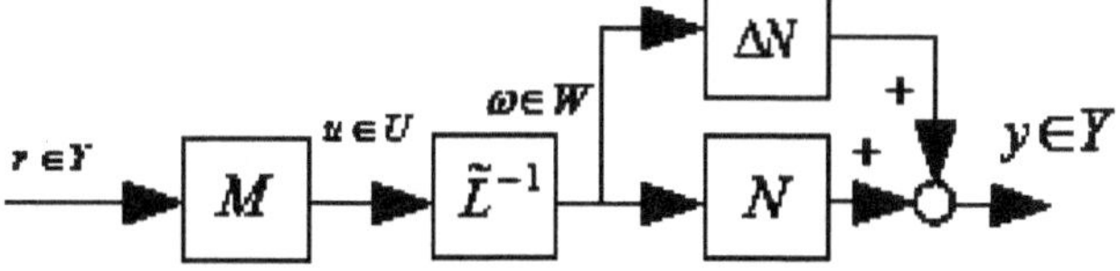

Fig. 7 The equivalent diagram of Fig. 6

In the following section, the proposed control scheme is confirmed by simulation.

4 Simulation Results

4.1 Model Checking of the Thermal Process

The thermal process with Peltier device is a typical nonlinear control system considering the endothermic amount of the Peltier device $u_d = ST_1 i - K(T_h - T_1) - Ri^2/2$. Due to the thermal radiation which is small relative to the other two kinds of transfer way under low temperature, we set a large initial temperature as 50℃. And the input current is limited between 0.0[A] and 2.2[A]. The parameters used in the simulation are shown in Table 2 (In order to see the simulation results clearly, we change the unit of the temperature "K" into "℃" in the simulation).

Table 2 Simulation parameters

Reference input (Desired temperature) $r=6$[℃]
Parameter $B=0.99$;
Current $i=2.2$ or 0.0 [A]
Initial Temperature: 50 [℃]
Simulation time: 600[s]
Sampling time: 100[ms]

In order to explain the heat radiation affecting the heat process, the simulations considering the heat radiation or not are carried out. The parameters used in the two simulations are the same. Fig. 8 shows the simulation result where the heat radiation is not considered to the process. And Fig. 9 shows the simulation result where the heat radiation had been considered to the process.

From Fig. 8, we can see the output temperature of the process is declined from 50℃ to 44℃ and finally falls in 44℃. But in Fig. 9, the output temperature of the process cannot decline to the 44℃, and it just falls in 44.3℃ at last. So it can be concluded that the thermal radiation on the thermal process of the aluminum plate influences the output temperature of the thermal process.

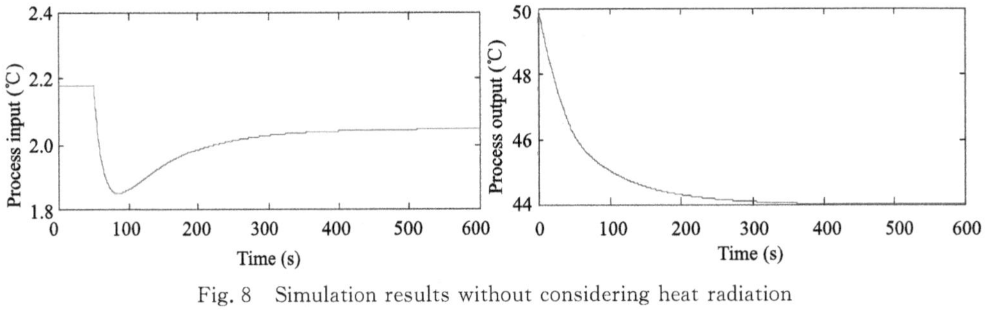

Fig. 8 Simulation results without considering heat radiation

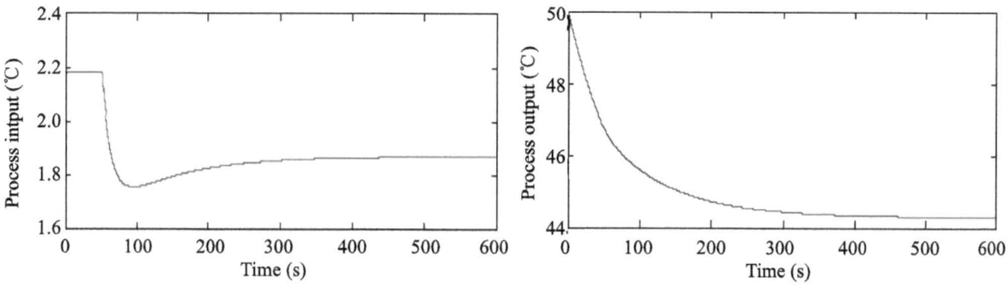

Fig. 9　Simulation results considering heat radiation

4.2　SVM-based Prediction for the Thermal Radiation

In this paper, the radial basis function is used to establish the M_{svm} model according to the support vector algorithm, where some parameters of support vector algorithm can be automatically determined, such as weight vector w, offset b, slack variables ξ_i, ξ_i^* and Lagrange multipliers α_i, α_i^*. Such that, SVM estimation accuracy depends on setting of parameters C, ε and σ. In this paper, we set $2\sigma^2$ as 1/10, C and ε are 100 and 0.1, respectively.

According to the relevant data of the process model, the model of the heat radiation of the thermal process can be obtained by using the SVM method. The obtained model which is estimated through the SVM method can be used to construct a new model. Fig. 10 shows the prediction results of the heat radiation using the built SVM model. The black line is the calculating result of the heat radiation based on the input and output data and the dot line is the predicting result of the heat radiation based on the SVM model.

It is clear to see from Fig. 10 that the two curves almost are coincident. So it can conclude that the SVM model of the heat radiation is efficient and exact.

4.3　Simulation on Robust Nonlinear Control for the Thermal Process

Robust right coprime factorization controllers are designed to make the thermal process reducing to 44℃ which considers the thermal radiation. And the simulation results are shown in Fig. 11. Fig. 12 shows the results of the aluminum plate thermal process considering the thermal radiation using the SVM model.

Fig. 11 shows that the output temperature of the thermal process which considering the thermal radiation can be fallen in 44℃ through adjusting the parameters of the controllers. Comparing with Fig. 11 and Fig. 12, it can be concluded that the SVM model of the heat radiation is efficient and exact.

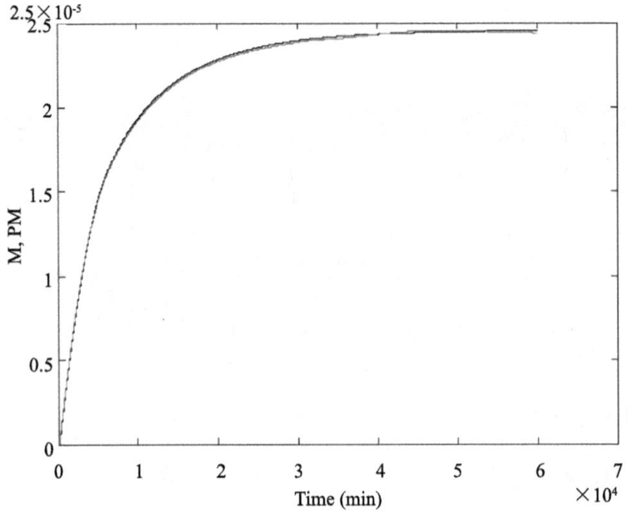

Fig. 10 The prediction result of the heat radiation

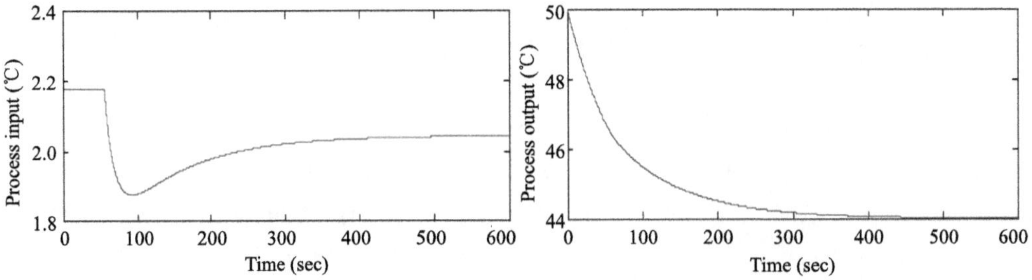

Fig. 11 Results of nonlinear control for the thermal process considering the thermal radiation

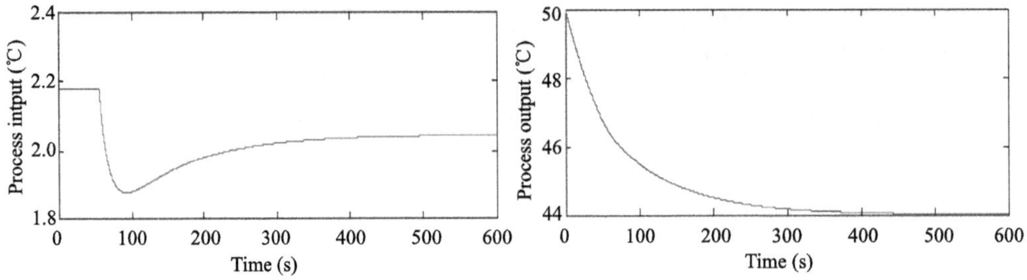

Fig. 12 Results using the proposed scheme

5 Conclusions

For the Peltier actuated aluminum plate thermal process, the heat radiation is certified to influence evidently on the temperature when temperature difference is large between the process and environment. In this paper, the model of the aluminum plate thermal process is built considering heat radiation, where the heat radiation is estimated by using a predictive model based on SVM. Furthermore, operator based robust nonlinear control with the SVM compensator is presented for the Peltier actuated thermal process. Finally, simulation results show the effectiveness of the proposed scheme.

Acknowledgement

This work is partially supported by Program for International Science Cooperation and Communication (2010DFA22770) and National Natural Science Foundation of China(61304115).

References

Bi S , Deng M. 2011. Operator based robust control design for nonlinear plants with perturbation. *International Journal of Control*, **84**(4):815-821.

Bu N, Deng M. 2011. System design for nonlinear plants using operator-based robust right coprime factorization and isomorphism. *IEEE Transactions on Automatic Control*, **56**(4):952-957.

Chavez J, Ortega A, Salazar J, *et al*. 2000. SPICE model of thermoelectric elements including thermal effects. *17th IEEE Instrumentation and Measurement Technology Conference*, pp. 1019-1023.

Chen G, Han V. 1998. Robust right factorization and robust stabilization of nonlinear feedback control systems. *IEEE Transactions on Automatic Control*, **43**:1505-1510.

Deng M, Inoue A, Edahiro K. 2007. Fault detection in a thermal process control system with input constraints using robust right coprime factorization approach. *Proc. of IMechE, Part I: Journal of Sys. and Control Eng.*, **221**(6):819-831.

Deng M, Inoue A, Edahiro K. 2010. Fault detection system design for actuator of a thermal process using operator based approach. *ACTA Automatica Sinica*, **36**:421-426, 2010

Deng M, Inoue A, Ishikawa K. 2006. Operator-based nonlinear feedback control design using robust right coprime factorization. *IEEE Transactions on Automatic Control*, **51**(4):645-648.

Deng M, Inoue A, Yanou A. 2007. Stable robust feedback control system design for unstable plants with input constraints using robust right coprime factorization. *Int. J. Robust Nonlinear Control*, **17**(18): 1716-1733.

Deng M, Jiang L, Inoue A. 2009. Mobile robot path planning by SVM and Lyapunov function compensation. *J. of the Institute of Measurement and Control*, **42**(8):234-237.

Drucker H, Burges C J C, Kaufman L. 1997. Support vector regression machines. *Advances in Neural Information Processing Systems. Cambridge:MIT Press*, 55-161.

Seifert W, Ueltzen M, Strumpel C, *et al*. 2001. One-dimensional modeling of a Peltier element. *Proc. 20th International Conference on Thermoelectrics*, 439-443.

Suykens J K A, Gestel T V, Brabanter J D. 2002. Least Squares Support Vector Machines. Singapore: World Scientific Publishing Co. Ltd..

Vapnik V N. 1998. Statistical Learning Theory. New York: John Wiley.

Vapnik V N. 1995. The Nature of Statistical Learning Theory. N Y: Springer O Verlag.

Wen S, Deng M, InoueA. 2012. Operator-based robust nonlinear control for gantry crane system with soft measurement of swing angle. *International Journal of Modelling, Identification and Control*, **16**(1):88-96.

InGaN/GaN MULTIPLE QUANTUM WELL SOLAR CELLS WITH GOOD OPEN-CIRCUIT VOLTAGE AND CONCENTRATOR ACTION①

LI Xuefei[1),2)], ZHENG Xinhe[2)]②, ZHANG Dongyan[2)], WU Yuanyuan[2)], SHEN Xiaoming[1], WANG Jianfeng[2)] and YANG Hui[2)]

[1)] Key Laboratory of New Processing Technology for Materials and Nonferrous Metals, Ministry of Education, College of Materials Science and Engineering, Guangxi University, Nanning 530004, China

[2)] Key Laboratory of Nanodevices and Applications, Suzhou Institute of Nano-tech and Nano-bionics, Chinese Academy of Sciences, Suzhou 215125, China

Abstract

The photovoltaic properties of large-chip-size (2.5×2.5 mm^2) InGaN/GaN multiple quantum well (MQW) solar cells grown by metal organic chemical vapor deposition were studied under concentrated AM1.5G sun irradiation. We demonstrate a high open-circuit voltage of 2.31 V for blue-light-emitting InGaN/GaN MQW solar cells under 1 sun. The higher open-circuit voltage is mainly ascribed to the extremely low reversed saturation current density of approximately 10^{-19} mA/cm^2. The open-circuit voltage and short-circuit current density were found to increase as sunlight intensity increases, with a peak value of 2.50 V observed at 190 suns, showing a great potential for concentrator applications.

Key words: multiple quantum well (MQW); solar cell; InGaN/GaN, energy band gap

1. Introduction

Since the revision of the energy band gap of InN, InGaN alloys have a tunable and direct energy band gap (varying from 0.7 to 3.4 eV), which almost matches the entire solar spectrum (Wu *et al.*, 2002; Wu *et al.*, 2003). This unique characteristic, in combination with the other advantages such as high absorption coefficient ($\sim 10^5$ cm^{-1}), large carrier mobility, and superior radiation resistance, makes InGaN alloys an excellent candidate for high-efficiency photovoltaic materials for both territorial and space applications (Wu *et al.*, 2003, Muth *et al.*, 1997; Geerts *et al.*, 1996).

A high background electron concentration that comes from native defects, however,

① The paper published in *Jap. J. Appl. Phys*, **51**(9), 092301(2012).

② E-mail address: xhzheng2009@sinano.ac.cn

makes the growth of p-type InGaN highly challenging (Singh *et al.*, 1997). Currently, many reports are still focused on p-GaN/i-InGaN/n-GaN double heterojunctions in the study of InGaN-based solar cells (Zheng *et al.*, 2008, Neufeld *et al.*, 2008; Jani *et al.*, 2007). To obtain good quantum efficiency at wavelengths of over 420 nm, an InGaN absorber with a high In content (>15%) is required. However, $In_xGa_{1-x}N$ ($x>0.15$) epilayers grown on GaN have a thickness limitation (generally less than 50 nm) because misfit dislocation density increases very rapidly with increasing x, which is harmful to the device performance once the layer thickness exceeds the critical one (Liu *et al.*, 2006; Holec *et al.*, 2006). The use of an InGaN/GaN multiple quantum well (MQW) as an active region could avoid the undesired dislocations, and thus, serves as an alternative way to develop InGaN-based solar cells (Dahal *et al.*, 2010; Bae *et al.*, 2011).

Reasonably high open-circuit voltage (V_{oc}) of 2 V has been achieved for $In_{0.3}Ga_{0.7}N$/GaN MQW solar cells that have a peak emission wavelength of 472 nm (2.64 eV) (Dahal *et al.*, 2009). Calculation shows the band gap to open-circuit voltage difference (E_g/q-V_{oc}) of 0.64 V, (Dahal *et al.*, 2009) in which E_g is the equivalent energy band and q is the elementary electric charge of an electron. For some InGaN/GaN heterojunction solar cells with thick bulklike intrinsic InGaN as an absorber, $E_g/q - V_{oc}$ is reported to be approximately 0.7-0.8 V (Zheng *et al.*, 2008; Ermolenko *et al.*, 2010). In addition, Dahel *et al.* (2010) reported an enhancement of both the open-circuit voltage and efficiency of the InGaN/GaN MQW solar cells with increasing light intensity up to 30 suns. Furthermore, it is expected that the efficiency under concentrated sunlight could be further enhanced by improving the interfacial quality through further growth and device processing optimization (Dahal *et al.*, 2010). Recently, Bae *et al.* (2011) have improved photovoltaic effects by using bottom reflectors and pyramid textured surfaces on vertical $In_{0.15}Ga_{0.85}N$/GaN MQW solar cells and obtained a reasonably high V_{oc} of 2.2 V. Sang *et al.* (2011) enhanced device performance with a decrease in dark current of more than two orders of magnitude by inserting a super thin AlN layer between the i-InGaN and p-InGaN layers. The leakage current density, which is the main component degrading the photovoltaic performance of devices, increases with the area/periphery ratio of the diode (Zeng *et al.*, 2010). Additionally, the chip size in most reports is generally 1×1 mm^2 or less (Zheng *et al.*, 2008; Bae *et al.*, 2011; Lang *et al.*, 2011; Farrell *et al.*, 2011). Here, we report a good open-circuit voltage of 2.31 V under 1 sun for large-chip-size (2.5×2.5 mm^2) InGaN/GaN MQW solar cells without any metal back reflector, antireflective coating, or passivation layer. The MQWs emit light with a peak wavelength of 449 nm (2.76 eV), indicating a corresponding E_g/q-V_{oc} of 0.45V for the corresponding solar cell structure. The difference is very close to that of typical GaAs or Si single-junction solar cells. Moreover, the good concentrator action of the MQW solar cells for open-circuit voltage and efficiency improvement up to 240 suns is demonstrated. A discussion is given regarding the experimental results.

2 Experimental Methods

Epitaxial layers of InGaN/GaN MQW were grown on (0001) sapphire by metal organic chemical vapor deposition (MOCVD). Twelve periods of $In_xGa_{1-x}N$ (3 nm)/GaN (15 nm) MQW were used as the absorbing region for the solar cells. The MQW samples were characterized by high-resolution X-ray diffraction (HRXRD), cross-sectional transmission electron microscopy (TEM), and scanning electron microscopy (SEM). The device consists of a 150 nm p-GaN region exhibiting a predominant electroluminescence (EL) emission peak of approximately 449 nm (data not shown here). The device with a size of 2.5×2.5 mm^2 was fabricated in the following steps: 1) amesa structure to expose n-GaN for contact was defined by standard photolithography and inductively coupled plasma etching with Cl_2/BCl_3 gas; 2) a current spreading layer of indium-tin oxide (ITO) (200nm) was deposited onto the entire device by optical coating and n-GaN was exposed through wet etching using diluted HCl solution, and then rapid thermal annealing was carried out at 500℃ in air for 10 min to obtain ohmic contact. 3) Ti/Al/Ti/Au (25/200/25/200 nm) electrodes were deposited on ITO/p- and n-GaN by electron beam evaporation, which was followed by a subsequent lift-off process. The devices were characterized using a microprobe station under an air mass 1.5 global (AM1.5G, 100 mW · cm^{-2}) solar simulator. Dark and illuminated current-voltage characteristics were measured by using a Keithley 2440 source meter. A standard silicon detector was used to calibrate the AM1.5G light spectra. The devices were placed on a heat sink with a closed-water circuit to optimize the thermal management when high concentration tests were performed.

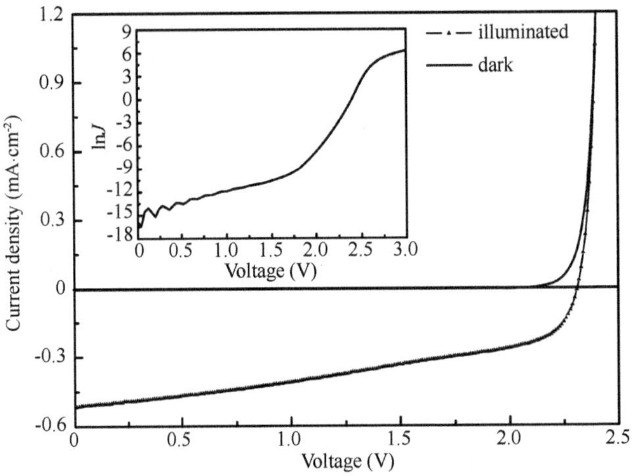

Fig. 1 *J*-*V* characteristics of a solar cell with InGaN/GaN MQW as the active region under dark and AM1.5G illuminated conditions. The inset shows ln*J*-*V* under dark condition for the same device

3 Results and Discussion

The current density (J) versus voltage (V) characteristics of the InGaN/GaN MQW solar cell under AM1.5G illuminated condition are shown in Fig. 1. The device displays a conversion efficiency of 0.59%, a short-circuit density (J_{sc}) of 0.52 mA · cm^{-2}, and a fill factor (FF) of approximately 45%. It is worth noting that a very high open-circuit voltage (V_{oc}) of 2.31 V, which is relatively higher than that of previously reported InGaN/GaN MQW solar cells with a similar In content in QWs or superlattices (Yang *et al.*, 2011; Zhang *et al.*, 2011; Nakao *et al.*, 2011) was obtained. If a combination of band gap (E_g) is considered, the band gap to open-circuit voltage difference (E_g/q-V_{OC}) could be regarded as a reference to judge material quality and device processing. That is, the lower the E_g/q-V_{oc}, the better the device performance. Here, the fabricated cells show an E_g/q-V_{oc} of approximately 0.45 V, in which E_g is calculated to be 2.76 eV on the basis of the peak emission wavelength of 449 nm from electroluminescence measurements. By defining E_g/q-V_{oc} and the circuit model of solar cells, one can see that E_g/q-V_{oc} is closely related to the short-circuit current density (J_{sc}) and reversed saturation current density (J_0) by $V_{oc} = (nk_BT/q)\ln(J_{sc}/J_0\text{-}1)$, where n is an ideality factor, k_B is the Boltzmann constant, and T is the temperature in Kelvin. To obtain n and J_0, the ln$J$ versus voltage curve is plotted in the inset of Fig. 1. We could infer that the ideality factor n is 2.11 and the corresponding reversed saturation current density J_0 is 1.5×10^{-19} mA · cm^{-2}, which is dependent on material quality (Steadman *et al.*, 1993). That is, the fewer the defects in the materials, the lower the J_0 for the device. Figure 2a shows the high-resolution X-ray diffraction (HRXRD) ω-2θ scanning spectra in which clearly resolved higher-order satellites and interference fringes are observed, indicating a figure-of-merit of high-quality MQW structures. The number of interference fringes between 0th- and 1st-order satellites [shown in the inset of Fig. 2a] corresponds well to the expected periods of the as-grown InGaN/GaN MQW. Also shown is further evidence of cross-sectional TEM (Fig. 2b) and top-view SEM (Fig. 2c) images of the InGaN/GaN MQW samples. It can be seen that twelve periods of InGaN/GaN MQW show almost no threading dislocations and no pits over them. These defects have been demonstrated to cause an increase in leakage current and finally a decrease in V_{oc} in solar cells (Kuwahara *et al.*, 2011). Therefore, these obtained results support the extremely low reversed saturation current density even for the cells with a large size of 2.5×2.5 mm^2. Also, note that by considering the diode equation in the case of series resistance and shunt resistance, a high shunt resistance could be beneficial for the high open-circuit voltage. Generally, a lower shunt resistance could be caused by device periphery and/or bulk leakage current owing to the high defect density for the larger chip size (Chen *et al.*, 2008). That is, the larger the chip, the lower the shunt resistance component. Here, we demonstrate a high shunt resistance (7.5 kΩ · cm^2), which is comparable to that in the case of smaller devices (Sang *et al.*, 2010; Stedman *et al.*, 1993). Therefore, owing to the high crystal quality of the MQW samples, a lower reversed saturation current density and a reasonably high shunt resistance in the equivalent circuit model of the solar cell could provide the solar cell a high open-circuit voltage.

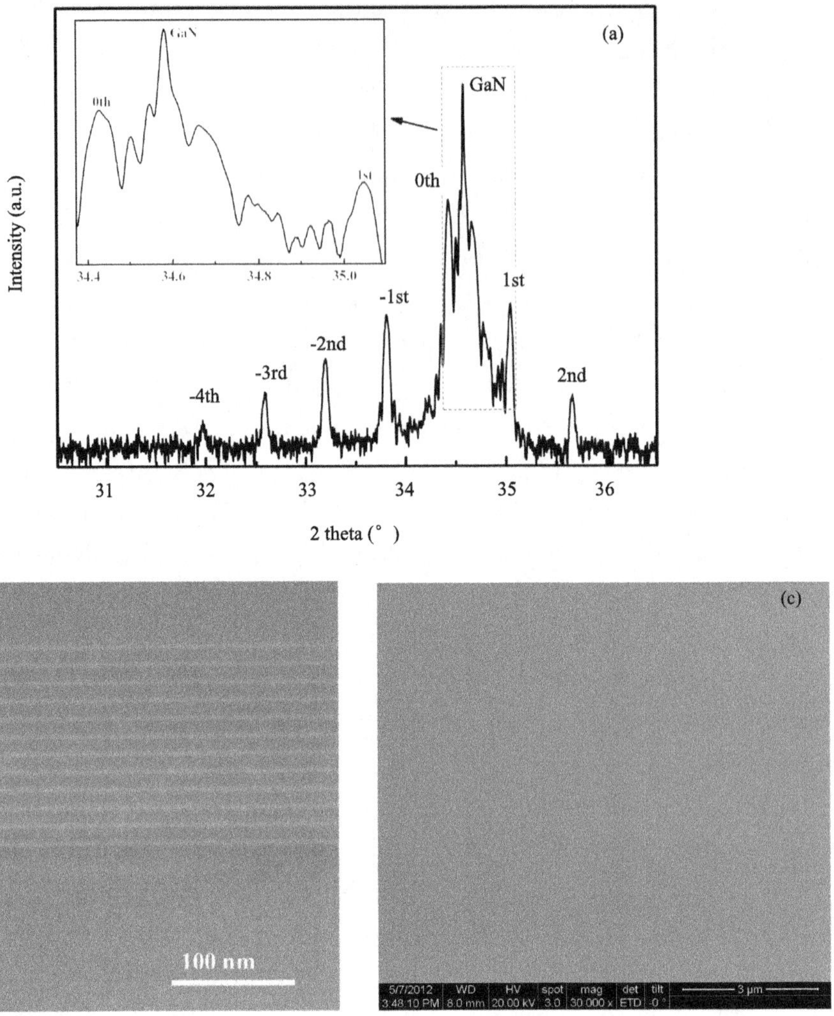

Fig. 2 (a) HRXRD ω-2θ scanning profile of the InGaN/GaN MQW epilayer. The inset graph shows the magnified profiles between 0th-and 1st-order satellite peaks. (b) Cross-sectional TEM image of InGaN/GaN MQW. (c) Top-view SEM image of the sample surface

Note that although a comparatively high V_{oc} is achieved, a lower FF of 45% is measured. In fact, besides the impact of V_{oc} on FF, FF could also be affected by series resistance. On the basis of the relationship between fill factor and series resistance, which was reported in the literature (Green *et al.*, 1983; Neufeld *et al.*, 2011), the decrease in FF is considered to be mainly due to the increase in series resistance. Here, our measured series resistance (R_s) of 4.1 kΩ, which is obtained via the forward *J*-*V* curve, could lead to a lower fill factor, possibly owing to the imperfect crystal quality of III-group nitride materials.

We also investigate the device performance under various concentrator conditions using a Fresnel lens to explore the potential use of InGaN/GaN MQW solar cells. The concentrator ratio was defined by the increase in I_{sc}/I_{sc}(1 sun) ratio. J-V curves of InGaN/GaN MQW solar cells under different illuminations are shown in Fig. 3(a). Because the number of carriers generated is proportional to the number of photons absorbed, J_{sc} was found to continuously increase with light intensity, which is as expected. To analyze the J-V characteristics under the concentrator case in detail, we focused on the equation in the case of $V \gg IR_s$. The J-V characteristics can be approximated by the following equation:

$$J = -J_{sc} + J_0[\exp(qV/nk_BT)-1], \quad (1)$$

where J_{sc}, J_0, q, n, k_B, and T have the same definitions as mentioned above. By using this equation, J_{sc} as a function of V_{oc} for different illumination intensities is plotted, as shown in Fig. 3b. A fitting to the experimental data by the equation with $J=0$ leads to a reversed saturation current density of approximately 1.2×10^{-19} mA/cm^2 and an ideality factor n of 1.83. These extrapolated values are slightly different from those based on the dark case, but we can safely conclude that J_0 is of the order of 10^{-19} mA/cm^2, which is much lower than the reported results.

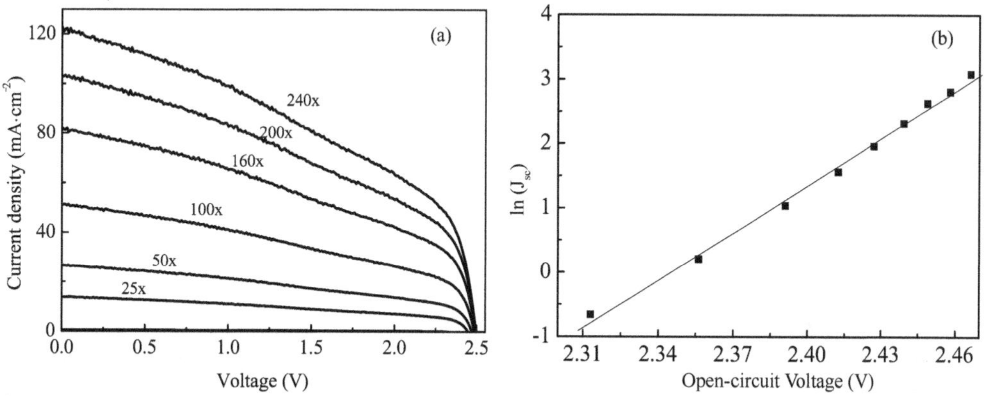

Fig. 3 (a) Current density versus voltage curve of the InGaN/GaN MQW solar cell under different illumination conditions. (b) lnJ_{sc} versus voltage curve of the InGaN/GaN MQW solar cell under different illumination conditions

The light intensity dependence of V_{oc} is shown in Fig. 4a. V_{oc} increased logarithmically with sun concentration up to 190, where a peak V_{oc} of 2.50 V was obtained, and then saturated up to the sun concentration of 240. As is well known, V_{oc} enhancement under concentrated condition is directly proportional to $nk_BT/q\ln C$, where C is the concentration ratio. If complex recombination mechanisms simultaneously act on a solar cell, V_{oc} will deviate from the linear relationship. In this work, the ideality factor n is close to 2, which indicates that the recombination current dominates in the space-charge region of the solar cell. When concentrated sunlight becomes stronger, the recombination in the interface region can be enhanced. Furthermore, the surface recombination loss increased with light intensity owing to the poor temperature conductivity of the sapphire substrate, although the heat sink was used.

Figure 4b also shows FF and η as functions of C. FF decreased because the series resistance of such large-chip-size devices cannot be ignored at higher concentration levels,

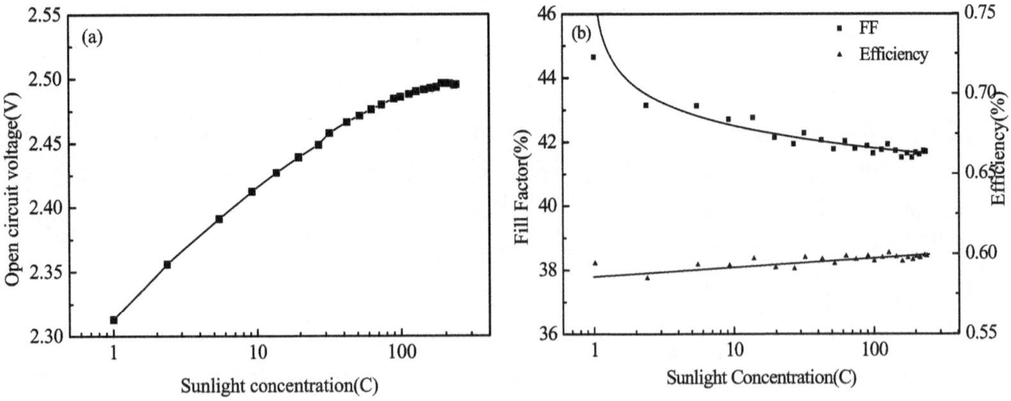

Fig. 4 (a) Open-circuit voltage as a function of sunlight concentration. (b) Fill factor and efficiency as a function of sunlight concentration. The solid lines are drawn as guides for the eyes

which suggests that the cell performance degrades slightly, which is consistent with the trend in Ge/GaAs/GaInP and GaInP/InGaAs/In GaAs multijunction concentrator solar cells(Geisz *et al.*, 2008; Guter *et al.*, 2009) and the results (Dahal *et al.*, 2009, 2010; Bae *et al.*, 2011; Ermolenko *et al.*, 2010; Sang *et al.*, 2011; Zeng *et al.*, 2010; Lang *et al.*, 2011). Although FF decreased with concentration ratio, η demonstrated a slight enhancement with concentration level. The results show that the performance of this large-chip-size solar cell is not markedly degraded under such a high sunlight concentration.

In summary, we successfully fabricated InGaN/GaN MQW solar cells with a large chip size. A high open-circuit voltage (2.31 V) was obtained under 1 sun illumination owing to the small reversed saturation current density and high shunt resistance component in the circuit model of solar cells, which could be attributed to the good crystal quality evidenced by TEM and HRXRD measurements. The open-circuit voltage and short circuit current density were found to increase as sunlight concentration increases up to 240 suns. The peak value of 2.50 V for open-circuit voltage was observed at 190 suns, and the conversion efficiency was enhanced slightly under all concentrator conditions.

Acknowledgments

This work was supported by the SONY-SINANO Joint Project (Grant No. Y1AAQ11001), the Suzhou Solar Cell Research Project (Grant No. ZXJ0903), the Ministry of Science and Technology of China (Grant No. 2010DFA22770), and the Guangxi Innovative Program of Graduate Students (Grant No. GXU11T32515).

References

Bae S Y, Shim J P, Lee D S, *et al.* 2011. Improved Photovoltaic Effects of a Vertical－Type InGaN/GaN Multiple Quantum Well Solar Cell. *Jpn. J. Appl. Phys.* **50**: 092301.

Chen X, Matthews K D, Hao D, *et al.* 2008. Growth, fabrication, and characterization of InGaN solar cells. *Phys. Status Solidi* A**205**: 1103-1105.

Dahal R, Li J, Aryal K, *et al.* 2010. InGaN/GaN multiple quantum well concentrator solar cells. *Appl. Phys. Lett.* **97**: 073115.

Dahal R, Pantha B, Li J, *et al.* 2009. InGaN/GaN multiple quantum well solar cells with long operating *wavelengths*. *Appl. Phys. Lett.* **94**:063505.

Ermolenko M V, Buganov O V, Tikhomirov S A, *et al*. 2010. Ultrafast all-optical modulator for 1.5 mu m controlled by Ti:Al_2O_3 laser. *Appl. Phys. Lett.* **97**:073113.

Farrell R M, Neufeld C J, Cruz S C, *et al*. 2011. High quantum efficiency InGaN/GaN multiple quantum well solar cells with spectral response extending out to 520 nm. *Appl. Phys. Lett.* **98**. 201107.

Geerts W, Mackenzie J D, Abernathy C R, *et al*. 1996. Electrical transport in p-GaN, n-InN and n—InGaN. *SolidState Electron*, **39**:1289-1294.

Geisz J F, Friedman D J, Ward J S, *et al*. 2008. 40.8% efficient inverted triple-junction solar cell with two independently metamorphic junctions. *Appl. Phys. Lett.* **93**:123505.

Green. M A. 1982. Accuracy of Analytical Expressions For Solar-Cell Fill Factors. *Sol. Cells.*, **7**:337-340.

Guter W, Schone J, Philipps S P, *et al*. 2009. Current-matched triple-junction solar cell reaching 41.1% conversion efficiency under concentrated sunlight. *Appl. Phys. Lett.* **94**:223504.

Holec D, Costa P M F J, Kappers M J, Humphreys C J. 2007. Critical thickness calculations for InGaN/GaN. *J. Cryst. Growth*, **303**:314.

Jani O, Ferguson I, Honsberg C, *et al*. 2007. Design and characterization of GaN/InGaN solar cells. *Appl. Phys. Lett.* **91**:132117.

Kuwahara Y, Fujii T, Sugiyama T, *et al*. 2011. GaInN-Based Solar Cells Using Strained-Layer GaInN/GaInN Superlattice Active Layer on a Freestanding GaN Substrate. *Appl. Phys. Express*. **4**:021001.

Lang J R, Neufeld C J, Hurni C A, *et al*. 2011. High external quantum efficiency and fill-factor InGaN/GaN heterojunction solar cells grown by NH_3-based molecular beam epitaxy. *Appl. Phys. Lett.* **98**:131115.

Liu R, Mei J, Srinivasan S, *et al*. 2006. Generation of misfit dislocations by basal-plane slip in InGaN/GaN heterostructures. *Appl. Phys. Lett.* **89**:201911.

Muth J F, Lee J H, Shmagin I K, *et al*. 1996. Absorption coefficient, energy gap, exciton binding energy, and recombination lifetime of GaN obtained from transmission measurements. *Appl. Phys. Lett.*, **71**:2572.

Nakao T, Fujii T, Sugiyama T, *et al*. 2011. Fabrication of Nonpolar a-Plane Nitride-Based Solar Cell on r-Plane Sapphire Substrate. *Appl. Phys. Express* **4**:101001.

Neufeld C J, Cruz S C, Farrell R M, *et al*. 2011. Observation of positive thermal power coefficient in InGaN/GaN quantum well solar cells. *Appl. Phys. Lett.* **99**:071104.

Neufeld C J, Toledo N G, Cruz S C, *et al*. 2008. High quantum efficiency InGaN/GaN solar cells with 2.95 eV band gap. *Appl. Phys. Lett.* **93**:143502.

Sang L W, Liao M Y, Ikeda N, *et al*. 2011. Enhanced performance of InGaN solar cell by using a super—thin AlN interlayer. *Appl. Phys. Lett.* **99**:161109.

Singh R, Doppalapudi D, Moustakas T D, *et al*. 1995. Phase separation in InGaN thick films and formation of InGaN/GaN double heterostructures in the entire alloy composition. *Appl. Phys. Lett.* **70**:1089-1091.

Steadman J W. 1993. in *The Electrical Engineering Handbook*, ed. R. C. Dorf (CRC Press, Boca Raton), 459.

Wu J, Walukiewicz W, Yu K M, *et al*. 2002. Unusual properties of the fundamental band gap of InN. *Appl. Phys. Lett.* **80**:3967.

Wu J, Walukiewicz W, Yu K M, *et al*. 2003. Superior radiation resistance of In1—xGaxN alloys: Full—solar—spectrum photovoltaic material system. *J. Appl. Phys.* **94**:6477-6482.

Yang C C, Jang C H, Sheu J K, *et al*. 2011. Characteristics of InGaN—based concentrator solar cells operating under 150X solar concentration. *Opt. Express* **19**:A695-A700.

Zeng S W, Cai X M, Zhang B P. 2010. Demonstration and Study of Photovoltaic Performances of InGaN p-i-n Homojunction Solar Cells. *IEEE J. Quantum Electron*. **46**:783-787.

Zhang X B, Wang X L, Xiao H L, *et al*. 2011. InGaN/GaN multiple quantum well solar cells with an enhanced open—circuit voltage. *Chin. Phys. B* **20**:028402.

Zheng X H, Horng R H, Wuu D S, *et al*. 2008. High—quality InGaN/GaN heterojunctions and their photovoltaic effects. *Appl. Phys. Lett.* **93**:261108.

EFFECTS OF THERMAL PRE-TREATMENT ON ANAEROBIC CO-DIGESTION OF MUNICIPAL BIOWASTES AT HIGH ORGANIC LOADING RATE①

GUO Jianbin, WANG Wei, LIU Xiao, LIAN Songjian and ZHENG Lei

Department of Environmental Engineering, School of Environment, Tsinghua University, Beijing 100084, China

Abstract

Anaerobic co-digestion of thermal pre-treated municipal biowaste (MBW) is a field of research that has had limited contributions to date. In this study, laboratory-scale semi-continuously fed anaerobic digesters treating thermally treated and non-treated MBW were operated at high organic loading rates (OLR). The results show that the methanogenesis process was inhibited by the accumulated volatile fatty acids when 30% (w/w) of dewatered activated sludge (DAS) was co-digested with food waste (FW) and fruit/vegetable residue (FVR) at high OLR≥ 10 kg volatile solid $m^{-3} \cdot d^{-1}$. Co-digestion with thermal hydrolysed DAS can significantly improve digester performance. In contrast to DAS, some adverse effects of thermal pre-treatment on the biodegradability of FW and FVR were observed. Therefore, co-digestion of FW, FVR with thermally treated DAS is suggested as an alternative to promote high methane production and process stability.

Key words: anaerobic co-digestion; municipal solid waste (MSW); dewatered activated sludge; thermal pre-treatment

1 Introduction

With the economic development and urbanisation of China, the production of municipal solid waste (MSW) has increased rapidly in recent years. In 2010, 158 Mt of MSW were collected and transported (National Bureau of Statistics of China, 2011). Of these, more than 60% is organic, that is, characteristically had a high water content and is biodegradable, for example, food waste (FW) and fruit/vegetable residue (FVR). In China, the annual production of dewatered activated sludge (DAS, 80% moisture content) from wastewater treatment plants has reached almost 30 Mt, and 80% of it has not obtained necessary stabilisation (Duan *et al.*, 2012).

Anaerobic digestion of the above municipal biowastes (MBW) is an effective way to

① The paper published in *Chemosphere*. 2013 Dec. 26. pii: S0045-6535 (13) 01667-6. doi: 10.1016/j.chemosphere.2013.12.007.

convert organic matter into renewable energy. Co-digestion of MSW with sewage sludge can offer a number of advantages: dilution of potential toxic compounds; improved balance of nutrients; synergistic effect of microorganisms; increased load of biodegradable organic matter and better biogas yields (Sosnowski *et al.*, 2002).

The hydrolysis stage is considered as a rate-limited step in the anaerobic digestion of solid wastes. Consequently, physico-chemical treatments are often used to promote solubilisation of organic matter. Thermal pretreatment has proved as a useful pretreatment method and it is thought that it can break cell structure to release organic matter inside the cells and consequently improve digestion performance. Therefore, most of the previous studies aimed to improve anaerobic degradability of activated sludge (Bougrier *et al.*, 2008; Christopher and John, 2009; Li and Noike, 1992) or activated sludge dewaterability (Neyens and Baeyens, 2003; Qiao *et al.*, 2010) via thermal pretreatment. Fewer studies have been undertaken on the effect of thermal pretreatment on mixed MBW. Qiao *et al.* (2012) found that the choice of feeding material (household sewage, sewage from mixture of domestic and industrial wastewater, and industrial and oil wastewater treatment plants) affects the performance of thermal pretreatment. Liu *et al.* (2012a) has studied the effect of thermal pretreatment on the physical and chemical properties of MBW and a batch test of each substrate was conducted to determine the biodegradability of the raw and thermally treated materials. The results showed that methane potential of thermally treated DAS increased by 35%. On the other hand, methane potential of thermally treated FW and FVR decreased by 8% and 12%, respectively. However, they had not considered that biogas production could possibly be stimulated by co-digested substrates in a semi-continuous system. Some conflicting results have been reported by Zhou *et al.* (2012) who has compared the performance of the semi-continuously fed digester processing thermally treated and non- treated MBW at a relatively low organic loading rate (OLR) of 1.5 and 3.0 kg VS $m^{-3} \cdot d^{-1}$. They found that the hydrolysis rate increased after thermal pretreatment. However, they did not find any obvious differences in volatile solids (VS) removal rate.

The objective of this study is to evaluate the effect of thermal pretreatment and different mixing ratios on the anaerobic co-digestion of FW, FVR and DAS at high OLR ($\geqslant$10 kg VS $m^{-3} \cdot d^{-1}$) via the stable operation of a semi-continuously fed digester.

2 Materials and Methods

2.1 Municipal Biowaste

FW used in this study was collected from a student canteen at Tsinghua University, FVR was collected from a wholesale market and DAS came from a municipal wastewater treatment plant in Beijing. Their differing characteristics are shown in Table 1. The inert materials in FW and FVR were manually separated (e.g., plastics, bone, wood and others). FW and FVR were shredded separately by a pulverizer to less than 3.0 mm. The mixed feedstock was kept at 4℃. The properties of the mixed substrates are shown in Table 2.

Table 1 Characteristics of raw materials used in this study (Liu *et al.*, 2012a)

	FW	FVR	DAS
Water content (%)	80.0	89.3	84.3
TS ($g \cdot kg^{-1}$)	200	107	157
VS ($g \cdot kg^{-1}$)	180	100	114
Carbohydrates ($g \cdot kg^{-1}$)	107	84	59
Crude protein ($g \cdot kg^{-1}$)	30	13	69
Crude fat ($g \cdot kg^{-1}$)	44	3	21
C/N	17	22	7

2.2 Thermal Pre-treatment

Thermal pre-treatment was applied at a temperature of 170—175℃ for 60 min. Steam was used as a heat source at a pressure of 0.5 MPa. A detailed description of the thermal pre-treatment process has been reported by Zhou *et al.* (2013). In the thermal pre-treatment process, MBW was inevitably diluted by the steam and some additional water, which resulted in around 1-fold dilution (Table 2).

Table 2 Properties of bio-waste with different mixing ratios used in this study.

Treatment	Substrates	Ratio (w/w/w)	TS ($g \cdot kg^{-1}$)	VS ($g \cdot kg^{-1}$)
1	FW, FVR, thDAS[a]	2 : 1 : 1	121±3	103±3
2	FW, FVR, dDAS[b]	2 : 1 : 1	126±7	107±6
3	FW, FVR, thDAS	2 : 2 : 1	108±3	91±4
4	FW, FVR, dDAS	2 : 2 : 1	112±4	93±3
5	FW, FVR, DAS	1 : 1 : 1	122±11	102±11
6	thFW, thFVR, thDAS	1 : 1 : 1	62±5	49±4

a: prefix th means thermal hydrolyzed.

b: In the consideration of dilution effect in the thermal treatment process, dDAS used in the experiment is the DAS diluted with water at a ratio of 1 : 1.

2.3 Digester

Six completely stirred anaerobic reactors (30 L) with an effective volume of 20 L were used. Each digester was sealed by a dull polished glass cap with a stirrer in the middle of the cap. Stirring was conducted automatically for 10 min $\cdot$ h^{-1} at a speed of 120 rpm throughout the entire experimental period. Two ports were fitted at the top and bottom of the digester walls for feeding and withdrawing liquid samples, respectively. The temperature was controlled at (38±1)℃ by water jackets surrounding the digesters.

2.4 Digester Start-up and Operation

Each digester was inoculated with 20 L inoculum (Total solids (TS): 3.9%; VS: 1.9%). The inoculum came from a pilot scale biogas plant (200 L) treating biowastes, which had run stably for more than 2 yr.

The OLR was increased aggressively until it reached 10 kg VS $m^{-3} \cdot d^{-1}$. Two high OLR levels (Table 3) were focused on in each treatment. The two OLRs were operated for at least 2-3 hydraulic retention time (HRTs). The nominal HRT is shown in Table 3. In treatment 6, the OLR cannot reach a nominal OLR of 10 kg VS $m^{-3} \cdot d^{-1}$ at fixed HRT because of the dilution effect of the thermal pretreatment process. The steady-state values of daily gas production and VS of the effluent were taken as the average of these consecutive measurements when the deviations between the observed values were less than 5% in all cases.

2.5 Monitoring

The TS and VS were analysed according to APHA (1995). Biogas production was monitored by using a drum type gas meter (LML-1, China). The quality of biogas was analysed with an infra-red sensor (OMBION 1.42 Biogas Check, Geotechnical Instruments (UK) Limited). The pH was measured during the experiments (FE20, Mettler-Toledo, Switzerland). pH is commonly used as a process indicator, but its effectiveness as a control parameter strongly depends on the alkalinity, i.e. the buffering capacity of the process. The most important buffer within the optimal pH for methanogenic organisms is bicarbonate. The ratio of volatile fatty acid (VFA) and total inorganic carbon (TIC) gives a good indication of the process stability of anaerobic digestion. Therefore, TIC and VFA content were analysed in this study using a titration method (Rieger and Weiland, 2006).

2.6 Statistical Analysis

At nominal OLR of 10 kg VS $m^{-3} \cdot d^{-1}$, independent-samples t-tests at the 0.05 significance level were conducted to determine whether the observed differences between thermally treated and non-treated treatments were significantly different.

3 Results

This section focuses the discussion (Section 4) on the effects of MBW thermal pretreatment and DAS content on digester performance. The discussions will be based on the experimental results below.

(1) pH, methane content and process stability

In treatments 1-4, a stable pH range of 7.6 — 7.8 was observed throughout the experiment. However, pH sharply decreased to 5.5—5.8 in treatments 5 and 6 when the OLR increased to 13.5 and 6.5 kg VS m^{-3} • d^{-1}, respectively (Fig. 1).

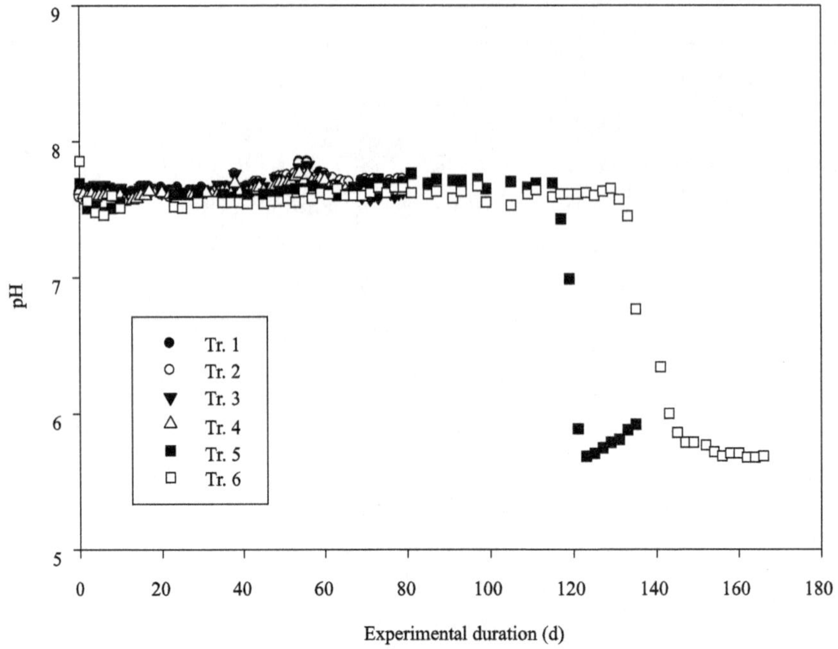

Fig. 1 Variations of pH values of all treatments (1-6) throughout the experiment: treatment 1 (●); treatment 2 (○); treatment 3 (▼); treatment 4 (△); treatment 5 (■) treatment 6 (□)

The ratio of VFA and TIC gives a good indication of the process stability of anaerobic digestion (Zhao and Viraraghavan, 2004; Sánchez *et al.*, 2005; Rieger and Weiland, 2006). As shown in Table 3, VFA/TIC ratio is very low in treatments 1—4 and 6 when the OLRs were around 10 and 5 kg VS m^{-3} • d^{-1}, respectively and no acidification was observed. However, in treatment 5, VFA/TIC ratio reached 0.64 at OLR of 10.2 kg VS m^{-3} • d^{-1}, which was much higher than the threshold of 0.3—0.4 recommended by Rieger and Weiland (2006). When the OLR increased to 13.5 kg VS m^{-3} • d^{-1}, the VFA/TIC ratio reached as high as 0.98. The digester acidified and the VFAs accumulated with a concentration of 6.30×10^3 mg HAc-eq L^{-1}. Similarly, in treatment 6, VFAs accumulated with a concentration of 5.20×10^3 mg HAc-eq L^{-1} when the OLR increased to 6.5 kg VS m^{-3} • d^{-1}.

(2) Methane content

The methane content determined during a 24 h feeding cycle (period between each feeding) at first target OLR level (nominal OLR of 10 kg VS m^{-3} • d^{-1} in treatment 1—5 and 5.0 kg VS m^{-3} • d^{-1} in treatment 6) of each treatment is plotted in Fig. 2. The feeding cycle works like a batch process in a semi-continuous system where the microbes would go through processes of hydrolysis, acidogenesis, acetogenesis and methanogenesis. During the process, methane content decreased after the first few hours due to hydrolysis of the organic matters, during which CO_2 was produced. After that, but CO_2 stripping off, with the consumption of CO_2 by hydrogenotrophic methanogens, CH_4 concentration recovered to the initial level. For the first targeted OLR level of each treatment, the average methane content was around 65%. The OLR did not affect the methane content in each treatment.

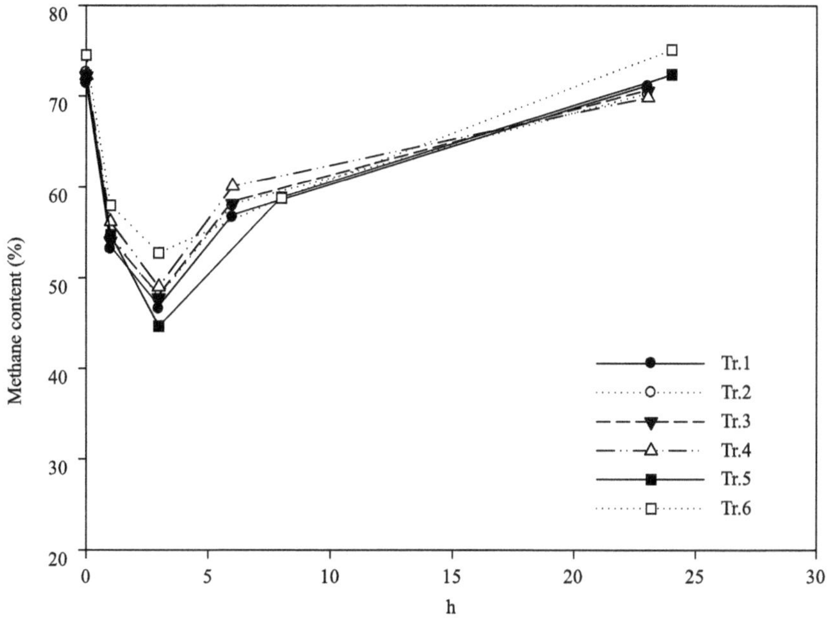

Fig. 2　Variations of methane content during feeding cycle at the first target OLR level: treatment 1 (●); treatment 2 (○); treatment 3 (▼); treatment 4 (△); treatment 5 (■) treatment 6 (□)

(3)Gas production

Biogas production increased with an aggressively increasing OLR (Fig. 3). The statistical difference of specific biogas production in treatments 1 and 2, 3 and 4 at nominal OLR of 10.0 kg VS $m^{-3} \cdot d^{-1}$ is indicated in supplementary material, respectively. In treatment 1 with thermal hydrolyzed DAS, specific biogas production was significantly higher than that in treatment 2 when the OLR was around 10.0 kg VS $m^{-3} \cdot d^{-1}$. A similar result was observed in treatments 3 and 4. The specific biogas production decreased when the OLR was increased to the next level (Table 3). On average, specific biogas production from treatments 3 and 4 is higher than that from treatments 1 and 2 because of the increased amount of FVR which has a higher biogas potential than DAS. In treatment 5, the specific biogas production was about 20%—30% lower than that in treatments 1—4 at a similar OLR of around 10.0 kg VS $m^{-3} \cdot d^{-1}$. In treatment 6, the specific biogas production of thermally treated MBWs was only 0.53 $L \cdot g^{-1}$ VS_{added} when the OLR was as low as 5.0 kg VS $m^{-3} \cdot d^{-1}$.

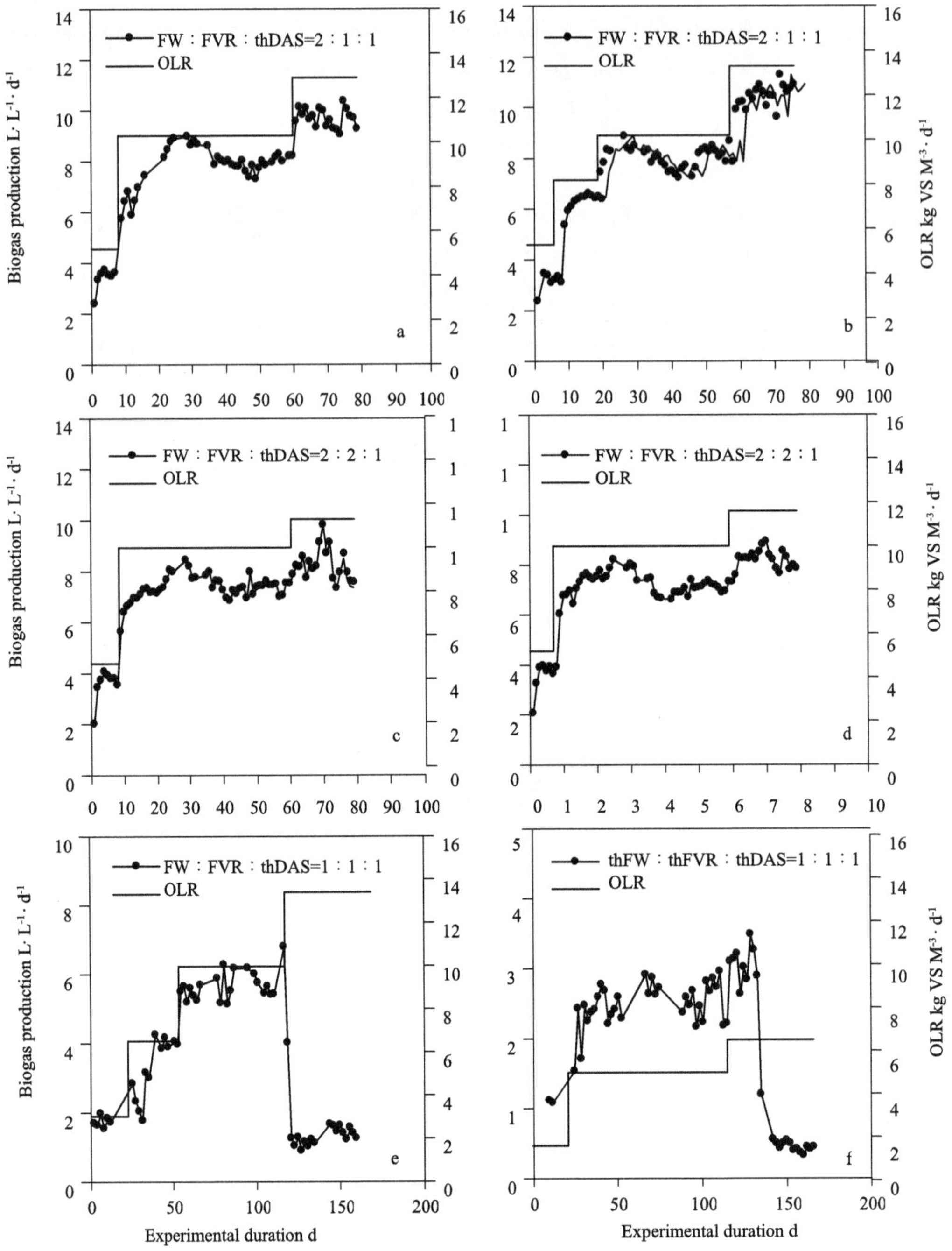

Fig. 3 Biogas production of all treatments (a-f) at aggressively increased OLR

Table 3　Steady-state performance of digesters in each treatment.

Treatment	OLR kg VS $m^{-3}\cdot d^{-1}$	Nominal HRT d	Specific biogas production $L\cdot g^{-1}VS_{added}$	VS removal efficiency %	VFA/TIC
1	10.3	10	0.76±0.03	64.2	0.13
	12.9	8	0.74±0.03	—	0.15
2	10.2	10	0.72±0.03	62.2	0.13
	13.3	8	0.71±0.03	—	0.15
3	10.0	10	0.81±0.03	64.1	0.13
	11.3	8	0.72±0.06	—	0.15
4	10.0	10	0.76±0.02	62.7	0.13
	11.6	8	0.70±0.03	—	0.15
5	10.2	10	0.55±0.03	59.5	0.64
	13.5	8	—	—	0.98
6	5.0	10	0.53±0.05	54.7	0.12
	6.5	8	—	—	0.89

(4)VS removal efficiency

In treatments 1—4, a similar VS removal efficiency was obtained in the range of 62% — 64%. Compared to that, at the first OLR level, the VS removal efficiencies in treatments 5 and 6 were 60% and 55%, respectively.

4　Discussion

4.1　Effect of Thermal Pretreatment on Digester Performance

Thermal pretreatment is considered a promising method to improve the properties of organic solid wastes, which can not only improve the hydrolysis rate of organic solids but also improve the dewaterability (Wang, 2010).

In consideration of the dilution effect in the thermal treatment process, the amount of thermally treated DAS that was actually co-digested in treatments 1 and 3 was only 13% and 10%, respectively. Therefore, the specific biogas production in treatments 1 and 3 is only a little higher than that in treatments 2 and 4, respectively. Statistical analysis also indicated that co-digestion with thermal hydrolysed DAS can significantly improve digester performance.

In treatment 5, the specific biogas production was 0.55 $L\cdot g^{-1}VS_{added}$ at OLR 10.2 kg VS $m^{-3}\cdot d^{-1}$. A very similar specific biogas production (0.53 $L\cdot g^{-1}VS_{added}$) was observed in treatment 6. However, the OLR was as low as 5.0 kg VS $m^{-3}\cdot d^{-1}$ because of steam dilution. Thus, on the assumption that there was a lesser steam dilution effect in a full scale thermal pretreatment process, the biogas production would be much lower when the OLR increased to the nominal OLR of 10 kg VS $m^{-3} \cdot d^{-1}$. The VS removal efficiency would be also much lower than 55%.

In previous studies, a 35% increased methane production was obtained from thermal treated DAS (thDAS) fermentation. However, a 8% and 12% decrease was obtained for thermal treated food waste (thFW) and thermal treated fruit and vegetable residue (thFVR), respectively (Liu *et al.*, 2012a). These results confirmed that thermal pretreatment of organic waste had been found to be responsible for the denaturation of organic matters (Eskicioglu *et al.*, 2006) and the formation of refractory or inhibitory dissolved organic compounds and most of them had been revealed as melanoidins (Zouari and Ellouzz, 1996; Penaud *et al.*, 2000).

Li and Noike (1992) attributed the decreased digester performance to thehigh concentration of soluble sugars and soluble protein in FW and FVR, which resulted in the formation of melanoidins during the thermal pretreatment process, and in turn, led to the decreased biodegradability. However, a smaller amount of melanoidins was produced in the thDAS due to the lower content of soluble sugar and soluble protein, and the higher methane production due to the increased soluble organics had lessened effect of melanoidins. Consequently, in treatment 6 of this study, when the three substrates had been thermally treated and co-digested in a semi-continuous feeding digester, an overall decreased performance was observed at OLR of 5.0 kg VS $m^{-3} \cdot d^{-1}$. When the OLR was increased to 6.5 kg VS $m^{-3} \cdot d^{-1}$ in treatment 6, the severely inhibitory effect was observed.

4.2 Effect of DAS Content on Digester Performance

A major limitation of anaerobic digestion of FW and FVR is the rapid acidification of these wastes resulting in a greater production of VFAs, which decreases the pH in the digester and inhibits the activity of methanogenic bacteria (Bouallagui *et al.*, 2009).

Introducing co-substrates with high nitrogen content offers a solution to adjust the nutrient content of FW and FVR. The unbalanced nutrients of DAS characterised by a low C/N ratio were also regarded as an important limiting factor to anaerobic degradation.

Bouallagui *et al.* (2009) reported that the addition of DAS with a ratio of 10% VS enhanced biogas yield by 43.8% and VS removal rate (73.1%—85.4%) by 11.7%, at a relative low range OLR 2.46—2.51 kg VS $m^{-3} \cdot d^{-1}$. In another study by Liu *et al.* (2012b), FW, FVW and DAS (mixing ratio=2 : 1 : 1, w/w/w) were co-digested in a continuous stirred-tank reactor for biogas production and stable operation was achieved at an OLR between 1.2—8.0 kg VS $m^{-3} \cdot d^{-1}$ and VS removal efficiency was between 61.7% and 69.9%.

As previously noted, in China, over 30 Mt of DAS (80% moisture content) is generated every year, and almost 80% of it does not achieve necessary stabilisation (Duan *et al.*, 2012). With the urgent need of DAS treatment in China in mind, in this study,

FW, FVR and DAS mixed as a ratio of 1 : 1 : 1 (w/w/w) were co-digested at a nominally high OLR of 10 kg VS $m^{-3}\cdot d^{-1}$ (Treatment 5). The results indicated that the digester could run stably at an OLR of 10.2 kg VS $m^{-3}\cdot d^{-1}$. However, the VFA/TIC ratio is as high as 0.64. According to Rieger and Weiland (2006), a VFA/TIC ratio above 0.3—0.4 may indicate acidification in the digester. This means that the digester needs to be monitored carefully as an accumulation of organic acids could ultimately acidify it. As expected, when the OLR increased to 13.5 kg VS $m^{-3}\cdot d^{-1}$, the VFA/TIC reached 0.98 and the accumulated VFA aggravated the acidification in the digester. These results indicate that a higher ratio (30%, w/w/w) of DAS resulted in a rather high viscosity of the substrates in the digesters (50000—80000 mPa • s). This high viscosity might restrict the mass transfer efficiency of the system and lead to acidification. Liu *et al.* (2012a) concluded that the viscosity of activated sludge (TS, 7.9%) decreased from 13500 to 1625 mPa • s after thermal pretreatment. Consequently, when thermal treated DAS was mixed in treatments 1-4, good process stability was observed when the OLR was in a range of 10.0—13.3 kg VS $m^{-3}\cdot d^{-1}$ (Table 3).

Acknowledgements

This study was supported by the China Postdoctoral Science Foundation funded project(2013M540965), the National Key Technology R&D program (2010BAC67B02), Ministry of Science and Technology of China (2010DFA22770) and the Produce-learn-research Project of Guangdong (20110903).

References

APHA. 1995. Standard Methods for the Examination of Water and Wastewater, 19th ed., American Public Health Association, Washington, DC.

Bouallagui H, Lahdheb H, Romdan E B, *et al*. 2009. Improvement of fruit and vegetable waste anaerobic digestion performance and stability with co-substrates addition. *J. Environ. Manage.* **90**: 1844-1849.

Bougrier C, Delgenes J P, Carrere H. 2008. Effects of thermal treatments on five different waste activated sludge samples solubilisation, physical properties and anaerobic digestion. *Chem. Eng. J.* **139**: 236-244.

Christopher A W, John T N. 2009. Hydrolysis of macromolecular components of primary and secondary wastewater sludge by thermal hydrolytic pretreatment. *Water Res.* **43**: 4489-4498.

Duan N, Dong B, Wu B, *et al*. 2012. High-solid anaerobic digestion of sewage sludge under mesophilic conditions: Feasibility study. *Bioresour. Technol.* **104**: 150-156.

Eskicioglu C, Kennedy K J, Droste R L. 2006. Characterization of soluble organic matter of waste activated sludge before and after thermal pretreatment. *Water Res.* **40**: 3725-3736.

Li Y Y, Noike T. 1992. Upgrading of anaerobic-digestion of waste activated-sludge by thermal pretreatment. *Water Sci. Technol.* **26**(3-4): 857-866.

Liu X, Wang W, Gao X B, *et al*. 2012a. Effect of thermal pretreatment on the physical and chemical properties of municipal biomass waste. *Waste Manage.* **32**: 249-255.

Liu X, Wang W, Shi Y C, *et al*. 2012b. Pilot-scale anaerobic co-digestion of municipal biomass waste and waste activated sludge in China: Effect of organic loading rate. *Waste Manage.* **32**: 2056-2060.

National Bureau of Statistics of China. 2011. *China Statistical Yearbook*, China Statistics Press, Beijing, China.

Neyens E, Baeyens J. 2003. A review of thermal sludge pre-treatment processes to improve dewaterability. *J. Hazard. Mater.* B **98**: 51-67.

Penaud V, Delgenes J P, Moletta R. 2000. Influence of thermochemical pretreatment conditions on solubilization and anaerobic biodegradability of a microbial biomass. *Environ. Technol.* **21**: 87-96.

Qiao W, Sun Y, Wang W. 2012. Treatment of 14 sludge types from wastewater treatment plants using bench and pilot thermal hydrolysis, *Water Sci. Technol.* **66**(4): 895-902.

Qiao W, Wang W, Wan X, *et al.* 2010. Improve sludge dewatering performance by hydrothermal treatment. *J. Residuals Sci. Technol.* **7**: 7-11.

Rieger C, Weiland P. 2006. Prozessstörungen frühzeitig erkennen. *Biogas J.* **4**: 18-20.

Sánchez E, Borja R, Weiland P, *et al.* 2001. Effect of substrate concentration and temperature on the anaerobic digestion of piggery waste in tropical climates. *Process Biochem.* **37**: 483-489.

Sosnowski P, Wieczorek A, Ledakowicz S. 2002. Anaerobic co-digestion of sewage sludge and organic fraction of municipal solid wastes, *Adv. Environ. Res.* **7**: 609-616.

Wang W, Hou H H, Hu S, *et al.* 2010. Performance and stability improvements in anaerobic digestion of thermally hydrolyzed municipal biowaste by a biofilm system. *Bioresour. Technol.* **101**: 1715-1721.

Zhao H W, Viraraghavan T. 2004. Analysis of the performance of an anaerobic digestion system at the regina wastewater treatment plant. *Bioresour. Technol.* **95**: 301-307.

Zhou Y J, Takaoka M, Wang W, *et al.* 2013. Effect of thermal hydrolysis pre-treatment on anaerobic digestion of municipal biowaste: A pilot scale study in China. *J. Biosci. Bioeng.* **116**: 101-105.

Zouari N, Ellouz R. 1996. Toxic effect of coloured olive compounds on the anaerobic digestion of olive oil mill effluent. *J. Chem. Technol. Biotechnol.* **66**: 414-420.